ADVANCED MATERIALS FOR
SEVERE SERVICE APPLICATIONS

Proceedings of a Japan–US Joint Seminar on 'Advanced Materials for Severe Service Applications' held in Tokyo, May 19–23 1986. Supported jointly by The Japan Society for Promotion of Science and The National Science Foundation.

ADVANCED MATERIALS FOR SEVERE SERVICE APPLICATIONS

Edited by

K. IIDA

*Department of Naval Architecture,
University of Tokyo, Japan*

and

A. J. McEVILY

*Department of Metallurgy,
University of Connecticut, USA*

ELSEVIER APPLIED SCIENCE
LONDON and NEW YORK

ELSEVIER APPLIED SCIENCE PUBLISHERS LTD
Crown House, Linton Road, Barking, Essex IGII 8JU, England

Sole Distributor in the USA and Canada
ELSEVIER SCIENCE PUBLISHING CO., INC.
52 Vanderbilt Avenue, New York, NY 10017, USA

WITH 32 TABLES AND 300 ILLUSTRATIONS

© ELSEVIER APPLIED SCIENCE PUBLISHERS LTD 1987
Softcover reprint of the hardcover 1st edition 1987

British Library Cataloguing in Publication Data

Advanced materials for severe service
applications.
1. Building materials   2. Hazardous
geographic environments
I. Iida, K.   II. McEvily, A. J.
624.1   TA403.6

Library of Congress Cataloging in Publication Data

Advanced materials for severe service applications.

"Proceedings of a Japan–US Joint Seminar on
'Advanced Materials for Severe Service Applications' held
in Tokyo, May 19–23, 1986. Supported jointly by the
Japan Society for Promotion of Science and the National
Science Foundation"—P.
Includes bibliographies and index.
1. Materials at high temperatures—Congresses.
2. Materials at low temperatures—Congresses.
3. Fracture mechanics—Congresses. I. Iida, K. II.
McEvily, A. J. III. Japan– US Joint Seminar on
"Advanced Materials for Severe Service Applications"
(1986: Tokyo, Japan) IV. Nihon Gakujutsu Shinkōkai.
V. National Science Foundation (U.S.)

TA418.24.A38   1987   620.1′121   87–590
ISBN-13: 978-94-010-8042-2   e-ISBN-13: 978-94-009-3445-0
DOI: 10.1007/978-94-009-3445-0

The selection and presentation of material and the opinions expressed are the sole
responsibility of the author(s) concerned.

Contents

v

Introduction

The material covered in the manuscripts published herein was subjected to public inquiry during the Japan–US Joint Seminar on Materials for Severe Service Conditions during 19–23 May 1986 at the Toranomon Pastral Guest House in Minato-Ku, Tokyo, Japan. This seminar was the latest in a series on advanced materials and applications initiated in the early 1970s by Professor T. Kanazawa of Japan and Professor A. S. Kobayashi of the United States. The 1986 seminar was organized by the undersigned with the able assistance of Professor H. Kobayashi and Dr H. Nakamura of the Tokyo Institute of Technology, and Dr K. Minakawa of the Nippon Kokan Technical Research Center. The seminar was sponsored by the US National Science Foundation and by the Japan Society for Promotion of Science.

This Proceedings volume is offered for its reference value in the enhancement of the understanding of the behavior of advanced structural materials for design applications involving adverse loading conditions and severe environments. During the seminar attempts were also made to extract priority issues of possible broad impact on science or technology, and to articulate possible guidelines for action plans.

The three overview contributions presented in the opening session clearly underscored the rewards of coordinated and concentrated scientific research in the rapid advance of technology. The Japanese Cryogenics Technology National Program includes strong materials involving the development of new materials and reliability and property evaluation. Included in the program is the development of appropriate testing protocol and equipment. The Japanese welding program illustrates the need at times for the development of new processes for specific applications. Economics and opportunities appear to be strong and proficient drivers in a move from wrought to cast alloys for advanced gas turbine and jet propulsion systems. Concentrated scientific research and development were given credit in lessening if not eliminating the adverse microstructures and nanostructures usually associated with castings.

The dynamic fracture intensity factor and crack growth rate relationship was isolated as a severe service parameter for presentation and discussion. Limited experimental results show that the arrest crack length of a propagating crack is governed by K_{1m}, the value of the dynamic stress intensity factor at low propagation velocities. It was also pointed out that the uniqueness or lack thereof in the K_I^{dyn} versus crack velocity relation is far from being settled.

The structural ceramics contributions pivot about the micromechanisms of their high temperature and fracture properties. Attention was focused on the role of stress-induced transformations in enhancing fracture toughness, and on the mechanism of creep crack growth in ceramics and ceramic composites. Consideration was also given to the nucleation of microcracks and the role of small flaws in the fracture of ceramics. The focus was on the development of ceramics of enhanced toughness and ductility. Coordinated research straddling solid mechanics developments of crack-tip mechanistics, elegant experimental demonstrations, and microstructural understanding were credited for the healthy state of knowledge in structural ceramics. There was a consensus of opinion that systematic research attention should be given to the understanding and enhancement of the high temperature strengthening and toughening effects of fibers in ceramic matrix composites.

The action items for high temperature oxidation and corrosion that were articulated included the effects of erosion and the perennial need to carry out dedicated research on mechanical properties as affected by hot-corrosion. Of particular concern is that erosion can greatly aggravate hot corrosion or oxidation attack of otherwise corrosion resistant alloys or coatings.

A contribution on materials behavior in the Tokamak simulators demonstrates the extreme complexity of materials needs in the adverse environments of controlled fusion-power reactors. It was beyond the scope of the seminar to advise specific action in this highly specialized area.

The contribution on the behavior of materials subjected to low cycle fatigue at elevated temperatures provided confidence that detailed understanding and broad impact information can be obtained in spite of the superposition of complexity of loading and the adversities of high temperatures. More directed research in the area of multiaxial loading was encouraged. There was a consensus that such research must be undertaken in a scientific manner so that improvements in the understanding of the mechanisms involved will continue to result.

The issue of whether or not there are really any creep–fatigue interactions in the mechanistic sense was often debated during the seminar. Some evidence for such interaction during the crack initiation stage is presented, but it concluded that the development of creep-fatigue constitutive relationships which incorporates the effects of frequency, anelastic effects, and the influence of the environment remains a formidable task. Action was recommended in terms of further studies into the mechanistics of creep–fatigue interaction in high-temperature engineering systems including the expeditious determination of creep–fatigue interaction effects in the higher temperature intermetallics, ceramics, and hybrid systems.

The presentations on satisfying the materials needs of Japan's proposed HTGR and in auditing the state of degradation of materials in operating fossil fuel power plants also underscore the fuller understanding of creep–fatigue–surface damage interactions. Action was suggested to develop constitutive relations for lifetime prediction that will take into account the interactions already uncovered. Alternative NDE techniques to determine the extent of temper embrittlement are also needed, as well as improved means for monitoring materials softening in order to audit the state of components.

Cryogenic creep was emphasized as a surprise problem in superconducting magnet applications. The action plan should include more concentrated study of understanding and defeating this particular creep phenomenon. The promotion of martensitic transformation in otherwise austenitic steels needs to be understood and suppressed in order to minimize permeability changes. The cost and magnetic stability aspects of the high manganese steels make them attractive for superconducting magnet applications. Action should be taken to further improve the cryogenic dynamic properties of these alloys.

Crack closure in fatigue crack propagation and the relative applicability of the different specific closure mechanisms were discussed. Air and vacuum tests were considered to be efficient means to separate material and environmental closure effects. Further action is needed to understand the kinetics of oxide formation at crack tips as well as the properties of the oxides formed on metal surfaces, particularly with respect to their role in affecting fatigue crack growth rates. It was also recommended that care should be taken in interpreting fatigue crack closure information. Consideration must be given to the nature of the closure process and the location of the contact points relative to the crack tip. A better understanding of the seeming insensitivity of the fatigue crack growth rate to materials strength and other properties with the exception of the modulus was deemed necessary for the sake of alloy design. Action should continue with respect to advances in the measurement as well as the conceptualization of crack opening levels, especially in the threshold region. The effect of a chevron notch as a crack starter was also discussed. The residual stretch at such a notch can lead to higher thresholds, and the stretch is to be avoided by removing the notch by EDM machining after precracking, or by eliminating the chevron notch entirely. Further work was proposed to determine whether fatigue crack growth and the closure level is path independent or not with respect to increasing ΔK as compared to decreasing ΔK tests.

Interest was also shown with respect to the understanding of the strong fatigue resistance enhancement associated with shot peening in terms of fracture mechanics parameters. Encouragement was also given to further studies of all aspects of the fracture mechanics of reinforced polymers, including the development of anisotropic driving intensity or energy release parameters as related to defects in the polymeric matrix. A most interesting presentation on dynamic and cyclic stress corrosion resistance of metals triggered interest in the determination of the mechanism involved in the promotion of hydrogen embrittlement in steels not known to be susceptible to this distress. The practical implications of these findings need to be analyzed fully with respect to relevant applications where safety is involved.

A presentation on the corrosion of weld metal and the correlation of test results with chemical composition underscored the importance of taking test results and analyses to the point where they will be useful for improvement of materials.

K. Iida
A. J. McEvily

1

A Review of Japanese Research on Materials Design for Severe Service Conditions

K. IIDA

Professor Emeritus, University of Tokyo; Department of Naval Architecture, Hongo 7-3-1, Bunkyo-ku Tokyo, Japan

ABSTRACT

This review deals with an outline of research projects, which are related to materials design for severe service conditions, and have been carried out in Japan under the sponsorship of the Science and Technology Agency.

The paper describes first an investigation on materials evaluation at cryogenic temperature, and then gives highlights of research works aimed at obtaining fundamental data about mechanical and metallurgical properties of degraded materials under severe service conditions.

1. INTRODUCTION

The Research Co-ordination Bureau of the Science and Technology Agency has every year sponsored several research projects for the purpose of promoting the advancement of science and technology. Two of the research projects carried out in these years are directly related to the present Seminar. The first one is a research project entitled 'Development of Superconductivity-Cryogenic Technology', which finished its first three year work period (1982–1984) at the end of March 1985. The second is a research project entitled 'Development of Reliability Assessment Techniques for Degraded Structural Materials', which completed its first term work of three years (1982–1985) at the end of March 1986.

The purpose of the first research project is to investigate basic manufacturing technology and practical application of superconductive and cryogenic materials and to develop super-fluidized helium circulation systems and baby cryogenic refrigerators of high efficiency. The final goal of the second research work was to establish reliable technical procedures for the prediction of the residual life of possibly degraded materials of structural components exposed to severe service conditions in thermoelectric power plants, chemical plants and others. The outline of the two research projects will be reviewed in the following.

2. RESEARCH PROJECT ON DEVELOPMENT OF SUPERCONDUCTIVITY-CRYOGENIC TECHNOLOGY

2.1. Research Subjects

The research items and budget in Japanese Yen in total for the first term of three years are as follows [1].

2.1.1. Development of high efficiency superconductive materials (161×10^6 *Jap. Yen*)
1. Manufacturing technology
   (a) Dissolution method.
   (b) Vaporizing method.
   (c) Powdering method.

2.1.2. Investigation of cryogenic materials (770×10^6 *Jap. Yen*)
1. Development of testing methods of cryogenic materials
   (a) Evaluation of notch toughness in high magnetic flux field.
   (b) Development of machine for multiple tension testing at cryogenic temperature.
2. Development and application technology of composite materials
   (a) Nonmetallic composite materials.
   (b) Carbon fibred composite materials.
   (c) Mechanical properties of FRP joint.

2.1.3. Investigation of cryogenic refrigerating system (81×10^6 *Jap. Yen*)
   1. Cooling system by super-fluidized helium.
   2. Development of high efficiency and baby refrigerator.
   3. Development of magnetic refrigerating system.

2.2. Main Results

Described in the following are abstracts of the results of each work of items 2.1.2.1(a) and 2.1.2.1(b), which are related to the present Seminar.

2.2.1. Evaluation of notch toughness in high magnetic flux field

In order to establish a method of evaluation of notch toughness of a nuclear fusion reactor material at cryogenic temperatures down to 4.2 K and under high magnetic flux field up to 10 T, a testing apparatus available to test a standard sized compact tension specimen was assembled on an experimental basis. A solenoid type superconductive magnet consisting of coils of Nb_3Sn and NbTi had a working room with 200 mm diameter and 540 mm height. The magnetic flux density in the working room was found to show a uniform flux distribution over the range of 100 mm diameter and 100 mm height. An acoustic emission measuring system and a fibrescope system developed for cryogenic use were utilized for the purpose of detecting crack initiation and observing fracture behavior, respectively. A specimen was axially loaded by a hydraulic cylinder of 10 ton capacity with the deformation rate ranging between 0·05 and 500 mm/min. Table 1 shows results for an Al-alloy A5083-O tested at 4 K and 77 K.

TABLE 1
Fracture toughness (J_{1C} in kJ/m^2) of Al-alloy A5083-O (H. Ogino *et al.*)

| Test temp. | Multiple tests | | AE method |
	Exp.	Ref.[a]	Exp.
4 K		25	22
77 K	20	21	20

[a]After Tobler, R. L. *et al.*, *ASME, J. Eng. Mat. Tech.*, Oct., 1977, p. 306.

2.2.2. Development of machine for multiple tension testing at cryogenic temperature

A system of fatigue testing machines and refrigeration circulation systems was developed for the purpose of accumulating fatigue data of materials of structural components for cryogenic temperature use. Principal specifications and technical problems investigated are shown in Table 2. Specimens of 304LN type stainless steel were fatigue tested by way of trial in liquid He using the developed testing systems, and the systems proved to be satisfactory. Fatigue tests are now under way on 304LN forged steel, high Mn steel and FRP composite material.

TABLE 2
Principal specifications of testing machines developed and technical problems
investigated (E. Izuno *et al.*)

Testing apparatus	Specifications	Problems
Bending fatigue of welded joints and strain cycling of base materials	Dynamic capacity: ± 10 t Reversed loading Cycling rate: up to 20 Hz Strain controlled	Strain measurement at cryogenic temp. Specimen holding tech.
Fatigue crack propagation rate	Dynamic capacity: ± 5 t Reversed loading Cycling rate: up to 40 Hz Computer controlled	COD measurement at cryogenic temp. Development of computer softs
Tension–torsion fatigue	Axial dynamic capacity: ± 5 t Dynamic torsional moment: 70 kg-m	High rigidity for higher response. Sealing to liquid He. Holding method of a specimen of composite material

3. RESEARCH PROJECT ON THE DEVELOPMENT OF RELIABILITY ASSESSMENT TECHNIQUES FOR DEGRADED STRUCTURAL MATERIALS

3.1. Research Subjects

The research items and budget in Japanese Yen in total for the first term of three years are as follows [2].

3.1.1. Elucidation of damage and fracture mechanism and development of prediction methods of residual life of structural metallic materials
1. Materials in high temperature environment (213×10^6 Jap. Yen)
   (a) Quantitative evaluation of damage and life by creep.
   (b) Quantitative evaluation of damage and life by creep–fatigue interaction
      — metallographical investigation;
      — dynamic investigation;
      — development of creep testing machine for very long duration test.

(c) Investigation of creep-fatigue crack propagation behavior
   — Experimental study;
   — Theoretical study.
(d) Investigation of metallographic structure change and embrittle-ment mechanism
(e) Evaluation methods of accumulated damage and retained strength of materials exposed to long duration of service environment
   — evaluation of accumulated damage;
   — evaluation of retained strength properties.
2. Materials in corrosive environment (139×10^6 Jap. Yen)
   (a) Evaluation method for stress corrosion cracking
      — electrochemical study of damage;
      — dynamic study;
      — development of residual life estimation method.
   (b) Quantitative evaluation of material damage due to corrosion fatigue.
   (c) Investigation of interaction effects by stress corrosion cracking and corrosion fatigue
      — metallographic study;
      — dynamic study.

3.1.2. Development of measuring techniques of material degradation and detection techniques of flaws
1. Measurement of damage of material (124×10^6 Jap. Yen)
   (a) Wave analysis in AE technique.
   (b) Ultrasonic inspection technique.
   (c) Magnetic inspection technique.
   (d) X-ray inspection technique.
2. Detection of crack (45×10^6 Jap. Yen)
   (a) Eddy current technique.
   (b) On-line crack detection system by AE technique.
3. Literature review on evaluation and detection techniques for estimation of material degradation and detection of flaws (6×10^6 Jap. Yen)

3.1.3. Service failure analyses and residual life evaluation systems
   1. Bridges (6×10^6 Jap. Yen)
   2. Ships and offshore structures (16×10^6 Jap. Yen).
   3. Thermoelectric power plants and chemical plants (6×10^6 Jap. Yen).

3.2. Main Results

The results of a few research works which seem to be closely related to the aims of the present Seminar are outlined in the following.

3.2.1. Quantitative evaluation of creep damage

Specimens of Cr–Mo–V, 2·25Cr–1Mo, SUS304, SUS316 and SUS321 steels were ruptured by creep after a time duration up to 100 000 h, and the rupture mode was analysed as a function of applied stress and test temperature. The results on a turbine rotor material of Cr–Mo–V steel are plotted in Fig. 1, which is divided into three categories of rupture mode that were investigated by scanning type and transmission type electron microscopes. Figure 2 shows the results for a boiler tube material of 304 stainless steel.

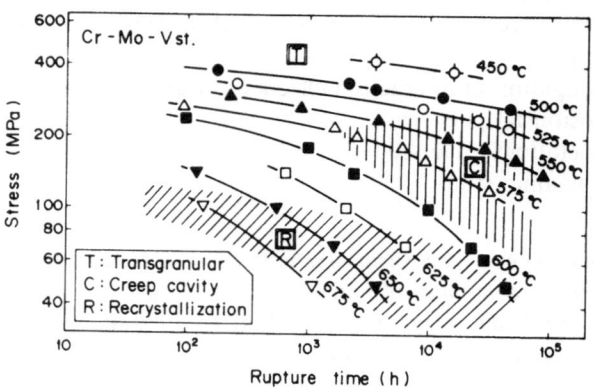

FIG. 1. Creep rupture mechanism diagram for Cr–Mo–V steel (S. Yokoi *et al.*).

With respect to Figs. 1 and 2, the characteristic change in density as a function of grain boundary cavity growth and microcracking was measured by a balance developed to have a high accuracy, in order to find a definite means of evaluating material degradation by creep. An examination by electron microscope showed the orientation of grain boundary cavities at a very early stage, that is less than 10% of the creep life. The change of density correlates well with the change in creep strain.

In Fig. 3, the half-value breadth by X-ray diffraction is plotted against a parameter which is a function of creep strain, time duration, applied stress and absolute temperature.

A method of evaluation of creep damage for the estimation of the

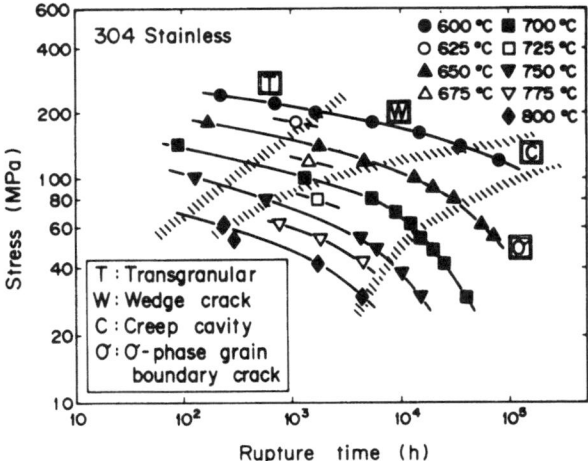

FIG. 2. Creep rupture mechanism diagram for 304 type stainless steel (S. Yokoi *et al.*).

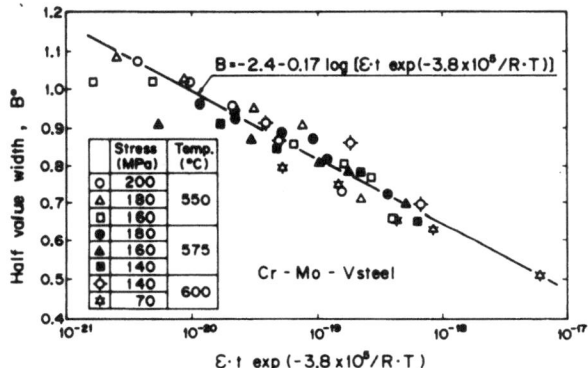

FIG. 3. Half-value breadth against creep rupture parameter (S. Yokoi *et al.*).

residual life to final creep rupture is suggested by Fig. 4 for a Cr–Mo–V steel. The figure indicates that the measurement of the half-value breadth and density change may be effective in the evaluation of the time spent in a certain creep condition.

3.2.2. Quantitative evaluation of damage by creep–fatigue interaction

For the purpose of obtaining a dynamical parameter, that is available for quantitative evaluation of material damage by creep–fatigue in-

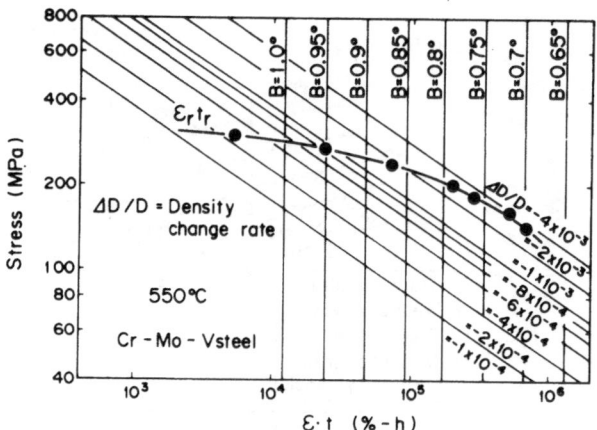

FIG. 4. Creep damage diagram for Cr–Mo–V steel (S. Yokoi *et al.*).

teraction in biaxial loading, a biaxial low cycle fatigue testing machine
was developed, which is capable of testing a cross-shaped specimen with
a biaxial strain ratio less than plus one and more than minus one, and at
an elevated temperature less than 650°C.

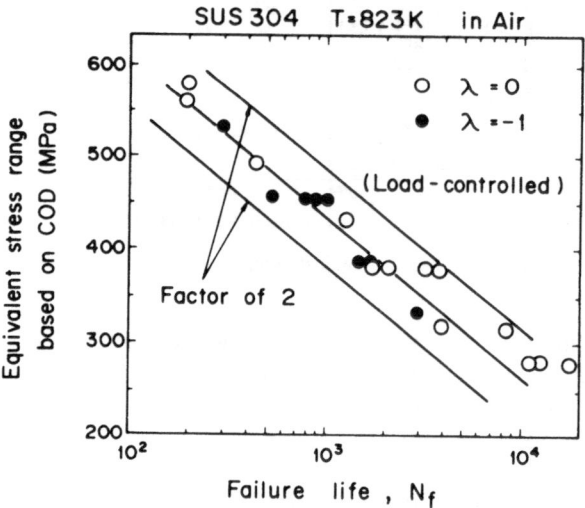

FIG. 5. Equivalent stress range based on COD vs. failure life under multi-axial fatigue
(M. Ohnami).

TABLE 3
Strain cycling fatigue test results of cross-shaped specimens (M. Ohnami)

Principal strain ratio	Failure life	Crack initiation life	Equiv. strain range (von Mises)	Equiv. strain range (COD)
1	1054	527	1·56%	1·40%
0	510	260	0·90%	1·17%
−1	1800	1150	0·90%	0·43%

As a fundamental investigation, elevated temperature and low cycle fatigue tests in uniaxial and biaxial loading conditions, creep tests and creep–fatigue interaction tests were carried out using 304 stainless and Cr–Mo–V steel specimens. An increase in micro-Vickers hardness of a 304 stainless specimen tested at 873 K showed an approximately linear relation against the increasing creep damage factor.

The results of push-pull fatigue tests (principal strain ratio = 0) and of reversed torsion (principal strain ratio = −1) were observed to form completely separated families, if both results are plotted in the relation of maximum principal stress range versus number of cycles to failure N_f. Both data fall into the same scatter band, if they are plotted in the relation of von Mises' equivalent stress range based on COD versus N_f, as shown in Fig. 5.

Shown in Table 3 are the results of biaxial, principal strain controlled low cycle fatigue tests of 1Cr–1Mo–0·25V steel specimens tested at 823 K and with a strain rate of 10^{-3}/s. As is evident from the table, the fatigue life is considerably influenced by the principal strain ratio.

3.2.3. Evaluation of accumulated damage due to duration of service environment

Pipes of 2·25Cr–1Mo, 9Cr–1Mo, 18Cr–8Ni and 18Cr–8Ni–Ti steels, which were used for a certain length of time in actual thermoelectric power plants, were examined by hardness check, tension tests at ambient and elevated temperatures, Charpy tests and creep rupture tests.

In the case of 9Cr–1Mo steel, no considerable changes in hardness and tensile properties were found as a result of service duration of 120 000 h, while the Charpy impact value at 0°C showed a remarkable decrease to 25% of the original value.

Figure 6 shows the results of accelerated creep tests of specimens cut out from a boiler tube, which was subjected to 50 000 hours in service.

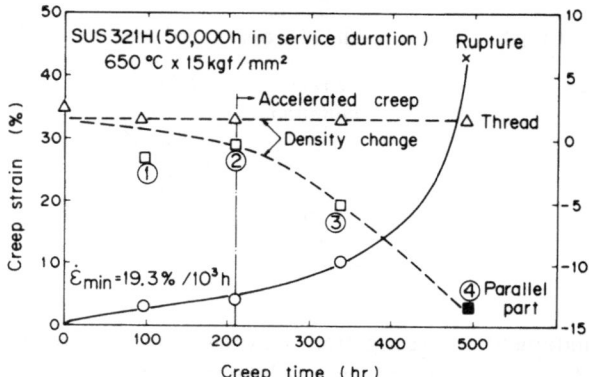

FIG. 6. Uniaxial creep strain as a function of creep time for SUS321H steel after service duration of 50 000 h in actual plant (K. Yoshikawa).

The observation that the density change agrees well with the creep strain change suggests the usefulness of creep cavity examination by the replica method.

The creep rupture master curves for the virgin and used materials of 9Cr–1Mo steel and 18Cr–8Ni–Ti steel are illustrated in Figs. 7 and 8. It is observed that the creep strength decreases as a result of service duration.

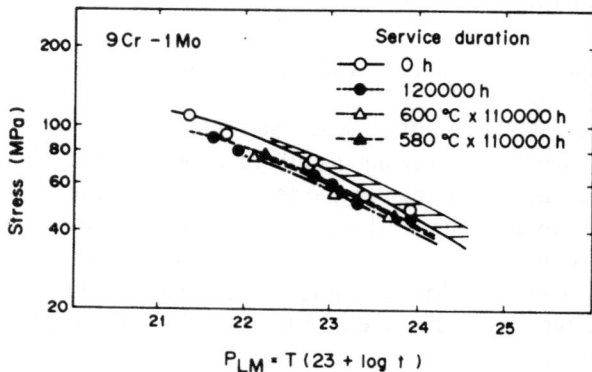

FIG 7. Creep rupture master curves for 9Cr–1Mo steel with a parameter of service duration (F. Masuyama *et al.*).

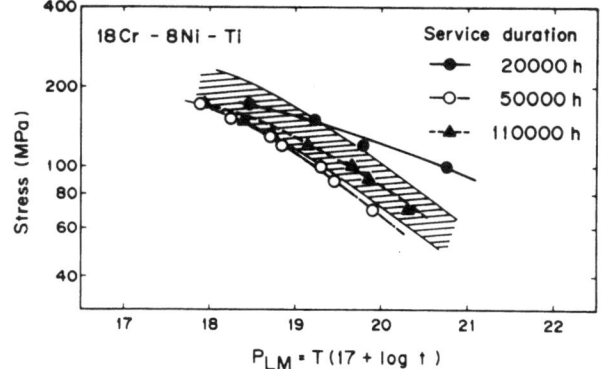

FIG. 8. Creep rupture master curves for 18Cr–8Ni–Ti steel with a parameter of service duration (F. Masuyama *et al.*).

3.2.4. *Service failure analysis of structural components in thermoelectric power and chemical plants*

An analysis of service failures in pressure vessels and piping systems in commercial thermoelectric power plants shows that 60% of failure incidents were experienced in pressure boundary components, while 30% and 10% were concerned with nonpressure boundary components and combustion system components, respectively.

The left-hand figure in Fig. 9 shows a classification of damaged

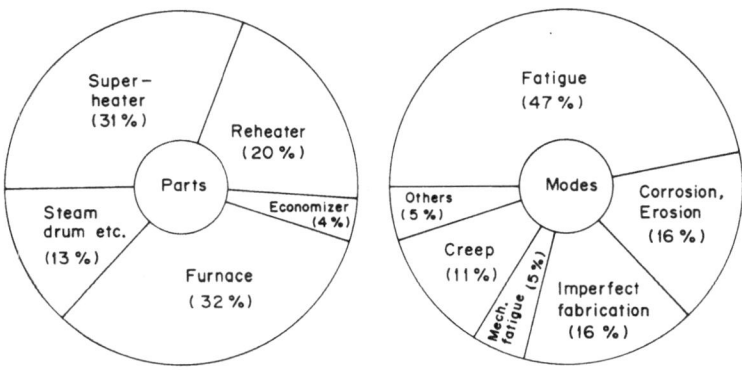

FIG. 9. Classification of damaged structural components and types of fracture in commercial power boilers.

components during service, and the right-hand one illustrates that the
fatigue failure has a majority in damage incidents. Most of the fatigue
failures occurred at the toe of butt-welded joints such as pipe-to-T and
pipe-to-reducer transition joints, which generate an additional stress
concentration due to the gross structural discontinuity.

A similar analysis was made of service failures during the past 5 years
from 1978 in chemical plants. The total number of incidents published is
315, which consists of 110 cases in oil refineries, 24 cases in petrochem-
istry, 23 cases in manure production, 25 cases in metal making and the
rest in other industries. Figure 10 illustrates classifications of damage

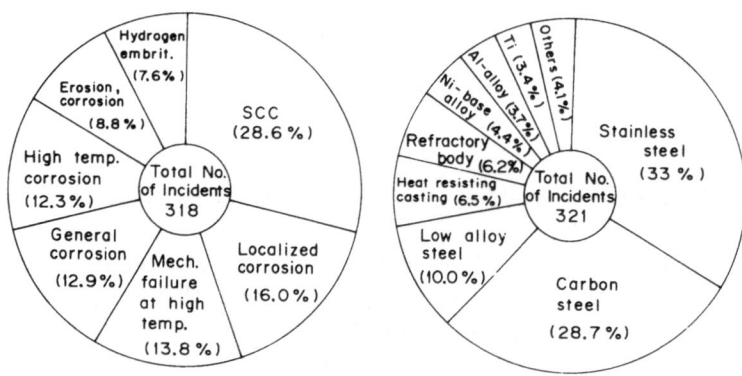

FIG. 10. Classification of damage types and damaged materials in chemical plants.

types and damaged materials in service failure cases including unpublished data. In the right-hand figure, the SCC includes cracking in
stainless steels by chloride, caustic potash and polythionic acid, cracking
in carbon steels by nitric acid and caustic potash, and cracking in brass
by ammonia. The local corrosion consists of pitting, crevice corrosion
and degalvanizing corrosion. And the oxidation, cementation, decarburization and molten salt corrosion are included in the high temperature
corrosion.

The highest frequency in terms of combination of material and failure
mode was experienced by the SCC in stainless steels (62 cases), which is
followed by local corrosion in rolled and cast steels (26 cases), the wear
or melt down of refractory bodies (20 cases), local corrosion in stainless
steels (19 cases) and general corrosion of rolled and cast steels (17 cases).

4. CONCLUDING REMARKS

Two research projects sponsored by the Science and Technology Agency were partly outlined. The second term of the projects is now in progress, and the results will be available by 1988.

REFERENCES

1. Research Co-ordination Bureau, Science and Technology Agency: Report on Development of Superconductivity-Cryogenic Technology, Dec. 1985 (in Japanese).
2. Research Co-ordination Bureau, Science and Technology Agency: Tentative Report of Investigation on Development of Reliability Assessment Techniques for Degraded Structural Materials, March 1986 (in Japanese).
3. Research Co-ordination Bureau, Science and Technology Agency: Report on Service Environment and Damage of Structural Materials in Thermoelectric Power Plants, Chemical Plants and Bridges, March 1984 (in Japanese).

2

Infringement of Defect Tolerant Castings on Traditional Wrought Superalloy Applications

J. K. Tien, J. C. Borofka and M. E. Casey

Center for Strategic Materials, Henry Krumb School of Mines, 918 Mudd Building, Columbia University, New York, NY 10027, USA.

ABSTRACT

The advancement of jet engine technology has traditionally given wrought superalloys eminence when design calls for both reasonable high temperature strength and dynamic fracture reliability. Despite the attractiveness of cast superalloys for their improved buy-to-fly ratios and higher strengthening γ' volume fractions, the expansion of polycrystalline castings into the more critical of the superalloy applications has been prevented by a lack of sufficient microstructural control in casting technology. Recently however, advances in casting technology are making possible the production of defect tolerant superalloy castings with improved fracture reliability. This new generation of superalloy castings now offers serious competition to wrought superalloys. Never before have wrought superalloys been so seriously threatened by their cast counterparts.

1. INTRODUCTION

Undoubtedly one of the worst environments encountered by a structural material is in the gas turbine or jet engine. High temperatures, high alternating stresses, and highly corrosive gases are superimposed to create a Dantean environment for these applications with human life liability.

The materials currently used for jet engine turbine disks and blades are the highly strengthened nickel-base superalloys. A listing of superalloy chemistries can be found elsewhere [1]. In some ways these materials can

15

be considered as a composite of about a dozen intentional elements, with two distinct phases—a solid solution matrix (γ) and a cuboidal array of precipitates of Ni_3Al (γ'). Monocrystalline superalloys are certainly designed to have just these two phases, since the need for the grain boundary strengthening carbide phases is precluded.

As any high strength material must, superalloys face the trade-off between strength and ductility or toughness, with strength winning out at high temperatures. Superalloys generally possess low fracture toughness and ductilities on the order of 10%. Accordingly, they are sensitive to defects, resulting in low and erratic fracture mechanics properties [2]. Wrought superalloys generally have better fracture and fatigue properties than cast superalloys, since defects and large grains are broken up and porosity is healed during hot working processes. Since the advent of gas turbine technology, wrought or hot worked superalloys have been the material of choice in critical structural applications requiring dynamic fracture reliability, such as turbine disks. Cast superalloys, including the monocrystals, have been used for turbine blades with their higher temperatures where a coarse grain size is needed to benefit creep resistance.

Cast superalloys have always been attractive from an economic and materials conservation standpoint. For example, a 400 lb ingot is required to produce a 22 lb finished wrought disk whereas only 88 lb of precision cast material is required for the same finished disk [3]. Despite this advantage, cast superalloys, as mentioned, had been hampered by poor microstructural control which prevented the manufacture of, for example, a reliable cast disk.

Today the situation is changing. Superalloy casting technology has acted to overcome the traditional problems inherent in superalloy castings—porosity, high inclusion contents, and large and nonuniform grain sizes—through process innovations. No doubt monocrystalline and directionally solidified superalloys have been a small although critical infringement on wrought superalloy applications, but castings are now being used for much larger components than just turbine blades.

2. PROBLEMS WITH CASTINGS AND RECENT SOLUTIONS

2.1. Cleanliness

Superalloys are highly defect sensitive due partly to the pre-eminence of planar slip. Low cycle fatigue properties are especially plagued since

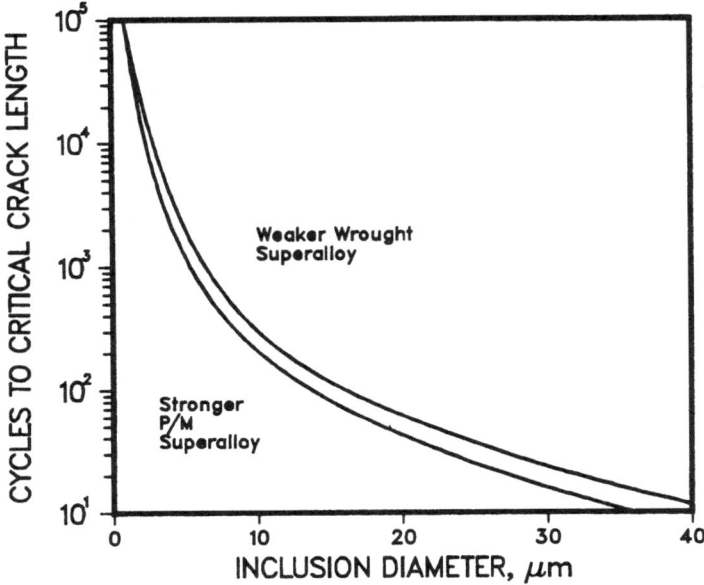

F_IG. 1. Calculated dependence of fatigue life on inclusion diameter for two different superalloys. One is a relatively strong P/M superalloy and the other is a relatively weak wrought superalloy [2].

impurity inclusions (carbides, carbonitrides, ceramics, oxides, etc.) can act both as crack initiation sites and crack propagation paths [4]. Figure 1 shows a calculated exponential drop in fatigue life for two superalloys as inclusion diameter increases. Stronger wrought superalloys, such as the P/M alloys are even more sensitive to inclusion contents than are weaker wrought superalloys. Figure 2 shows a similar trend for K_{IC} as a function of the number of inclusions per grain. This is the number density of inclusions in the 'dirty' areas of the material, since inclusions do not occur as a homogeneous distribution. Although both diameter and number density of inclusions are related to volume fraction, the size of inclusions by itself must be considered, especially near the surface where oxidation effects related to inclusions are important. Figure 3 represents actual and typical experimental data showing the sensitivity of a typical superalloy to inclusions [5].

Hot and redundant working of superalloys always disperses the inclusions and makes them smaller, less shapely, and more mechanically inert. Accordingly, castings were not considered for fatigue-limited

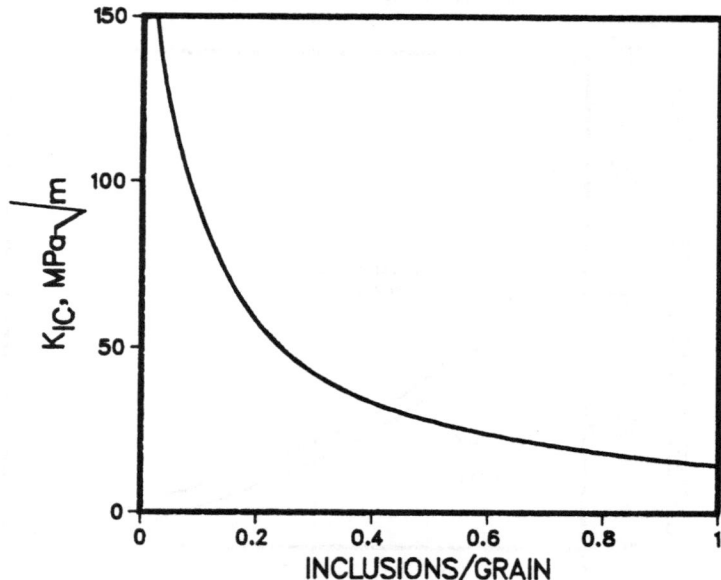

FIG. 2. Calculated dependence of K_{1C} on inclusion number density [2].

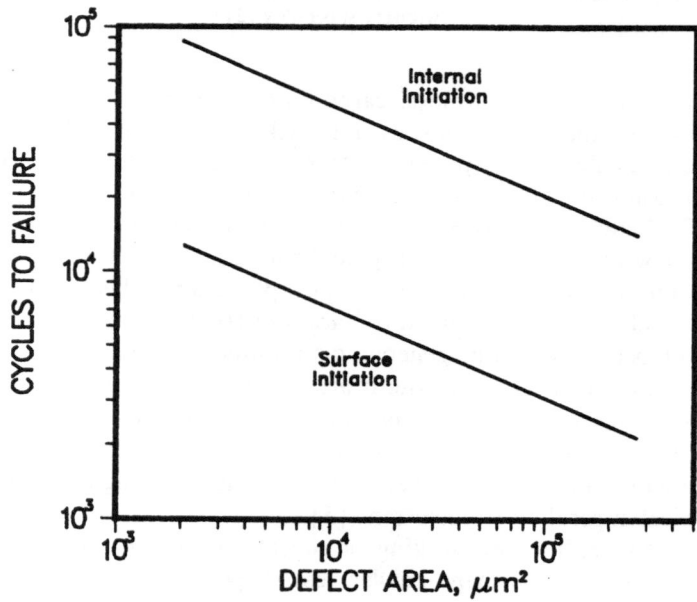

FIG. 3. The effect of defect size and initiation site on LCF life of a P/M superalloy at 538°C [5].

engine applications until after a superalloy was cleaned of the inclusions, many of which are spallations from the ceramic crucibles used during the initial vacuum induction melting (VIM) process. Recently, the process innovation of electron beam processing has resulted in promisingly clean superalloys.

Electron beam copper hearth remelting (EBCHR) of superalloys has yielded cleaner metal than ever available before. The superalloy is melted in a water-cooled copper hearth and electron guns are used to locally heat and somewhat agitate the melt so that inclusions may float to the top or sink to the bottom of the hearth to be skimmed off later by the use of skimmer dams or weirs [6]. Besides EBCHR, which is still under development, commercially cleaner alloys are also resulting from the electrode slag remelt process (ESR). The slag pool at the top of the ESR mold can essentially getter or collect much of the inclusions [6]. Foam-like filters used during VIM process also help to eliminate ceramic inclusions [7]. Table 1 shows an example of the reduction in both oxide size and content of electron beam refined VIM/VAR material and ESR refined VIM material [8]. It is important to note that the incentive or payoff for the production of clean metal is not linear. Much room for improvement remains.

TABLE 1

Ceramic inclusion sizes and contents for VIM/VAR processed material and ESR and EBCHR refined VIM material

Alloy and process	*Largest oxide particle (μm)*	*Total oxide content (ppm)*
A high γ' fraction P/M superalloy		
VIM/VAR	163 ± 64	2.8 ± 2.3
ESR	102 ± 71	0.6 ± 0.2
EBCHR	24 ± 3	0.2 ± 0.02
A wrought iron-nickel base superalloy		
VIM/VAR	281 ± 218	1.4 ± 0.9
ESR	68 ± 12	0.6 ± 0.4
EBCHR	37 ± 28	0.1 ± 0.1

2.2. Porosity

Once a superalloy is 'free' of inclusions, porosity then becomes the next weak link. Porosity occurs in castings due to the change in volume associated with the liquid–solid phase transformation and fine pores are often found between the dendrites in castings. Casting pores have long been recognized as adversely affecting the mechanical behavior of the barely ductile superalloys [9,10]. Fatigue cracks nucleate at voids, decreasing the LCF life of a porous alloy relative to that of fully dense material. In addition, levels of porosity vary from casting to casting, decreasing the reliability with which fatigue lives can be predicted and used in life predictive designs. Porosity is also detrimental to both tensile and creep ductility since the link up of pores and voids provides an easy path for a crack to follow as it propagates to failure. This is especially the case for the polycrystalline superalloys.

Hot Isostatic Pressing (HIP) is now routinely used as a post-solidification treatment to eliminate porosity in cast superalloys for advanced applications [11]. Figure 4 demonstrates the healing of porosity by HIP through a side by side comparison of the macrostructures of an as-cast and a cast + HIP superalloy. The mechanism by which

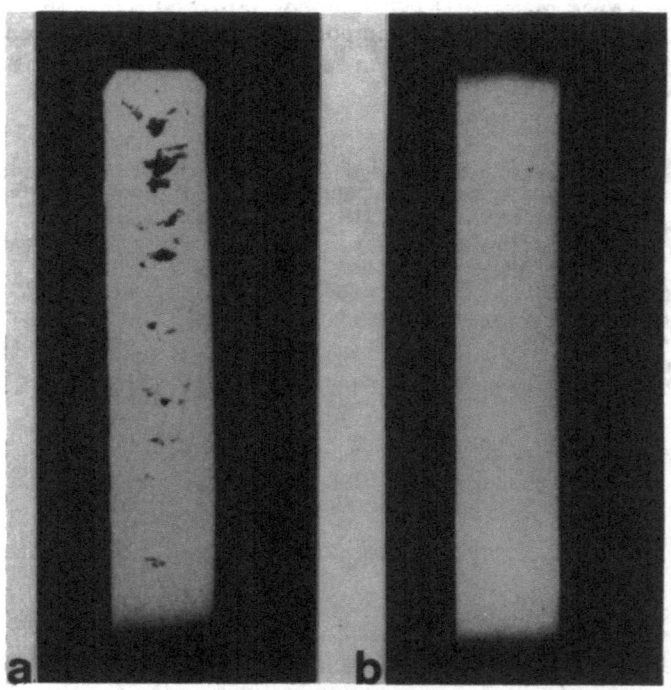

FIG. 4. Macrostructures of (a) as-cast and (b) cast + HIP superalloy bar.

TABLE 2

Cycles to failure for a typical cast superalloy. RO (run-out) indicates that the test was terminated after 1.0×10^7 cycles. Start indicates that the test failed during the first cycle

Mean stress	70			100			120		
Alternating stress (ksi)	5	10	20	5	10	20	5	10	20
As-cast	RO	RO	RO	1.0×10^7 1.6×10^4	9.0×10^6	Start 2.2×10^4	9.4×10^4	6.8×10^4	Start Start 1.1×10^4
Cast + HIP	RO	RO	RO	RO	RO	1.7×10^4	1.1×10^6	3.0×10^5 1.1×10^6	2.3×10^4

casting pores are eliminated has been postulated to be a two step process. The first step requires a HIP temperature and pressure combination which exceeds the yield stress of the material resulting in void collapse due to the deviatoric stress states about the voids. After void collapse or convolution has occurred, a temperature and hold time must be maintained which will result in the eventual sintering together of the void surfaces. The extent of void closure therefore is extremely sensitive to the HIP parameters of temperature, time and pressure. However, once HIP densification parameters are established the process appears to be insensitive to the size or shape of the casting defect or its location in the material (except for surface connected voids), and void closure appears to be uniform [10].

HIP has been shown to increase tensile and creep ductility as well as the LCF life of many cast superalloys. The scatter in LCF life is also reduced as a result of HIP. Table 2 shows the effect of HIP on the LCF life of a typical cast superalloy [12]. The number of cycles to failure for the HIP alloy is compared to that of the as-cast alloy. The HIP material demonstrated performance superior to that of the as-cast alloy under all fatigue loading conditions studied.

The use of HIP to correct for casting porosity does have some limitations, however. Surface pores can not be closed by this method since they are exposed to the high pressure gases of the HIP press. In addition, the temperature window over which this process can be conducted effectively is narrow for superalloys. If the HIP temperature is too high, grain growth can occur as γ^1 precipitates begin to dissolve. If, on the other hand, the temperature is too low it may not be possible to exceed the yield stress of the material—a most necessary condition for complete void collapse—due to limitations of the attainable pressure in the HIP press. It should also be kept in mind that the improvement in mechanical properties brought about by the elimination of internal voids through HIP always depends upon the starting porosity of the material.

2.3. Grain Control

Microstructural coarseness or nonuniformity of grain size are the predominant features of conventionally investment cast material. The columnar grains and chemical inhomogeneities associated with conventional castings greatly reduce the fatigue performance and reliability of cast turbine hardware.

Today, new casting techniques in various stages of development exist which produce very fine grained and homogeneous microstructures.

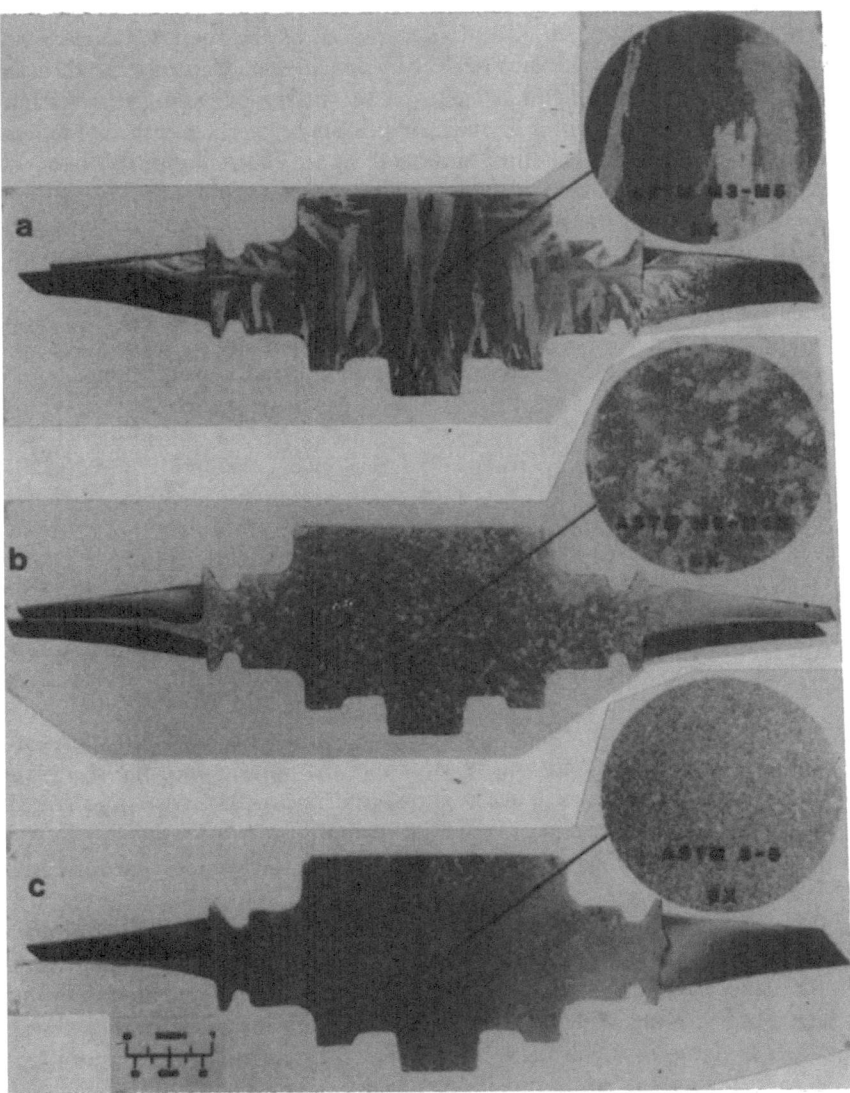

FIG. 5. Macrostructure of (a) a conventionally cast and (b), (c) fine grain precision cast integral wheels.

Most details of these casting processes are proprietary, but three general principles for controlling microstructures are used [13]: mechanical vibration, thermal control, and inoculation of the melt with seeds to promote nucleation of grains. The last method is rarely used with superalloys due to their high inclusion sensitivity. Mechanical vibration acts to break up dendrites as they grow to increase the number of nuclei in the melt, resulting in a finer grain size [11]. Thermal control involves the arrangement of heat sinks and insulation around the mold to control thermal gradients, the reduction of melt superheat time and temperature to retain the refractory metal carbides in the melt to serve as nucleation sites and the reduction of the pour temperature to increase the cooling rate in the mold thus restricting grain growth.

The first generation of fine grain casting techniques involved continuous agitation of the mold during solidification. The castings produced by this technique exhibited an equiaxed dendritic structure with grain sizes ranging from ASTM M9 to M13, close to ASTM 0. Figure 5 shows the refinement in macrostructure of fine grain castings compared to conventionally cast material.

The newest generation fine grain casting process involves a very rapid solidification rate thereby producing a cellular structure. It has long been known that cellular solidification occurs on a much finer scale than dendritic solidification and greatly restricts the propensity for microsegregation [14]. The grain size produced by the cellular solidification type process ranges from ASTM 3 to 5 and is comparable to that produced by wrought processes.

The benefits of the fine grain casting processes include increased uniformity in properties, chemistry and precipitate morphology over conventional castings. Figure 6 shows the improvement in mechanical properties [15]. Fine grain casting techniques still result in microporosity, so that HIP is still necessary for soundness and performance reliability.

Fine grain casting techniques are being used to produce complex geometries such as integral wheels as well as fine grain cast disks, blades and vanes for small turbine engines (Fig. 7). Another application is in the fine grain casting of forging preforms. Fine grained and homogeneous billets as large as 500 lb have been produced. Several commercial alloys have been cast and used directly as forging stock. Successful forgings on these materials have been conducted using upset ratios as high as 80% [16]. Fine grained casting techniques could eventually replace primary

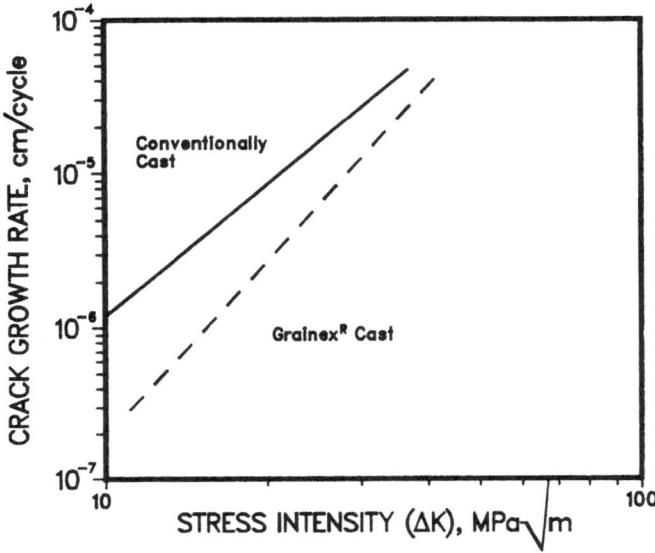

FIG. 6. A comparison of crack growth rates for a conventionally cast and a fine grain cast superalloy at 427°C [15].

FIG. 7. Typical precision cast superalloy integral wheels.

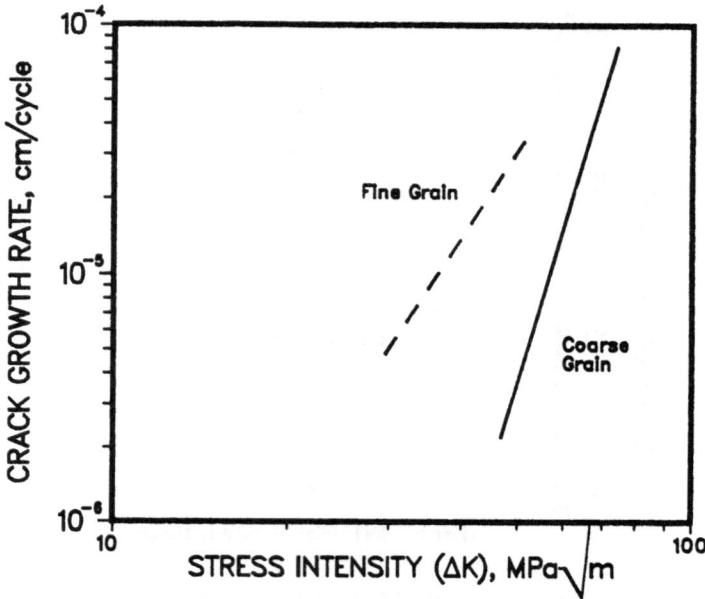

FIG. 8. The effect of grain size on crack growth rate for a typical wrought superalloy. The fine grain size is ASTM 9 and the coarse grain size is ASTM 3 [17].

hot working methods as a means of refining superalloy microstructures to the point where they are suitable for forging.

On the subject of grain size, a possible dilemma needs to be discussed. With the presumption of fatigue lives in turbine disk applications being crack initiation controlled, wrought turbine disks have been traditionally fine-grained, and have performed well over the years. Naturally, the fine grain size and the attendant Hall–Petch effect also resulted in high strengths without any loss of the already marginal toughness. Recent work [17,18] shows that crack propagation resistance, on the other hand, may be enhanced by having coarse grains (Figs. 8 and 9). A dilemma may occur if fracture mechanics design policies are recommended as the life-design protocol for advanced engines. Based on life-designs from da/dn curves, and a mandated presumption of crack propagation being the life controlling mode, coarser grained disks will result. However, what if initiation and not propagation is really the most common rate controlling step? Should one trade a quarter of a century of experience for the privilege of applying fracture mechanics design concepts in this case?

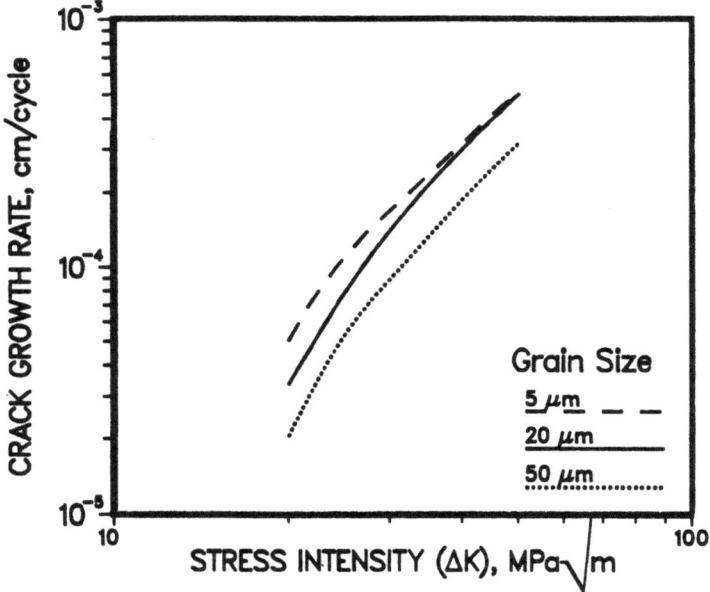

FIG. 9. The effect of grain size on crack growth rate for a typical P/M superalloy in air at 650°C [18].

3. DISCUSSION AND CONCLUDING REMARKS

During recent years, many entrepreneurial companies in the USA have in great measure integrated the science and technology of clean alloy processing, porosity elimination and grain size control to result in precision casting of microstructurally sound and dynamically reliable cast-to-shape turbine components.

An exciting development that keynotes the rapid advances made by the precision near-net-shape superalloy casting industry is their current ability to commercially compete even with the sheet products industry. Thin walled superalloy shrouds or casings are now routinely cast to near-net-shape (Fig. 10). This would not have been possible without the availability of clean superalloys, since some inclusions in conventional castings are on the order of the thickness of the shrouds.

Near-net-shape superalloy castings also have the potential of competing well against near-net-shape P/M superalloy components. This is because making powders and then consolidating them to shape will

FIG. 10. A thin-walled precision cast turbine exhaust case.

always be expensive relative to any cast-to-shape process. The savings gained from precision cast-to-shape process is loudly advertised by examples where components made from this process have replaced previously wrought, heavily machined and assembled components (see Fig. 11).

At this juncture of time, the cast-wrought replacements are still more with respect to the less critical components in engines, and many are in

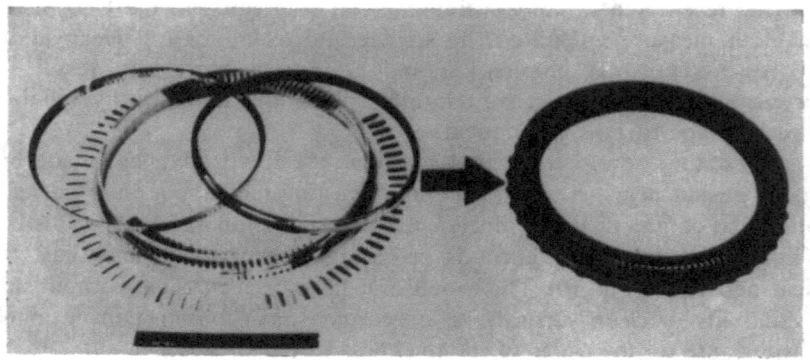

FIG. 11. A comparison of a wrought and a precision cast compressor vane and shroud. The wrought component requires the assembly of 125 separate parts while the precision cast component is a single unit.

non-flight propelling gas turbine engines like the auxiliary power units on board aircraft. However, as the cast components gather rotating time and experience, they also gather credibility. Accordingly, we believe that it is only a matter of time before, say, the critical jet engine disks are precision cast-to-shape. Interestingly, ESR cast-to-shape jet engine disks are a reality in Chinese (PRC) jet engines [19].

The question is whether the casting gains are triggering constructive defensive and/or offensive reactions from the wrought superalloys industry. The answer is yes. Already the forgers are capitalizing on the fine grain and clean metal innovations integrated by the superalloys precision casting industry [16]. Cast fine grained preforms can be much cheaper than fine grain powder metallurgy preforms. That fine grains will result in superplasticity of the superalloys during hot working is well known [20–22]. Fine grains can also result from vacuum arc remelt processes. The primary example of this is the VADER™ process (Special Metals Corporation Trademark) [23]. Enhanced workability through grain control and through solid-state solvi control [24] is somewhat removing the strength and composition restrictions with respect to just which superalloys are workable. Castings, of course, do not have the limitations

FIG. 12. Bronze castings, unearthed at Ban Chiang, Thailand, circa 2000 BC.

on what compositions are or are not castable, since superalloys (strong or weak) share similar melting ranges and castability.

The emergence of castings (and not just the fabled monocrystals) for advanced applications in the extremely adverse turbine engine environments shows that once there is scientific understanding and articulation of the problems, technological solutions will result when there are proper economic incentives. However, lest we become too proud with the achievements discussed above, we offer for contemplation Fig. 12. These are precision bronze castings from the Orient, with geometries similar to integral turbine wheels, dated circa 2000 BC (1500 Before Confucius) [25]. The circle closes.

ACKNOWLEDGEMENTS

The authors acknowledge NSF, AFOSR, NASA, ONR, Westinghouse, and CBMM for support of some of the work cited in this paper. The authors thank J. Mihalison of Howmet, M. Gell of United Technologies, and J. Snow of Garrett Turbine Engines for providing figures used in this paper.

REFERENCES

1. *Inco Metals Handbook*, 1977.
2. TIEN, J. K. and SCHWARZKOPF, E. A. *Proc. Conf. on Electron Beam Melting and Refining, State of the Art 1983*, Reno, NV, Nov. 1983, (R. Bakish ed.) Bakish Materials Corp., Englewood, NJ, 1983, p. 6.
3. Metal Powders Report, **35** (1980), 507.
4. SHAMBLEN, C. E. and CHANG, D. R. *Met. Trans.*, **16B** (1985), 775.
5. CHANG, D. R., KRUEGER, D. D. and SPRAGUE, R. A. *Superalloys 1984. Proc. of the Fifth Int. Conf. on Superalloys*, Seven Springs, PA, Oct. 1984 (M. Gell et al. eds), AIME, Warrendale, PA, 1984, p. 245.
6. TIEN, J. K. and JARRETT, R. N. *Proc. of the Conf. on Electron Beam Melting and Refining, State of the Art 1984*, Reno, NV, Nov. 1984 (R. Bakish ed.), Bakish Materials Corp., Englewood, NJ, 1984, p. 321.
7. MAURER, G. E. *Proc. of the Japan-U.S. Seminar on Superalloys*, Fuji Kyoiku Kenshusho, Susono, Japan, Dec. 1984, (R. Tanaka et al. (ed.), The Japan Institute of Metals and The High Temperature Alloys Committee TMS of AIME, p. 409.
8. SHAMBLEN, C. E., CHANG, D. R. and CORRADO, J. A. op. cit., *Superalloys 1984*, p. 509.
9. TIEN, J. K. and GAMBLE, R. P. *Mater. Sci. Eng.*, **8** (1971), 152.

10. WASIELEWSKI, G. E. and LINDBLAD, N. R. *Superalloys-Processing, Proc. of the Second Int. Conf. on Superalloys*, Seven Springs, PA, Sept. 1972, MCIC, Columbus, OH, 1972, p. D1.
11. EWING, B. A. and GREEN, K. A. *op. cit., Superalloys 1984*, p. 33.
12. SCHWARZKOPF, E. A. and TIEN, J. K. Unpublished Research, Columbia University, 1985.
13. WOULDS, M. and BENSON, H. *op. cit., Superalloys 1984*, p. 3.
14. TIEN, J. K. and GAMBLE, R. P. *Proc. of the Int. Conf. on the Strength of Metals and Alloys*, ASM, Vol. 3 (1970), p. 1037.
15. Howmet Turbine Components Corp., Technical Bulletin TB3000.
16. BRINEGAR, J. R., NORRIS, L. F. and ROZENBERG, L. *op. cit., Superalloys 1984*, p. 23.
17. ANTOLOVICH, S. D. and JAYARAMAN, N. *Fatigue Environment and Temperature Effects*, Sagamore Army Materials Research Conf. Proc., Bolton Landing, NY, July 1980 (J. J. Burke and V. Weiss, eds), Plenum Press, New York, p. 119.
18. GAYDA, J. and MINER, R. V. *Met. Trans.*, **14A** (1983), p. 2301.
19. CHANG, J. F., HU, G. P. and GAO, L. *Superalloys 1980. Proc. of the Fourth Int. Conf. on Superalloys*, Seven Springs, PA, Sept. 1980 (J. K. Tien *et al.* eds), ASM Metals Park, OH, p. 245.
20. EDINGTON, J. W., MELTON, K. N. and CUTLER, C. P. *Prog. Mater. Sci.*, **21** (1976), 61.
21. HOWSON, T. E., COUTS, W. H. and COYNE, J. E. *op. cit., Superalloys 1984*, p. 275.
22. BOROFKA, J. C. and TIEN, J. K. to be published.
23. BOESCH, W. J., TIEN, J. K. and HOWSON, T. E. *J. Metals*, **122** (1982), No. 5, 49.
24. JARRETT, R. N. COLLIER, J. P. and TIEN, J. K. *op. cit., Superalloys 1984*, p. 457.
25. RAYMOND, R. *Out of the Fiery Furnace: The Impact of Metals on the History of Mankind*, MacMillan, South Melbourne, 1984, p. 39.

3

Welding of Materials for Use in Severe Service Conditions

F. MATSUDA

Welding Research Institute, Osaka University, Mihogaoka 11–1, Ibaraki, Osaka 567, Japan.

ABSTRACT

This author has introduced state-of-art welding technology to several construction materials used in severe environmental service in Japan.

The materials involved are HY-type high yield strength steel for submarine, Invar (Fe–36%Ni) metal for LNG containers, advanced Cr–Mo steel for heat-resistant vessels, fully austenitic and duplex stainless steels for chemical plants, aluminum alloys for vacuum vessels and molybdenum for lining metal of nuclear fusion reactor vessels. Welding difficulties in the above materials have been mainly attributed to cold and hot crackings during welding and to the lack of ductility in welding joint.

1. INTRODUCTION

Efforts in many fields are being expanded to develop new materials, which are used in severe environment. As a result, new utilizations of conventional materials are frequently being proposed. Most of these materials are fabricated by the process of industrial welding. Welding technology for these newly developed materials is also being investigated as new problems are encountered.

The author has introduced the state-of-art of welding technology in Japan for use of these materials in severe conditions. This report will discuss the metallurgical aspects of weld cracking during fabrication and weld ductility after welding.

2. GENERAL TREND IN WELDING TECHNOLOGY DEVELOPMENT

Materials which have been specifically developed for welding recently with typical application in industrial use, are collectively listed in Table 1. Also principal severe conditions encountered are marked.

The principal welding methods used for these materials are listed in Table 2. The use of gas tungsten arc welding (GTAW) and electron-beam welding (EBW) in fusion welding processes and resistance and brazing processes have prevailed recently. Narrow gap welding (NGW) is gradually being applied in the conventional arc welding field.

In the following, the author will briefly mention some unresolved welding problems encountered during and after welding.

(1) Hot cracking: hot cracking is divided into two categories of supersolidus and subsolidus cracking. Note that here the solidus means the real solidus under which no liquid exists between dendrites. These crackings are classified as,

Hot cracking
$\begin{cases} \text{Supersolidus crack:} & \begin{cases} \text{Solidification crack (in weld metal)} \\ \text{Liquation crack (in HAZ and/or} \\ \quad \text{weld metal)} \end{cases} \\ \text{(caused by low} \\ \quad \text{melting liquid film)} \\ \text{Subsolidus crack:} & \text{Ductility-dip crack (in weld metal} \\ \text{(caused by precipitation} & \quad \text{and or HAZ)} \\ \quad \text{and/or sliding of grain boundaries)} \end{cases}$

Ductility changes during and after solidification are illustrated in Fig. 1. Some typical sketches of hot cracks which are often observed are shown in Fig. 2. These cracks are intergranular cracks.

(2) Cold cracking: most cold cracking in high strength steel weldment is hydrogen-induced-delayed type fracture. This crack susceptibility increases with increase in hardness of the HAZ or weld metal, in diffusible hydrogen and in restraint or external stress. Mechanism of cold cracking is illustrated in Fig. 3. Classification according to mechanism for preventing cold crack susceptibility is; (a) decreased hardenability of steel by reducing carbon content and cooling rate after welding; (b) decreased diffusible hydrogen content using ultra low hydrogen electrode, higher preheating temperature and atmosphere of argon or vacuum; and (c) decreased restraint stress by improving joint configuration.

Typical cold cracks experienced are sketched in Fig. 4. The cold cracks are sometimes intergranular but most often transgranular cleavage with dimple pattern.

(3) Reheat cracking: in order to release the welding residual stresses,

TABLE 1

Materials used under severe service conditions with welding technologies

Material	Severe service condition						Industrial example
	Low temp.	High temp.	High vac.	High press.	High strength	Special	
(1) HY-type steel				◎	◎		Submarine shell, pressure vessel
(2) Maraging steel				◎	◎		Rocket casing
(3) High Mn steel					○	○	Non-magnetic structure
(4) High Cr–Mo steel						○	Reactor tube, chemical vessel
(5) Ferrite/M stainless steel		○	○			○	Fusion reactor vessel, chemical plant
(6) Austenitic stainless steel		○	◎			○	Low temp., vacuum, chemical plants
(7) Invar	◎						LNG vessel wall
(8) Ni-heat alloy	◎	◎				○	JT-60 vessel wall, engine, vessel
(9) Al alloy	◎		◎			○	Low temp. vessel, vacuum vessel, aircraft
(10) Ti alloy				◎		○	Submarine shell, aircraft, chemical plant
(11) Ta metal						◎	Chemical plant
(12) Mo metal		◎					JT-60 wall, furnace
(13) Ceramics		◎				○	Ultra high temp. material, chemical plant

TABLE 2

Main welding processes used

	Fusion welding[a]							Pressure welding			Brazing
	Manual (MMAW)	Submerged (SAW)	MIG[b] (GMAW)	MAG[b] (GMAW)	TIG[b] (GTAW)	Plasma arc	Electron beam (EBW)	Resistance	Friction	Explosion	Brazing diffusion TLP bonding
(1) HY-type	O			O	O		O				
(2) Maraging					O		O				O
(3) High Mn	O			O		O	O	O	O		
(4) High Cr–Mo	O	O		O							
(5) Fe/M Steel	O	O			O		O	O	O		
(6) Aust. Steel	O	O		O	O		O	O	O		O
(7) Invar	O			O	O			O			
(8) Ni-heat Alloy	O	O	O	O	O	O	O	O	O		O
(9) Al Alloy			O		O	O	O	O	O	O	O
(10) Ti Alloy			O		O	O	O	O	O		O
(11) Ta Metal			O		O		O	O	O	O	O
(12) Mo Metal					O		O	O	O		O
(13) Ceramics								O	O		O

[a]Include narrow gap welding (NGW).
[b]Include pulsed arc welding.

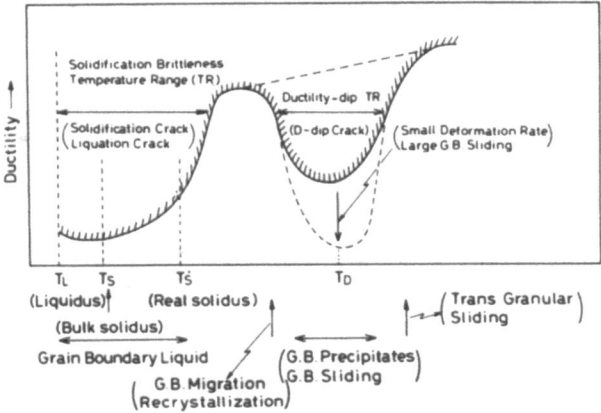

FIG. 1. Ductility curve and hot cracking susceptibility during cooling after welding.

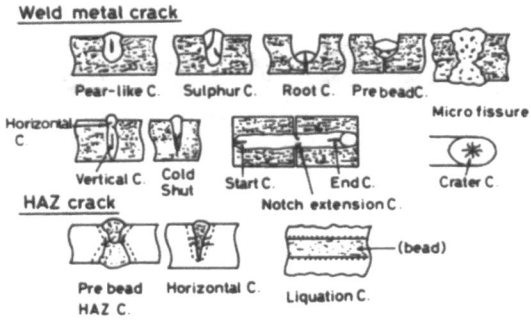

FIG. 2. Typical examples of hot crack experienced.

the welded joints of some metals and alloys are heat-treated in a furnace after fabrication. Reheat cracking may occur in the weldments of Cr–Mo steels, some austenitic stainless steels and heat resistant nickel alloys after heat treatment. In order to prevent reheat cracking improved purity of material, joint configuration and heat treatment are used for special metals and alloys.

(4) Porosities: porosities are very often seen in weld metal of some materials. Porosities are mainly caused by the following gases; N_2, H_2, CO and CO_2 for steel, N_2 and Ar for stainless steel, H_2 for aluminum, N_2 for nickel, O_2 and H_2 for titanium, and O_2 for tantalum and molybdenum.

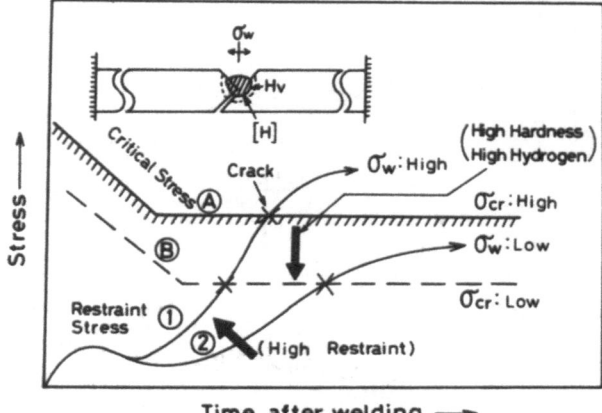

FIG. 3. Mechanism of cold cracking (hydrogen-delayed crack).

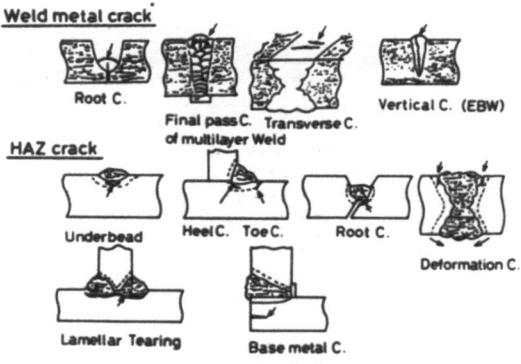

FIG. 4. Typical examples of cold crack experienced.

(5) Brittleness by oxidation and nitrification: reactive and refractory metals, such as Ti, Zr, Hf, V, Nb, Ta, Cr, Mo and W, are easy to react with O_2 and N_2 in air at high temperature. Therefore, brittle mechanical properties are observed in weld metal and HAZ of these metals. Brittleness can be inhibited by welding in complete argon shielding or in an argon atmosphere, or in a vacuum chamber.

(6) Coarse-grained brittleness: coarse-grained brittleness occurs in weld metal and HAZ of molybdenum, tungsten, nickel, aluminum and

titanium and their alloys. Reducing weld heat input is the only technique for preventing this brittleness, but there is some limitation in applying this process.

(7) Ductility of weld joint: ductility at low or high temperature is required for welded joint of materials used in severe environment. Increased ductility in joints has been required recently to assure the safety of products. Ductility in joints subjected to long-time exposure of radioactive environment has been required recently for some special materials.

(8) Static and dynamic strength: ultra low or high temperature strength is also required for welded joint of some materials. Static and dynamic strength in special environment should be evaluated.

(9) Stress corrosion cracking (SCC) and corrosion characteristics: for ultra high strength steel and some alloys, SCC feature is very important in industrial use. Corrosion in a special environment must be considered for special materials.

3. ADVANCED WELDING TECHNOLOGY FOR INDUSTRIAL MATERIALS

3.1. HY-type High Strength Steel

High strength steel of HY110 to HY180 has been developed for submersibles in Japan. Submersible 'Sinkai 2000' which can dive 2000 m deep was fabricated in 1981 using HY-130 (0·08C–5Ni–0·5Cr–0·5Mo–V) high strength steel.

3.1.1. Welding problems

(a) *Cold cracking* [1]. In general, diffusible hydrogen in weld metal in mercury measurement, which has a detrimental effect on cold cracking, can not be reduced less than 2 ml/100 g Fe on an average in manual metal arc welding (MMAW), submerged arc welding (SAW) and gas metal arc welding (GMAW), but can be reduced to 0·1 ml/100 g Fe in GTAW with argon shielding. Usually steel of lower grade than HY-110 can be welded with about 100°C preheating by MMAW, SAW and GMAW, but more grades higher than HY-130 are difficult to weld soundly even with 100°C preheating. Recently, HY-130 steel was soundly welded with GTAW of low hydrogen level. 'Sinkai 2000' was completely circumferentially welded with GTAW.

Increasing strength of base and weld metal shows increasing crack susceptibility in general. However HY-type steel HY-180 containing 10% Ni shows less crack susceptibility than HY-130 containing 5%Ni. This is due to the reduction of restraint stress of the welded joint by transformation expansion to martensite at lower temperature during cooling after welding. Therefore, among Hy-type steels HY-130 steel is most susceptible to cold cracking.

(b) *Hot cracking.* Crack susceptibility to hot cracking has been investigated with the Varestraint test [2]. As a result, increasing Ni and C, and S increases hot crack susceptibility, as shown in Fig. 5. The longer the total length of cracks, L_T, the worse the crack susceptibility in HY-type steel.

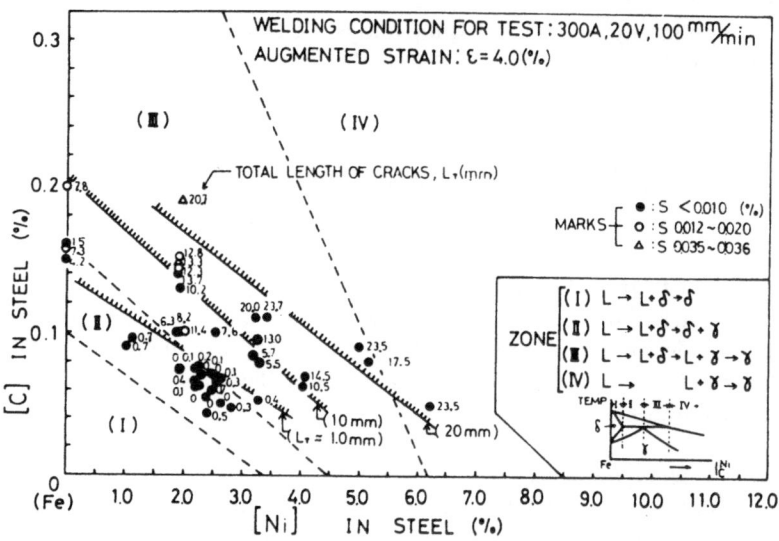

FIG. 5. Equivalent crack length at 4% augmented strain in the Varestraint cracking test.

(c) *Ductility.* In general, increasing strength of weld metal and HAZ decreases their ductility at room temperature and −70°C. However, steels up to HY-180 have no special ductility problem at present (requirement of ductility in weldment, $_vE_{-70} \geqq 7$ kgf m).

(d) *SCC in sea water.* Increased strength in weld metal and HAZ decreases their SCC resistance in sea water. Improvement of SCC resistance for higher strength HY-type steel should be further investigated in future.

3.2. Maraging Steel

Some maraging steels are used in Japan for high speed rotating vessels, and rocket chambers for space research. Thin maraging steels with strength levels of 200–300 kgf/mm^2 are mainly welded using EBW. Joint efficiency in the tensile strength of these steels is about 95–80% with aging treatment of 500°C × 4 h after EBW [3]. Notch tensile strengths and notch sensitivity ratios of welded joints are also investigated. Maximum notch tensile strength is reached at a strength level of about 270 kgf/mm^2 in both base and weld metals.

3.3. High Manganese Steel

Steels of 12–24% Mn containing high carbon are used for base frames of experimental nuclear reactor vessels and guide frames of railroads for high speed motor cars, because they are nonmagnetic, of high strength and high ductility at low temperature. Many different types of high manganese steel have been developed, but 15Mn and 18Mn type steels have been completely welded with MMAW, GMAW and EBW. Welding electrodes have been developed for MMAW and GMAW [4]. Hot cracking is one of the welding problems in these steels. The influence of phosphorus on hot cracking is much severer than that of sulphur in 0·7C–15Mn–1Ni steel. Reducing P to less than 0·015% improves hot crack susceptibility of this steel in 0·008–0·01%S [4].

3.4. Cr–Mo Steel

Many different Cr–Mo steels have been used for high temperature, high pressure vessels and tubes. Recently 5Cr–0·5Mo and 9Cr–Mo(1Mo or 2Mo) type steels, and welding electrodes for MMAW and SAW have been developed. For 9Cr–Mo steel low carbon type welding electrodes of 0·1C–9Cr–1Mo–V–Nb and 0·06C–9Cr–2Mo–Nb–V are recommended in order to prevent cold cracking and to improve ductility of weldment.

3.4.1. Welding problems for 9Cr–Mo steel

(a) *Cold cracking.* As diffusion of hydrogen in weld is delayed due to high content of chromium, hydrogen delayed cracking in HAZ is observed several days after welding. Reduced hydrogen and carbon in electrodes, and cooling rate after welding and higher preheating are recommended for preventing cold cracking.

(b) *Hot cracking.* Figure 6 shows variation in solidification crack susceptibility of various Cr–Mo steels with carbon content using the Varestraint test of 4·0% augmented strain level [5]. Longer L_T results in

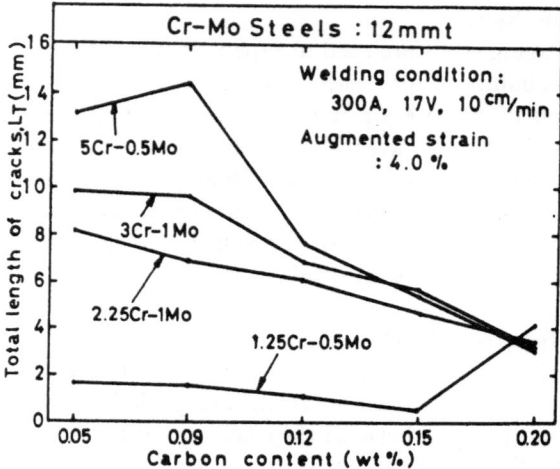

FIG. 6. Effect of carbon content on equivalent crack length, L_T, at 4% augmented strain in the Varestraint cracking test.

worse crack susceptibility in Cr–Mo steel. Increasing carbon content reduces solidification crack susceptibility for 5Cr–0·5Mo, 3Cr–1Mo and 2·25Cr–1Mo steels. However, it is reported [6] that for 9Cr–Mo steel, increasing carbon contents increases crack susceptibility. The total crack length in the Varestraint test is represented by $(50C + 6Nb - Mo - 1)$.

(c) *Hydrogen attack.* Resistance from hydrogen attack at high temperature and high hydrogen pressure is very important for Cr–Mo steel. Much research in this field is under way. Increasing chromium and reducing carbon improve the resistance to such attack in atmospheres of $300 \, \text{kgf/cm}^2 \, H_2$ and $600°C$ [7].

3.5. Ferrite/Martensite and Ferrite Stainless Steels

As these stainless steels have advantageous heat resistant, high strength, magnetic properties with low brittleness after irradiation, they are used for chemical plants and pressure vessels and will be used for experimental nuclear fusion reactor vessels.

3.5.1. 12–15Cr F/M stainless steels

Cold cracking and brittleness of welded joints are important problems in the welding of these stainless steels. In order to prevent cold cracking, reducing (C + N) content [8] and hydrogen content and preheating to more than 250°C are recommended when electrodes of the same steel are used. Electrodes of austenitic stainless steel are often used for these steels. For improved ductility, postweld heat-treatment (PWHT) of 700–750°C × 1 h is recommended after welding [9].

3.5.2. *18, 26, 30Cr ferrite stainless steel*

Lower ductility of weld metal and hot cracking are encountered in welding these steels. Increased carbon, nitrogen and oxygen lower ductility and increase hot cracks [10]. Therefore the shielding process in air during welding is necessary for improved weld soundness. Moreover reducing sulphur and alloying rare earth metals are useful for preventing hot cracking.

3.6. Austenitic Stainless Steel

3.6.1. *Fully austenitic stainless steel (SUS310S, 25Cr–20Ni)*

Fully austenitic stainless steel is very crack susceptible (solidification crack) in welding, although the welded joints of fully austenitic stainless steels are superior in low temperature ductility, high temperature strength and have superior corrosive properties in comparison with austenitic-ferrite stainless steel. SUS310S (AISI 310S correspondence) is one of these typical fully austenitic stainless steels. The mechanism for the crack susceptible property during welding of SUS310S steel has been investigated recently, and the following counterplans have been discussed [11]; (a) Reduced P and S contents to less than 0·01% in P + S content in SUS310S and (b) addition of alloy rare earth metals (REM) or La element in SUS310S steel. The amount of La added is represented by the equation, La(in wt%) = 4·5[P] + [S]. When condition (a) or (b) was satisfied, SUS310S steel was successfully welded without cracking when GTAW or EBW were used. Following these results, weldable SUS310S will be developed in Japan as low P and S steel.

3.6.2. *α + γ dual phase stainless steel*

α + γ dual phase stainless steel is the center of attention now because of the excellent SCC resistance and mechanical properties. Development of welding electrode or wire with chemical contents corresponding to the base metal is required. Hot cracking and ductility of the welded zone are the most important problems in welding these steels. About 20% ferrite is optimum for reduced crack susceptibility. The reason for a beneficial dual phase for crack susceptibility is also being investigated using a new concept [11].

3.6.3. Hydrogen embrittlement and hydrogen-induced cracking of stainless steel weld metal

In oil refineries and chemical plants, weld metals of stainless steels are often subjected to high temperature, high pressure of hydrogen at 200–450°C, 10–30 atm. Cracking due to hydrogen embrittlement has been seen in weld metal of SUS347 of fillet welded joints and disbonding in HIC has been observed in overlaid weld metal with SUS309 on Cr–Mo steel. This embrittlement and HIC are accelerated by the formation of chromcarbide and sigma phase by stress relieving heat treatment after welding. Many counterplans for these problems have been investigated. For hydrogen embrittlement, austenite stable weld metal and of heat treatment after welding have been used. For disbonding in HIC of overlaid weld metal, high speed welding conditions, pre-overlaying of low carbon Cr–Mo weld metal on Cr–Mo steel, growth control of solidification ferrite structure, special heat treatment using high frequency electric power have been used.

3.7. Invar (Fe–36%Ni) Alloy

Invar alloy has mostly been used recently as a lining metal of LNG container in a carrier utilizing its low expansion coefficient and its excellent low temperature ductility and strength. Thin invar alloy is welded with GTAW without filler wire. However the weld bead is very crack susceptible when it is reheated with repeat welding for repair of weld defect. This cracking at about 600–1000°C is caused by combined precipitation of S, O, Al and Mn along the grain boundary and serration of grain boundary during grain boundary sliding [12]. An example of such a crack is shown in Fig. 7. This problem was solved by reducing weld heat input, reducing impurities such as S, O, N, Al and P and completion of shielding in Ar gas during welding [12].

3.8. High Nickel Heat Resistant Alloys

High nickel alloys are prevalently used for vacuum vessels of JT-60 reactors, engines for aircraft and so on. Welding processes in industrial production have been extended recently to EBW, TLP bonding and friction welding, in addition to GTAW, GMAW, SAW and MMAW. Hot cracking is one of the most serious problems in welding. Solidification and liquation cracking are often seen in weld metal and HAZ due to low temperature melting of P and S and eutectics of Al, Ti, Nb compounds in alloys. Reducing P and S is beneficial for prevention of the cracking. However there are many problems still in existence.

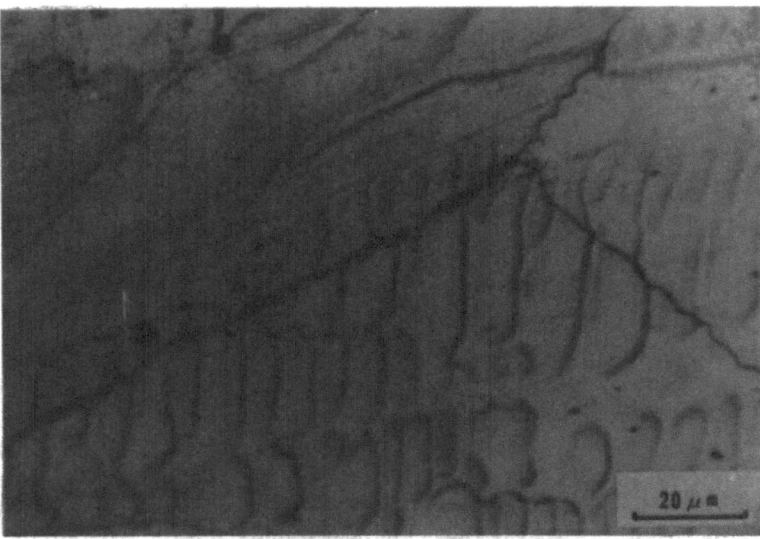

FIG. 7. Typical example of reheat crack in Invar weld metal (polished and etched surface).

3.9. Aluminum and its Alloys

Aluminum alloys are being used more widely because of their many advantages as being lighter, of excellent ductility in low temperature, good conductivities for heat and electricity, non-magnetic, excellent formability and anti-corrosive. Moreover it was established recently that aluminum alloys are beneficial for use in vacuum vessels of high energy accelerators due to its lower degasification from the inner surface of Al–Mg alloy in vacuum and its larger decay factor of radioactivity in comparison to steels. The obvious uses of aluminum alloys are as follows:

(a) Vessels and equipment for low temperature gas and liquid. Many LNG containers and pipes on shore and ship are fabricated by A5083 (Al–4·5Mg–0·7Mn) alloy (10–180 mm in thickness) in Japan. Welding of these structures is mostly performed with high current GMAW and DCSP–GTAW using 5183 filler wire.

(b) Vacuum chambers of high energy accelerators and ultra vacuum equipments. Mg in aluminum alloy is reported to be beneficial for low degasification in vacuum chambers [13]. In the Japanese National Laboratory for High Energy Physics (Tsukuba), the high energy accelerator whose total circumferential length is about 3 km was successfully constructed using A6063 (Al–0·5Mg–0·4Si) with DCSP–GTAW, pulsed-GMAW and high current GMAW.

(c) Railroad vehicles. Vehicles of new 'Shin Kansen' were fabricated by A7N01, (Al–4·5Zn–1·5Mg–0·5Mn), 7003 (Al–6Zn–0·7Mg) and 5083. Many vehicles of subways, private and national railroads are being fabricated by A6N01 (Al–0·6 Mg–0·7 Si). Moreover, experimental ultra high speed motor-cars are fabricated by A7N01 and 2024 (Al–4·5Cu–1·5Mg–Mn) alloys.

(d) Aircraft. Fuel tanks of N–II rockets were fabricated by A2219 (Al–6·3Cu–0·3Mn) with DCSP–GTAW.

(e) Ships and boats. Many ships and boats, naval and civilian, are fabricated by A5083 and 5052 (Al–2·5Mg–0·25Cr) with many different welding processes.

3.9.1. Welding problems in aluminum welding

(a) *Strength of welded joint.* Hardness of HAZ in cold rolled or aged alloys is usually reduced to that of annealed alloy and the strength of the welded joint is also reduced relative to the high strength base metal. Increasing weld heat input in welding reduces joint strength of aluminum alloys.

(b) *Welding cracking (solidification and liquation cracks).* Aluminum alloys of Alxxx, 3xxx, 4xxx and 5xxx (except 5005) are usually insusceptible to cracking during welding, but alloys of A2xxx, 7xxx, 6xxx and 5005 are in general much more susceptible. A2xxx alloys of filler wire with higher copper content including aluminum (Al–6Cu for example) are being developed to prevent cracking. For 7xxx and 6xxx, filler wire with higher Mg content (5–6Mg–Al for example) in addition to Zr or Ti element has been developed [14]. However weld cracking is not solved completely.

(c) *Porosities and lack of fusion.* Hydrogen which is dissociated from water, grease or hydro-oxides generates weld porosities in aluminum weld metal. Moreover high melting point aluminum oxide, such as Al_2O_3, often lacks fusion of welded joint during welding. These problems have not been solved.

(d) *SCC.* Tensile stresses with external and internal weld residual stresses and severe environmental conditions such as salty environment, results in SCC in A2xxx, 7xxx and some 5xxx weld metals and HAZ. In order to avoid SCC, many processes have been tried such as: (i) increasing recrystallizing temperature by addition of small amount of Zr,

Mn or Cr; (ii) addition of small amount of Cu or Li; (iii) improvement of rolling or heat treatment condition; and (iv) use of over-aged material.

3.9.2. Welding processes

The most popular welding processes for welding aluminum alloys are GMAW (high current, pulsed, fine electrode, etc.), GTAW (DCSP, pulsed, AC etc.), plasma welding, resistance welding and EBW, brazing.

3.10. Titanium and its Alloys

Titanium and its alloys have been used for materials of chemical plants and aircraft in Japan because of their excellent corrosive resistance and high relative strength. The new scientific submarine 'Sinkai 6000', which can dive to 6000 m deep into Ocean, will be using 100 mm thick Ti–6Al–4V alloy. Welding of spherical shell of about 2 m in diameter will be done mainly with EBW besides arc weldings. Titanium is one of the most reactive metals and thus gas contamination such as oxygen and nitrogen, must be considered carefully in welding of these metals. Complete argon shielding GTAW or GMAW with trailing nozzle is recommended for welding in air environment. Arc welding in a controlled atmosphere welding chamber, and EBW in vacuum are also strongly recommended. Another welding problem is porosities in weld metal due to oxygen and/or hydrogen. Advance cleaning of the weld is strictly required in order to avoid porosities. Extra low interstitial grade (ELT) titanium metal is commercially used for ductility improvement of welded joints. Titanium welding is completed with GTAW, GMAW, plasma welding, EBW, laser beam welding, diffusion bonding and explosive welding. However welding of steel with titanium is still difficult.

3.11. Tantalum

Because of its excellent corrosive resistance, tantalum is gradually being used for chemical equipment. However tantalum has a high density ($16.6 \, g/cm^3$) high melting temperature ($3000°C$) and is reactive with oxygen and nitrogen at temperatures over $400°C$. Therefore the shielding problem during welding is one of the most important points for sound tantalum welding. Thin sheet, less than about 1 mm in thickness, is being welded with GTAW and for thickness exceeding about 1 mm, it is welded with EBW in Japan.

3.12. Molybdenum

Molybdenum is used for lining metal of experimental nuclear fusion vessels. Mo sheet and plate are made by both powder metallurgy and EB melting process. However weldability of both these is not very good because of the coarsened grain size of weld metal and HAZ resulting in brittlement. The ductile–brittle transition temperature (DBTT) of the welded zone of pure molybdenum is generally elevated to more than 150°C even with EBW [14]. Figure 8 shows for example the DBTT of base and welded joint. At present, DBTT of welded joint of pure molybdenum is not met by the designer of the reactor. However recently Re with added molybdenum was found to yield excellent DBTT even welds. Weldability of several percent of Re added molybdenum will be investigated in the future.

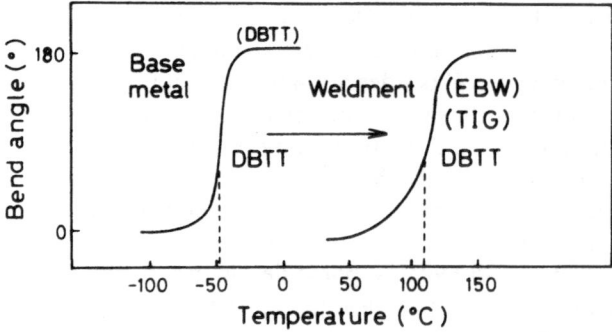

FIG. 8. DBTT of base and weld metal of pure molybdenum.

3.13. Ceramics

Welding various ceramics is being actively investigated in many fields. They are divided into two categories, that is, welding of ceramics–ceramics and ceramics–metal. The welding process is roughly divided into four groups, that is; (a) liquid phase bonding using low melting point insert metal, glue, low melting coating with CVD, PVD or heat treatment; (b) solid phase bonding by means of solid phase reaction, or high pressure at high temperature; (c) fusion welding by laser beam, and (d) others utilizing gas reaction. Although experimental results have been successful few industrial uses have been reported. Welding of fine optical fibres (125 μm in diam.) is now completed and widely used with a tungsten arc heat source. A welding machine with automatic controller is commercially available. Optical fibres are now used in Japan for communication on a nationwide scale.

REFERENCES

1. SHINOZAKI, K. Weld cracking of HY-type steel, PhD thesis, Osaka University, 1984, 12.
2. KIHARA, K. *et al. Trans. JWRI*, **2** (2) (1973), 83–95.
3. FUJITA, M. *et al. J. Iron Steel*, **70** (14) (1964), 120–7.
4. SAWA, S. *et al. J. JIM*, **18** (8) (1979), 573–81.
5. MATSUDA, F. *et al. Trans. JWRI*, **6** (2) (1977), 53–8.
6. ABE, N. *et al.* WM-956-84, JWS.
7. CHIBA, R. Hydrogen attack of pressure vessel in high temperature high pressure hydrogen environment, Ph.D. thesis, Osaka University, 1983.
8. ANPO, H. *et al. J. WES*, **24** (5) (1976), 57–61.
9. STAINLESS INSTITUTE. SAS 801, 1979 (in Japanese).
10. NAKAO, Y. *et al.* JWES-SM-8301, 1983, 5, JWES, 465–477.
11. KATAYAMA, S. Weld cracking of SUS310S, Ph.D. thesis, Osaka University, p. 5, 1981.
12. ZHANG, Y. Weldability of Invar, Ph.D. thesis, Osaka University, p. 3, 1986.
13. KATO, Y. *et al. J. JILM*, **35** (4) (1985), 228–33.
14. MATSUDA, F. *et al. J. JWS*, **4** (1) (1986), 115–20.

4

Dynamic Stress Intensity Factor versus Crack Velocity Relation

A. S. KOBAYASHI and K. H. YANG

University of Washington, Department of Mechanical Engineering FU-10, Seattle, WA 98195, USA.

ABSTRACT

The dynamic stress intensity factor versus crack velocity relations of Homalite-100, polycarbonate, hardened 4340 steel, reaction bonded silicon nitride and alumina, which were obtained by the first author and his colleagues, are reviewed and discrepancies with published data are discussed. The experimental results show that the dynamic stress intensity factor versus crack velocity relation and the dynamic crack arrest stress intensity factor are not unique material properties.

1. INTRODUCTION

The dynamic fracture community's preoccupation with the dynamic stress intensity factor, i.e. the crack driving force, versus the crack velocity relation is demonstrated by the fact that six out of the seven review papers dealt with the experimental aspects of dynamic fracture mechanics in the recent issue of the *International Journal of Fracture* [1]. The survey paper by Dally *et al.* [2] describes the major findings to date and indicates possible sources of experimental error which may have led to the current controversies on this subject. The purpose of this paper is to present the authors' experimental results on the dynamic stress intensity factor versus crack velocity relations of polycarbonate, 4340 steel, reaction bonded silicon nitride and alumina in addition to the well-studied Homalite-100 [3–5] and their view on the uniqueness or lack thereof in the dynamic stress intensity factor versus crack velocity relation and/or in the dynamic crack arrest stress intensity factor.

2. MATERIALS AND TEST RESULTS

2.1. Photoelastic Polymers

Literature on dynamic stress intensity factor, K_I^{dyn}, versus crack velocity, $\dot{a}$, relations, which have been obtained through dynamic photo-elasticity and caustics analyses of polymers [3–7], show that the terminal crack velocity is test specimen dependent while the 'near vertical stem' of these relations is either a unique [2, 3] or a nonunique [5–7] material property. In order to support their nonuniqueness claim, the authors and their associates re-evaluated the dynamic photoelastic data using an updated data processing procedure which incorporates higher order terms of the dynamic crack tip stress field. Figures 1 and 2 show the K_I^{dyn} versus $\dot{a}$ relations for Homalite-100 and polycarbonate fracture speci-mens. No attempt was made to fit an average K_I^{dyn} versus $\dot{a}$ curve through the wide scatter of data generated from various batches of Homalite-100 and polycarbonate sheets tested over a period of ten years. Figure 1 shows that the scatter bands about the imagined vertical stems of the dynamic tear test (DTT), single edge notched (SEN), modified compact (M–CT) and wedge-loaded rectangular double cantilever beam

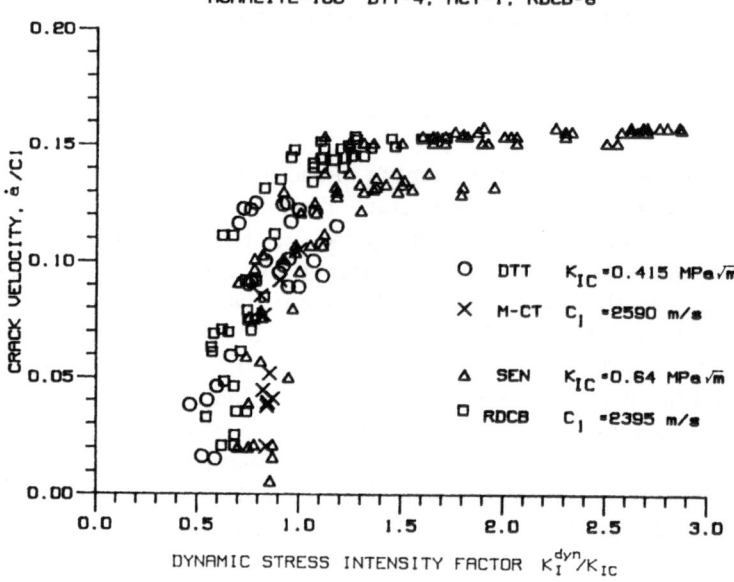

FIG. 1. Dynamic stress intensity factor versus crack velocity relation for Homalite-100.

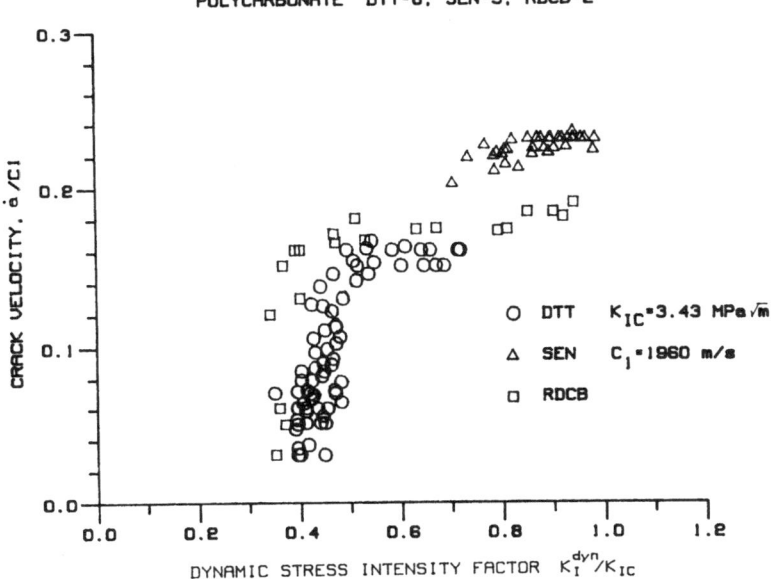

FIG. 2. Dynamic stress intensity factor versus crack velocity relation for polycarbonate.

(WL–RDCB) Homalite-100 specimens are similar to those shown in Ref. 5. However, differences in the minimum dynamic stress intensity factor, K_{Im}, of the vertical stems of the DTT and SEN specimens are larger than that reported in Ref. 6. The difference in K_{Im} for the more ductile polycarbonate WL–RDCB and DTT specimens is about 10% and is in agreement with the general observation by Rosakis *et al.* [8].

Figure 3 shows the K_I^{dyn} versus crack extension relations of four SEN specimens subjected to different fixed grip loading conditions [7]. Also shown are the corresponding static stress intensity factors. This figure, which is similar to the well-publicized results of Kalthoff *et al.* [9], demonstrates that the dynamic crack arrest stress intensity factor, K_{Ia}^{dyn} is a constant for the same specimen while the static crack arrest intensity factor, K_{Ia}^{stat}, varies with the crack initiation condition.

Since the above photoelastic results are in general agreement with the caustic results, the published discrepancies in the K_I^{dyn} versus $\dot{a}$ results cannot be attributed to the differences in the experimental procedures alone. The discrepancies could be attributed in part to experimental errors and inherent errors in the data reduction procedures used by

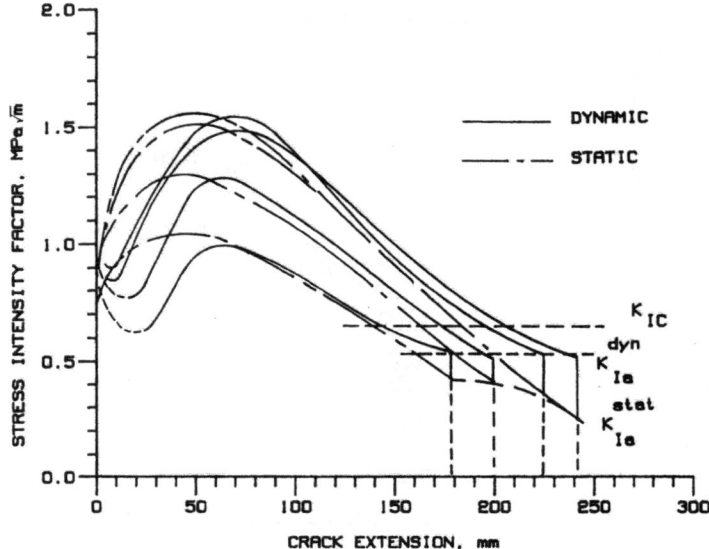

FIG. 3. Stress intensity factors versus crack length for Homalite-100 SEN specimens.

various investigators. One such experimental error is induced in the
crack velocity data obtained from the discrete and somewhat fuzzy ultra-
high speed photographs of the rapidly propagating crack tip. Accurate
crack velocity measurements have been made by the more precise
ultrasonic fractography [10] in CT, SEN, three-point bend specimens
and are similar to those reported in Refs 12–14. The existence of small
but sharp changes in crack velocities, which are comparable to those
reported in Ref. 16, was observed in the Charpy specimens [11] which
were subjected to severe stress wave effects. As will be shown later, such
discontinuous crack velocities were also observed in small hardened 4340
steel and ceramic specimens where the stress wave effect is pronounced.

2.2. 4340 Steel

The hybrid experimental–numerical procedure [17] was used to determine the K_I^{dyn} versus $\dot{a}$ relation for 4340 steel hardened to Rockwell
C44. The dynamic crack extension histories in four wedge-loaded,
modified-tapered double cantilever beam specimens (WL–MTDCB) with
a chevron starter notch were measured by a KRAK-GAGE and
FRACTOMAT (both TTI Division, Hartrum Corp., Chaska, MN,

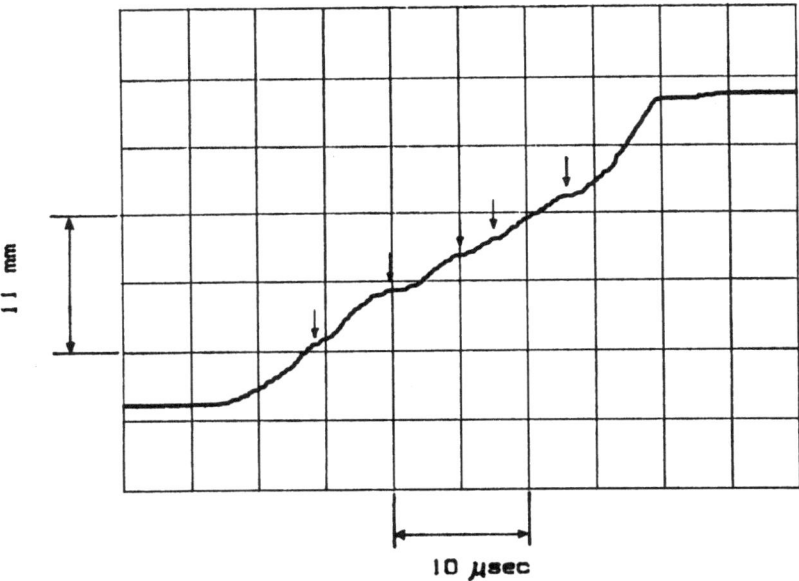

FIG. 4. Crack extension versus time for blunt notch 4340 steel WL–MTDCB specimen.

USA). The result shows that initial and slower crack propagation in the chevron notch specimens is followed by rapid crack propagation and subsequent deceleration. The latter crack deceleration is interrupted by a number of short intervals of crack arrest. Figure 4 shows a typical crack extension record of a fracturing blunt notch 4340 WL–MTDCB specimen. Temporary crack arrests in the crack extension history are indicated by arrows, where the average time between each crack arrest coincides with the average transit time of shear wave from the crack tip to the lateral edge of the specimen and back. Such intermittent crack arrest, as long as 20 µs, was reported by Van Elst [18] and de Graaf [19], who used streak photography to record continuous crack extension in Robertson type low-carbon steel specimens. Ravi-Chandar and Knauss [5] and Rosakis *et al.* [8] also reported the presence of discontinuous crack velocities in their highly dynamically loaded specimens.

Returning to the hybrid experimental-numerical procedure, an average of the measured crack extension histories, without crack arrest, of four 4340 steel WL–MTDCB specimens was then used to drive a dynamic finite element code in its generation phase and the dynamic stress intensity factor was computed. Figure 5 shows the K_I^{dyn} versus $\dot{a}$ relation

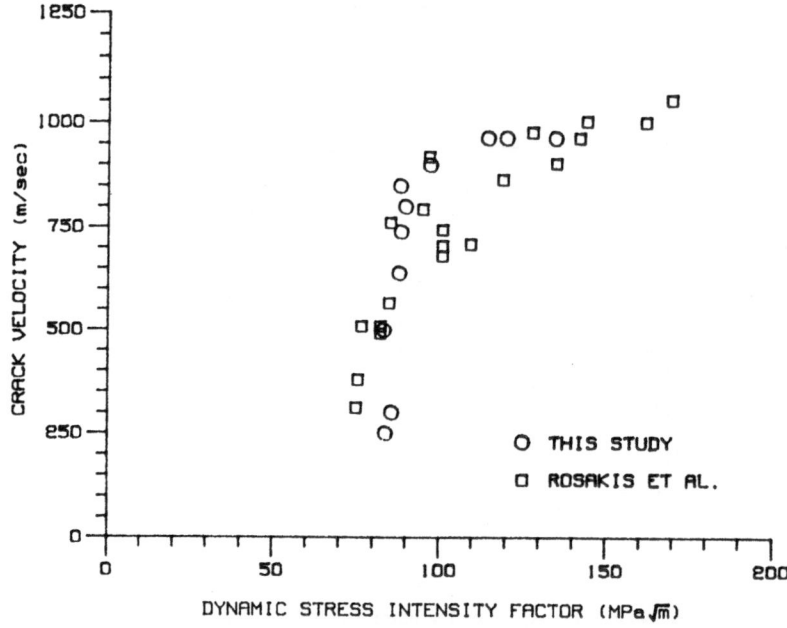

FIG. 5. Dynamic stress intensity factor versus crack velocity relation for 4340 steel.

for this study as well as that of Rosakis *et al.* [8]. The remarkable agreement between the two independent results is noted despite the dissimilarities in specimen geometries.

2.3. Reaction Bonded Silicon Nitride (RBSN)

Despite the differences in K_I^{dyn} versus $\dot{a}$ relations, a vertical stem in these relations always existed in the photoelastic polymers and 4340 steel specimens discussed so far. However, limited dynamic fracture studies of extremely brittle materials, such as glass and structural ceramics [20–22], show that K_{Im} and hence the vertical stem in the K_I^{dyn} versus $\dot{a}$ curve does not exist in some materials. Figure 6 shows K_I^{dyn} versus $\dot{a}$ relation of reaction bonded silicon nitride (RBSN) WL–MTDCB specimens loaded to fracture under both static and dynamic conditions. The specimen geometry is identical to that of the 4340 steel WL–MTDCB specimens. Since the static stress intensity factor in this specimen decreased with increasing crack length, a rapidly propagating crack in a statically loaded, annealed 4340 WL–MTDCB specimen naturally arrested after

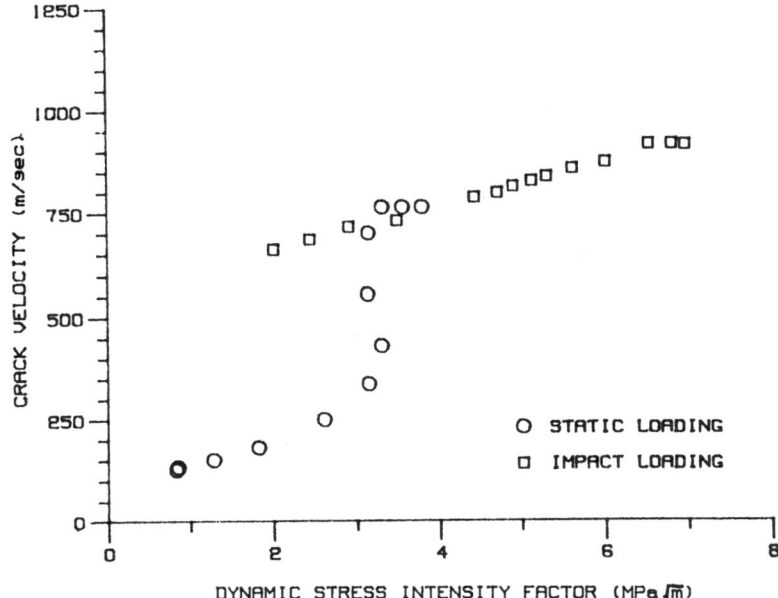

F<small>IG</small>. 6. Dynamic stress intensity factor versus crack velocity relation for blunt notch RBSN WL–MTDCB specimens.

propagating through much of the specimen length. For a statically loaded RBSN specimen, however, Fig. 6 shows that the propagating crack attempted unsuccessfully to arrest. A crack propagating under dynamic loading, on the other hand, shows little tendency for arresting. Figure 6 thus shows the lack of any unique vertical stem in the K_I^{dyn} versus $\dot{a}$ curves for reaction bonded silicon nitride.

2.4. Alumina

Alumina, prenotched three-point bend specimens were subjected to impact loading at room temperature and at 1500°C. For the room temperature tests, the remaining ligament in the bend specimen was instrumented with a KRAK–GAGE and the crack extension history was recorded together with the impact and reaction loads. The average time varying boundary conditions of five specimens were used to drive a dynamic finite element code in its generation phase and the dynamic stress intensity factor was computed. Figure 7 shows the resultant dynamic stress intensity factor versus crack velocity relation. While the

A. S. Kobayashi and K. H. Yang

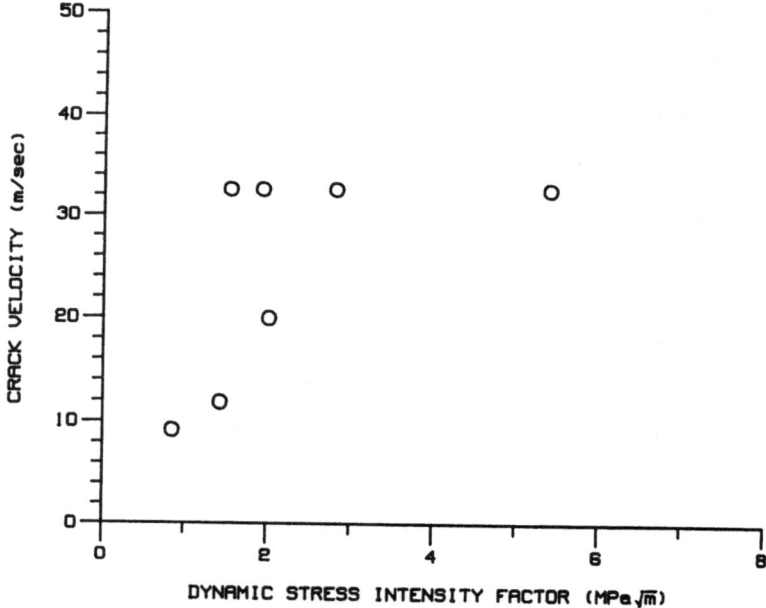

FIG. 7. Dynamic stress intensity factor versus crack velocity relation for alumina three-point bend specimen under impact loading.

data are not extensive as that of RBSN specimens, the same trend of no crack arrest is also observed in these alumina specimens.

3. CONCLUSIONS

As profoundly stated by many authors in [1], the controversy regarding the uniqueness or lack thereof in the K_I^{dyn} versus $\dot{a}$ relation is far from being settled. Available experimental results indicate that in the absence of stress wave effects, such as in infinitely large fracture specimens under benign loading, K_I^{dyn} versus $\dot{a}$ relations may possess a unique K_{Im} or a vertical stem. Such a unique vertical stem is not observed in dynamic fracture specimens of smaller size and/or under dynamic loading. A study of various experimental data shows that the consistency in data scatter cannot be totally attributed to experimental errors and that the intermittent crack arrest and the discrete changes in crack velocity are caused by the reflected stress waves.

4. DISCUSSION

In the pursuit of the above uniqueness controversy, we pose the question 'for what reason?' The end use of the sought K_I^{dyn} versus $\dot{a}$ relation is as the fourth constitutive equation for estimating the dynamic fracture response of an elastic solid. Limited numerical experiments show that the arrest crack length of a propagating crack is obviously governed by K_{Im} [22]. For a dynamically loaded specimen or in the presence of severe stress wave effects, however, small differences in K_{Im} may not cause large differences in the arrest crack length. On the other hand, the same difference in K_{Im} may cause large differences in the arrest crack length in the absence of stress wave effects.

ACKNOWLEDGEMENTS

Most of the work reported here was obtained under ONR contract Nos. N00014-76-C-0060 NR 064-478 and N00014-85-K0187. The authors wish to acknowledge the support and encouragement of Dr Yapa Rajapakse, ONR, during the course of this investigation. The 4340 steel and ceramic fracture results were obtained under NASA contract NAGW-199.

REFERENCES

1. *Int. J. Fract.*, **27** (3–4) (1985).
2. DALLY, J. W., FOURNEY, W. L. and IRWIN, G. R. *Int. J. Fract.*, **27** (3–4) (1985), 159.
3. DALLY, J. W., *Exp. Mech.*, **19** (10) (1979), 349.
4. KOBAYASHI, A. S. and MALL, S., *Exp. Mech.*, **18** (1) (1978), 11.
5. RAVI-CHANDAR, K. and KNAUSS, W. G., *Workshop on Dynamic Fracture*, California Institute of Technology 1983, p. 119.
6. KALTHOFF, J. F., BEINERT, J. and WINKLER, S., Institut fuer Festkoepermechanik report prepared under Electric Power Research Contract RP 1022-1 IKFM 40412 (1978).
7. KOBAYASHI, A. S. and MALL, S., *J. Poly. Engng Sci.*, **19** (2) (mid-February 1979), 131.
8. ROSAKIS, A. J., DUFFY, J. and FREUND, L. B., *Workshop on Dynamic Fracture*, California Institute of Technology, 1983, p. 100.
9. KALTHOFF, J. F., BEINERT, J. and WINKLER, S., *Fast Fracture and Crack Arrest*, ASTM STP 627 (G. T. Hahn and M. F. Kanninen (eds), 1977, p. 161.

10. KERKHOF, F. *Proceedings of 3rd International Congress on High-Speed Photography*, Butterworths, London, 1957, p. 194.
11. TAKAHASHI, K., MATSUSHIGE, K. and SAKURADA, Y., *J. Mater. Sci.*, **19** (1984), 4026.
12. BRADLEY, W. B. and KOBAYASHI, A. S., *Exp. Mech.*, **10** (3) (1970), 106, March.
13. KOBAYASHI, A. S., SEO, K., JOU, J. Y. and URABE, Y., *Exp. Mech.*, **20** (9) (1980), 301.
14. MALL, S., KOBAYASHI, A. S. and URABE, Y., *Exp. Mech.*, **18** (12) (1978), 449.
15. ROSAKIS, A. J. and ZEHNDER, A. T., *Int. J. Fract.*, **27** (1985), 169.
16. KNAUSS, W. G. and RAVI-CHANDAR, K., *Int. J. Fract.*, **27** (1985), 127.
17. KOBAYASHI, A. S., *Nonlinear and Dynamic Fracture Mechanics* (N. Perrone and S. N. Atluri (eds), ASME AMD-35, 1979, p. 19.
18. VAN ELST, H. C., *Trans. Metall. Soc. AIME*, **230** (1964), 460.
19. DE GRAAF, J. G. A., *Appl. Optics*, **3** (11) (1964), 1223.
20. KOBAYASHI, A. S., EMERY, A. F. and LIAW, B. M., *Fracture Mechanics of Ceramics, Vol. 6* (R. C. Bradt, A. G. Evans, D. P. H. Hasselman and F. F. Lange (eds)), Plenum Press, New York, 1983, p. 47.
21. KOBAYASHI, A. S., EMERY, A. F. and LIAW, B. M., *J. Am. Cer. Soc.*, **66** (2) (1983), 151.
22. LIAW, B. M., KOBAYASHI, A. S. and EMERY, A. F., *Fracture Mechanics, Vol. 17*, J. H. Underwood, R. Chait, C. W. Smith, D. P. Wilhem, W. A. Andrews and J. C. Newman (eds), in ASTM STP 905, 1986, p. 95.
23. CHEVERTON, R. D., GEHLEN, P. C., HAHN, G. T. and ISKANDER, S. K., *Crack Arrest Methodology and Applications*, ASTM STP 711, G. T. Hahn and M. F. Kanninen, 1980, p. 392.

5

Quantitative Evaluation of Microcracking in Alloys and Ceramics

T. KISHI and M. ENOKI

Faculty of Engineering, Institute of Interdisciplinary Research, Komaba 4-6-1, The University of Tokyo, Meguro-ku, Tokyo, Japan.

ABSTRACT

An advanced acoustic emission technique was developed to obtain a source function tensor (seismic moment) for microcracks, that allows evaluation of the fracture mode, size and nucleation rate of microcracks, in addition to location. At first, the theoretical background of acoustic emission source characterization is summarized and the meaning of the source function is discussed. This source function can be evaluated by a deconvolution integral of the detected signals, using the transfer function of the measuring system and a dynamic Green's function of the medium. The transfer function of the measuring system can be estimated by means of a simulated signal, while the dynamic Green's function was estimated by computer simulation using the finite difference method. The source function can be evaluated by time-domain deconvolution.

In the case of alumina, microcracks were located at the front of precracks, the size estimated, and compared with direct observations in the scanning electron microscope. Microcracks less than 50 μm contribute to the enhancement of fracture resistance and those larger than 70 μm become the origin for unstable fracture.

1. INTRODUCTION

Evaluation of the nucleation and coalescence of microcracks in materials is an essential step to understand the fracture mechanism, because in almost all cases of fracture these microcracks become the motive source for final unstable fracture. Meaningful parameters for the understanding of the characteristics of these microcracks are:

(1) location;
(2) inclination of fracture plane;
(3) fracture mode;
(4) size;
(5) rise time (mean nucleation rate).

Acoustic emission (AE) can detect growing microdefects and is a technique both to improve the fundamental understanding of fracture and to evaluate structural integrity [1–4].

However, there exists a serious problem, because AE signals detected by ordinary equipment are influenced both by the shape and size of the medium and measuring system, as shown in Fig. 1.

On the other hand, the source function of the AE signals $D(t)$ can give quantitative information about microcracking [5]. This paper is concerned firstly with the theoretical background of AE source character-

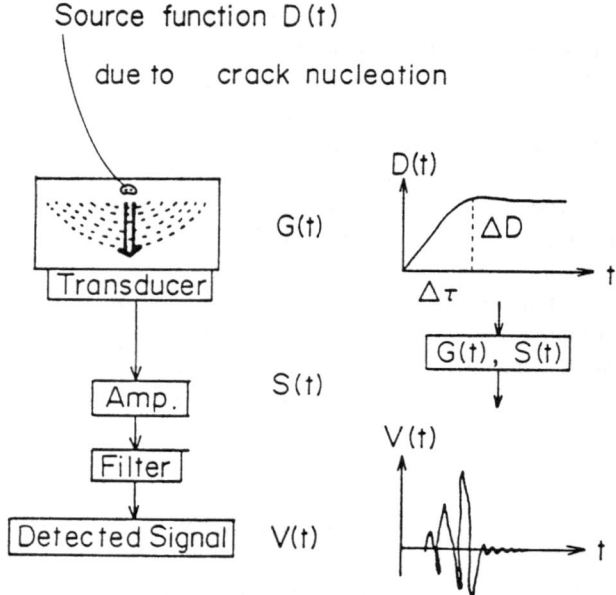

FIG. 1. The relation between source function $D(t)$ and detected signal $V(t)$ in acoustic emission signal processing, where $G(t)$ is a dynamic Green's function and $S(t)$ is a response function of measuring system.

ization, secondly with the hardware and software to estimate the source function. Lastly the source functions due to microcracking are obtained in metals and ceramics, and their microfracture processes are discussed.

2. THEORETICAL BACKGROUND OF THE AE SOURCE CHARACTERIZATION

2.1. Linear Response Theory

The displacement due to elastic waves radiated by an incremental extension of a microcrack can be formulated by models used in seismology [6], dynamic fracture mechanics [7] and dislocation theory [8]. In this treatment, displacement at the position x is given by [9]

$$u_i(x, t) = G_{ij,k'} * D_{jk} \tag{1}$$

where $*$ is a convolution integral, $G_{ij,k'} = \partial G_{ij}/\partial x_{k'}$ is a dynamic Green's function of the medium and D_{jk} is a source function (seismic moment). In the case where the measuring system, including the transducer, is linear, the output signal $V(t)$ is related to $u_i(x, t)$ by the transfer function of the transducer $S_i(t; x, \tau)$ [10]

$$\begin{aligned} V(t) &= S_i * u_i \\ &= S_i * G_{ij,k'} * D_{jk} \end{aligned} \tag{2}$$

2.2. Physical Meaning of Source Function D_{jk}

In eqn. (2), a source function due to a microcrack, D_{jk}, is expressed by

$$\begin{aligned} D_{jk}(x', t) &= C_{lmjk} \cdot \phi_l(t) \cdot \Delta A_m(t) \\ &= \{\lambda \delta_{lm}\delta_{jk} + \mu(\delta_{ij}\delta_{mk} + \delta_{lk}\delta_{jm})\} \cdot \phi_l(t) \cdot \Delta A_m(t) \end{aligned} \tag{3}$$

where λ and μ are the Lamé elastic constants. The discontinuity of the displacement $\phi_l(t)$ and the surface area $\Delta A_m(t)$ correspond to the Burgers vector and the vector loop area in dislocation theory. By evaluating the source function tensor of D_{jk}, we can define the fracture mode, crack volume and rise time (life time) of crack nucleation.

In the case of a tensile crack shown in Fig. 2, $l = m = 3$, then the only nonzero terms in eqn. (3) are

$$D_{11}(x', t) = D_{22}(x', t) = \lambda \phi_3(t) \cdot \Delta A_3 \tag{4}$$

and

$$D_{33}(x', t) = (\lambda + 2\mu)\phi_3(t) \cdot \Delta A_3 \tag{5}$$

Tensile Crack   (at l = m = 3)

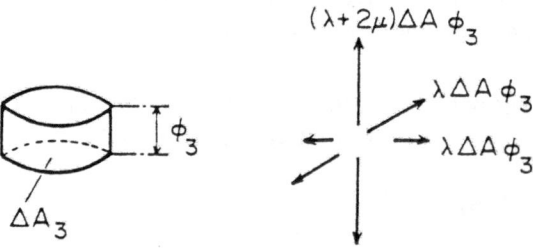

FIG. 2. Source function D_{jk} in the simple case of tensile crack.

From eqn. (5), the cracking area ΔA_3 and crack opening displacement $\Delta \phi_3$ in microcrack formation can be estimated by using stress–strain analysis [15].

2.3. The Transfer Function of the Measuring System, Including Transducer, $S_i(t)$

We developed a transducer calibration system on the basis of reference signals simulated by breaking a pencil or a glass capillary, using capacitive transducers [11]. The calibration system, as shown in Fig. 3, consists of an aluminum alloy plate, width 500 mm and thickness 60 mm,

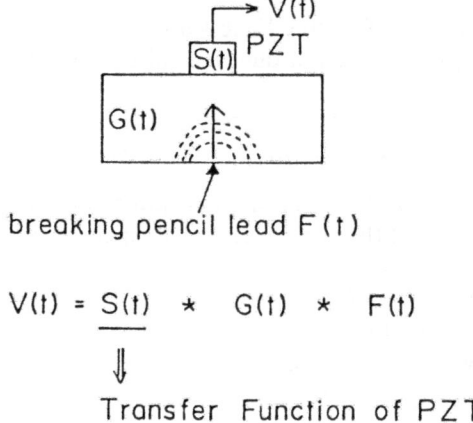

$$V(t) = \underline{S(t)} \;\; * \;\; G(t) \;\; * \;\; F(t)$$

$$\Downarrow$$

Transfer Function of PZT

FIG. 3. Calibration system to obtain a transfer function of PZT transducer by simulated reference signals of breaking pencil lead.

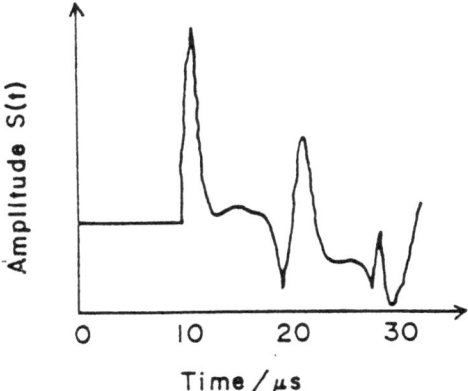

F IG. 4. Transfer function of wide band piezo transducer calibrated by simulated signal of breaking pencil lead.

which acts as one plate of a parallel plate capacitor. The other plate is a highly polished brass cylinder, 6 mm diameter and 10 mm length.

The source function of a breaking pencil lead, $F(t)$, was obtained by deconvolution of eqn. (2) from the detected signal, $V(t)$, using both Knopoff Green's function $G(t)$ [12], and the transfer function of a capacitive transducer, which reacts to displacement. The result showed that a rise time of a breaking pencil lead was about 0·5 to 0·7 μs and a release force 5 N. The frequency response range was 3–5 MHz.

An example of the transfer function $S(t)$ of a wide band transducer (AET, FC-500), is shown in Fig. 4. The frequency response of a transducer was flat between 50 kHz and 2 MHz, and had components up to 13 MHz.

2.4. The Dynamic Green's Function $G_{ij,k'}$

The dynamic Green's function, $G_{ij,k'}$, can be evaluated by three different methods; analytical, experimental and numerical simulation. The analytical method [6, 12] is valid only for infinite or semi-infinite media. Hence we developed both the experimental method and simulation by finite difference.

The experimental method [11, 13] is simple and meaningful results of source functions were obtained by using this Green's function. But the pencil breaking signal is a monopole force, so the Green's function for the dipole source, related to fracture, has to be replaced by $(\partial G_{33}/\partial t)/V_L$, for the first arrival of a longitudinal wave having velocity V_L [15]. This

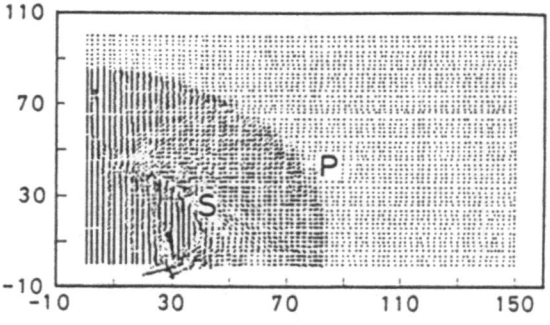

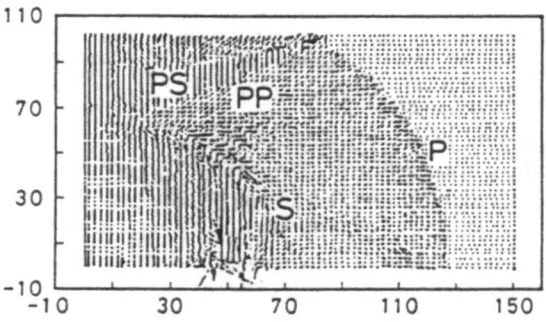

FIG. 5. Elastic wave propagation in a finite plate simulated by finite difference method, where P shows longitudinal wave and S shear wave.

experimental method is valid only for a source for which the simulated signal can be given.

We have also developed a three-dimensional numerical simulation of elastic wave propagation [14]. Figure 5 represents the result of the wave propagation in a finite block. From these results $G_{ij,k'}$ at an arbitrary position can be estimated.

2.5. Time Domain and Frequency Domain Deconvolution Technique

Time domain and frequency domain deconvolution is adopted to obtain the source function D_{jk}. We digitized these functions at sampling intervals of 25–50 ns. Then eqn. (2) can be rewritten as

$$V(I) = \sum_{K=1}^{I} g(I - K + 1) D(K) \qquad (6)$$

where $g(t) = S(t) * G(t)$. $D(I)$ is given by

$$D(I) = \{V(I) - \sum_{K=1}^{I-1} g(I-K+1)D(K)\}/g(I) \qquad (7)$$

3. EXPERIMENT

Figure 6 shows a block diagram of the data acquisition and analysis system for source characterization. The transducers used are wide band AET-FC500 (flat till $\sim 2\,\mathrm{MHz}$, maximum $\sim 13\,\mathrm{MHz}$) and PAC P-50 ($\sim 2\,\mathrm{MHz}$ maximum). Through an impedance converter, AE signals are amplified by a BX-31 amplifier, having a gain 0–40 dB, and a frequency response, DC-70 MHz. For small signals, two sets of BX-31 amplifiers are used in tandem. Output signals are directly digitized using the 8 channel AE-9620 wave memory, with a 25 ns sampling interval, a maximum precision of 10 bits and data length of 1 M words. These data are stored in the memory of a HP-216 desk top computer. The decon-

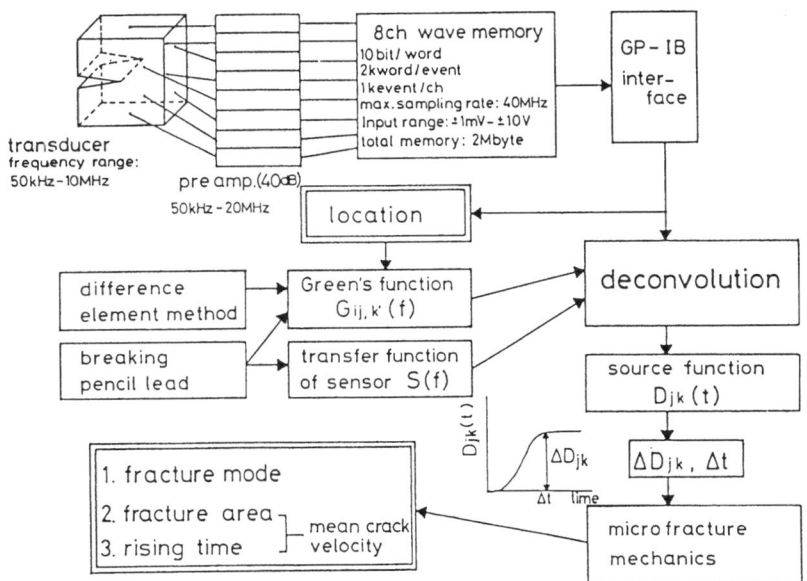

FIG. 6. A block diagram of data acquisition and analysis system for AE source characterization.

volution analysis is carried out using HP-9845 or HP-216 computers, with the dynamic Green's function $G_{ij,k'}$ and response function $S_l(t)$.

4. EVALUATION OF MICROCRACKING [16]

4.1. Three-dimensional Location of Microcracks

Three-dimensional location was carried out by measuring the time difference of initial longitudinal waves between two independent transducers. From the arriving time of waves t_1, t_2, t_3, ..., time difference $\Delta t_{ij} = t_i - t_j$ is estimated. Location, r, is obtained by solving the following nonlinear equation

$$\alpha \Delta t_{ij} = |r - r_i| - |r - r_j| \tag{8}$$

where α is the longitudinal wave velocity, and i and j represent the number of transducers. Figure 7 indicates the location of microcracks which nucleate before macrocrack growth during fracture toughness testing in A470 steel.

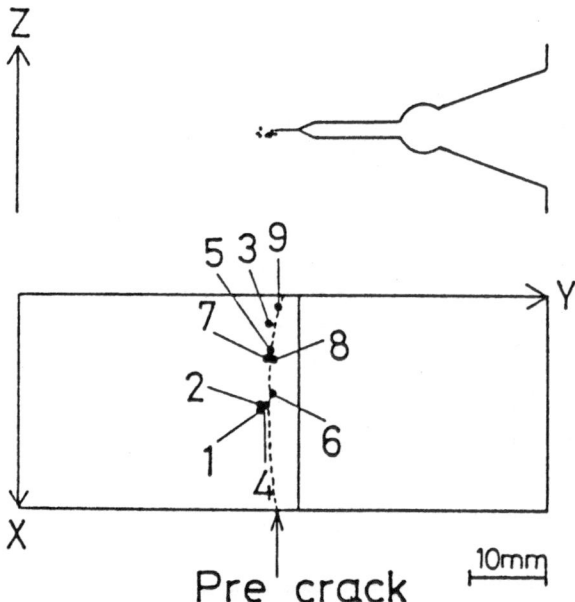

FIG. 7. The location results of microcracks during fracture toughness testing in A470.

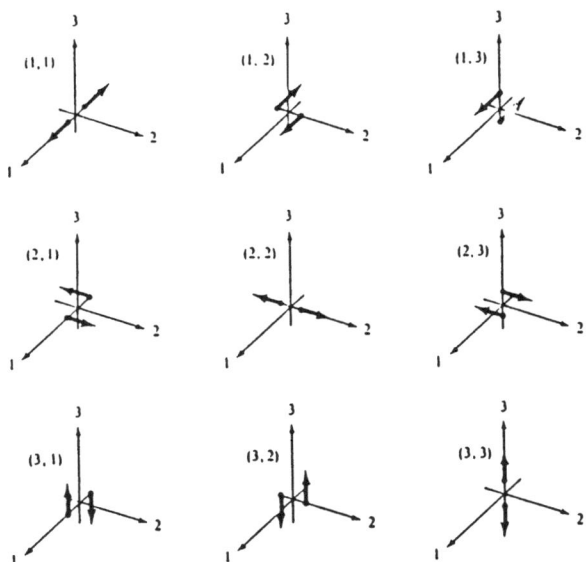

F IG. 8. Nine tensor components input at a located position for finite difference method.

4.2. Dynamic Green's Function $G_{ij,k'}$ by Finite Difference Method

In order to obtain dynamic Green's functions by finite difference methods, the nine tensor components (such as shown in Fig. 8) were input at a located position (e.g. position No. 1 in Fig. 7). Displacement at the surface of a transducer simulated by using these inputs is given in Fig. 9. Nine components are obtained for each of the six transducers (a total of 54 signals).

4.3. Source Function Tensor D_{jk}

Using the transfer function of the measuring system S_i and the dynamic Green's function, the source function tensor D_{jk} due to the nucleation of quasi-cleavage in A470 steel was estimated by the deconvolution integral in eqn. (7). A typical result is

$$D_{jk}(t) = \bar{D} \begin{pmatrix} 0.41 & 0.04 & -0.22 \\ & 0.17 & 0.01 \\ \text{sym.} & & 1 \end{pmatrix} T(t)$$

$$\bar{D} = 4.02 \times 10^{-4} \text{ Nm}$$

where $T(t)$ is a time function, given in Fig. 10. This time function shows that a microcrack nucleates in $0.3\ \mu$s.

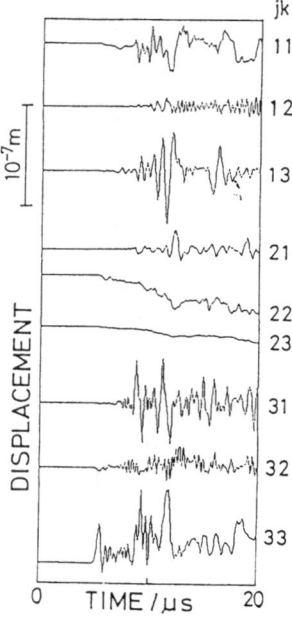

FIG. 9. Displacement at the surface position of a transducer simulated by finite difference method.

4.4. Fracture Mode and Cracking Size

As shown in Fig. 10, the inclination of the microcrack plane to the main crack surface was 15·0° and the displacement direction 58°, indicating that this microcrack formed subject to mixed mode conditions.

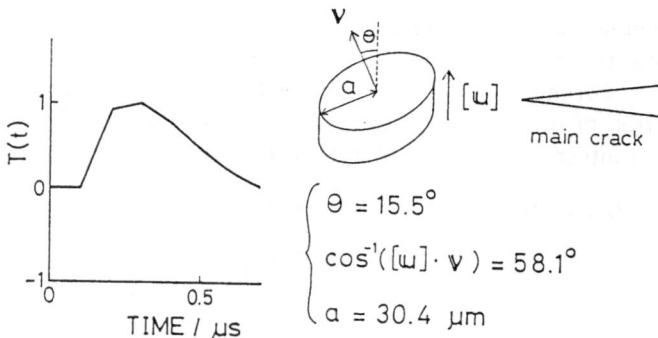

$$\Theta = 15.5°$$
$$\cos^{-1}([u] \cdot v) = 58.1°$$
$$a = 30.4 \ \mu m$$

FIG. 10. The obtained result of a source time function and the inclination of cracking plane and displacement direction of the microcrack.

Under the assumption of a constant stress, the fracture radius a can be evaluated by [11]

$$a = \left(\frac{3}{8\pi} \frac{1-2v}{1-v^2} \frac{D_{jj}}{\sigma \cos^2 \theta} \right)^{1/3}$$

In this case a was estimated as $30.4\,\mu m$.

Table 1 shows the obtained radius of a quasi-cleavage facet. The radius ranged from 20 to $40\,\mu m$, which corresponds to the cleavage facets in Fig. 11.

TABLE 1
The results of nucleation rate and radius of quasi-cleavage facet

No.	$\bar{D}(10^{-3}\,Nm)$	$\Delta t(\mu s)$	$a(\mu m)$
1	0·50	0·2	30
2	0·39	0·3	27
3	0·43	0·2	28
4	1·2	0·4	40
5	1·5	0·7	43
6	0·22	0·2	22
7	0·33	0·2	26
8	1·1	0·4	38

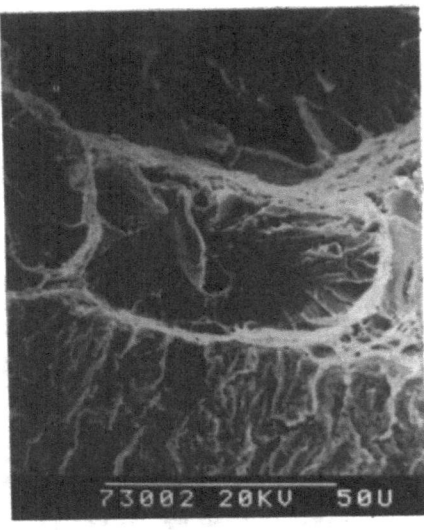

FIG. 11. The cleavage facet of A470.

5. APPLICATION TO THE MICROFRACTURE PROCESS IN ALUMINA CERAMICS [17, 18]

During fracture toughness testing of alumina using 1 in compact tension specimens, many AE signals were detected before macroscopic crack extension (Fig. 12). At the initial stage, microcracks nucleate at the

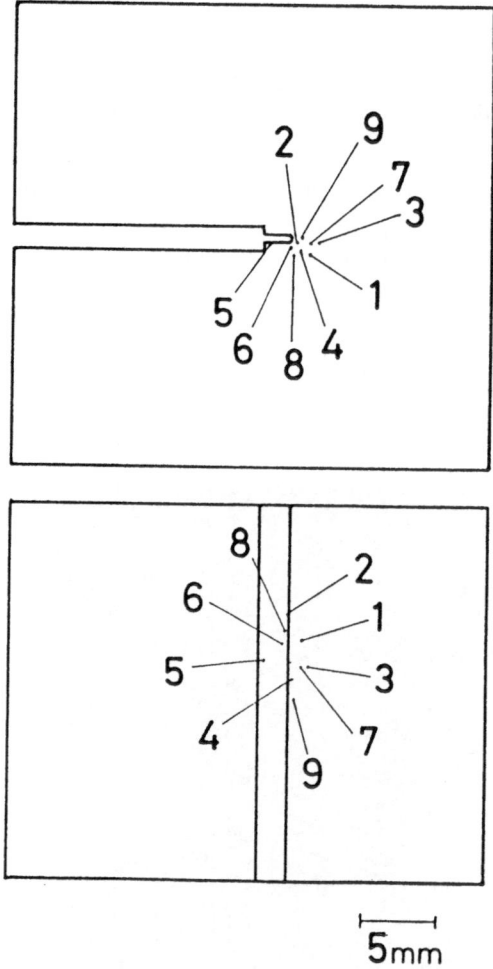

5mm

FIG. 12. The results of three dimensional location of AE sources detected during fracture toughness testing using CT (1 inch thickness) specimen.

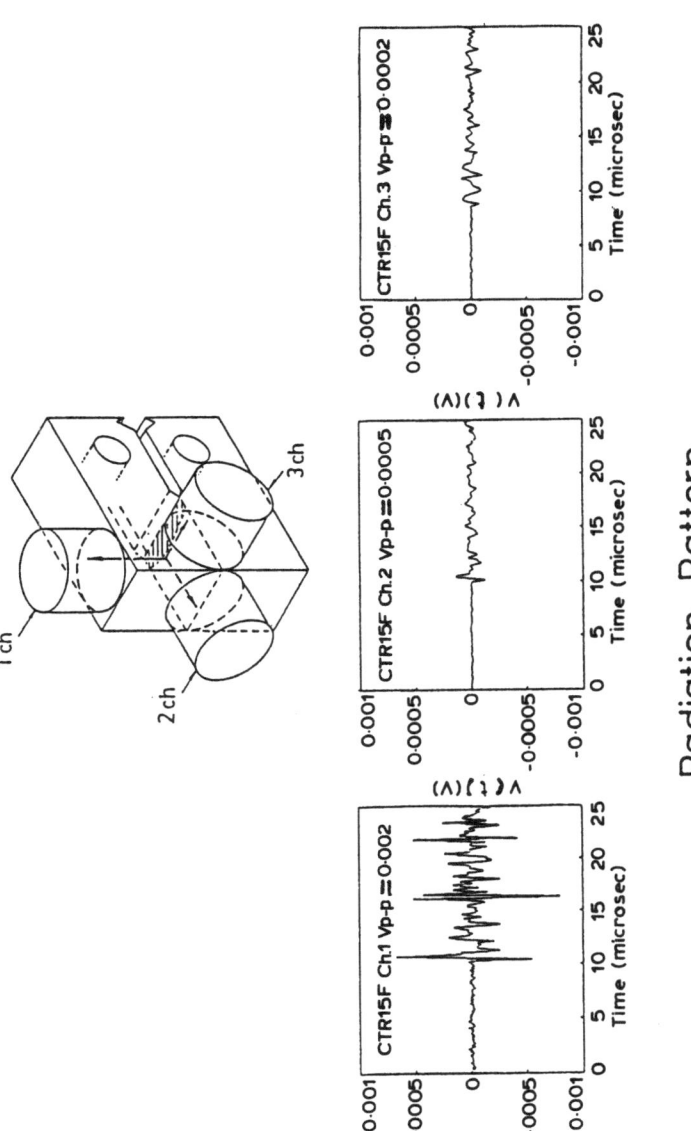

Radiation Pattern

FIG. 13. AE waveforms detected by three transducers.

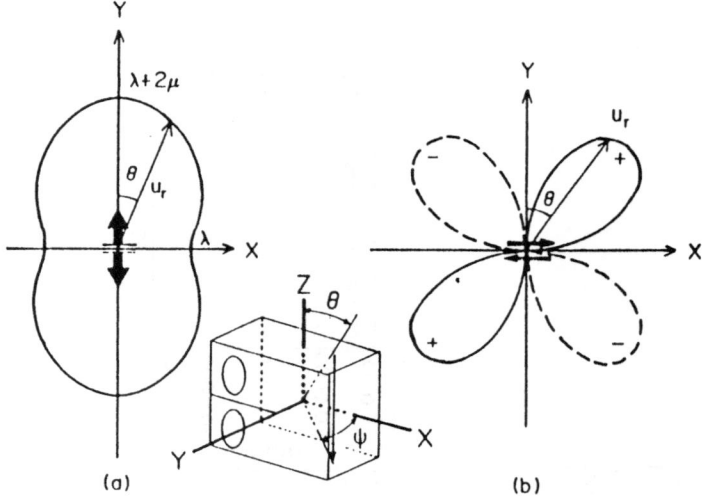

FIG. 14. Radiation pattern of AE signal of (a) Mode I tensile cracking and (b) Mode II, III shear cracking.

middle of the specimen thickness and at the latter stage near the specimen surface. Figure 13 gives AE waves of an event detected at three transducers, which indicate that this microcrack consists of Mode I cracking, because the radiation pattern in Mode I always has the same phase, as shown in Fig. 14 (obtained by eqn. (1)).

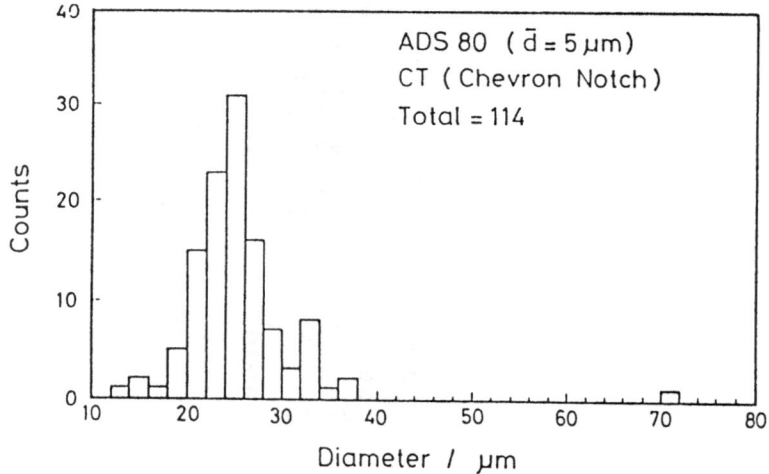

FIG. 15. The distribution of crack diameter evaluated by AE source characterization.

FIG. 16. Microfracture process which is observed directly in scanning electron microscope.

Figure 15 shows the distribution of microcrack diameters: 10–50 μm intergranular cracks were quantitatively evaluated and these microcracks correspond to the Mode I cracks directly observed during fracture toughness testing in scanning electron microscopy (Fig. 16).

At the latter stage larger 70–100 μm cracks are observed. These are due to the intergranular cracking of coarse grains and/or the coalescence of smaller intergranular cracks. These larger cracks become the origin of final fracture because the stress intensity factor K due to these cracks ($K = \sigma \sqrt{\pi a} \simeq 12 \, \text{kg/m}^{3/2}$ at $\sigma = 30 \, \text{kg/mm}^2$, $a = 50 \, \mu\text{m}$) reaches a K_{IC} value evaluated by DT testing.

It can be concluded that in ceramics there exist two types of microcracking. Smaller ones which contribute to the enhancement of fracture toughness and larger ones which are directly connected with unstable fracture.

6. CONCLUSIONS

A system to obtain the source function of microcracking is established. Using this system of AE wave analysis, a source function tensor was obtained and then the fracture mode, and fracture rise time in metals and ceramics were evaluated, in addition to the three-dimensional location of these microcracks.

REFERENCES

1. KISHI, T. and KURIBAYASHI, K., *Fundamentals of Acoustic Emission*, Joint Meeting of Acoustical Societies of America and Japan 1978, p. 105.
2. KISHI, T., *Proc. Joint JSME-SESA Conf. Exp. Mech.*, Honolulu, 1982, p. 773.
3. KISHI, T. and MORI, Y., ASTM STP 697 1979, p. 131.
4. SAKAKIBARA, Y., KISHI, T. and YAMAGUCHI, K., *Progress in Acoustic Emission II*, Japan Soc. NDI, Zao, 1984, p. 278.
5. KISHI, T., *Z. Metallkunde*, **76** (1985), 512.
6. AKI, A. and RICHARD, P. G., *Quantitative Seismology 1*, W. H. Freeman and Company, San Francisco, 1980, p. 599.
7. BURRIGE, R. and WILLIS, J. R., *Proc. Camb. Phil. Soc.*, **66** (1969), 433.
8. MURA, T., *Bull. Am. Phys. Soc.*, **9** (1961), 521.
9. CERANOGLU, A. N. and PAO, Y. H., *J. Appl. Mech.*, **48** (1981), 125.
10. HSU, N. N., SIMMONS, J. A. and HARDY, S. C., *Mat. Eval.*, **35** (1977), 100.
11. OHISA, N. and KISHI, T., *Proc. Joint JSME–SESA Conf. on Exp. Mech.*, Honolulu, 1982, p. 359.
12. KNOPOFF, L., *J. Appl. Phys.*, **29** (1958), 661.
13. KISHI, T., OHNO, K. and KURIBAYASHI, K., *HIHAKAI–KENSA*, **30** (1981), 911.
14. KIMURA, T., WAKAYAMA, S., OHIRA, T. and KISHI, T., *Proc. 4th AE Symp.*, Japan Soc. NDI, Tokyo, 1983, p. 19.
15. KISHI, T. and OHIRA, T., *Trans. Japan. Inst. Metals*, **24** (1983), 255.
16. ENOKI, M., Thesis of Master Course, the University of Tokyo, 1986.
17. KISHI, T. and WAKAYAMA, S., *4th Int. Symp. on Frac. Mech. Ceramics*, Virginia, 1985.
18. WAKAYAMA, S., Private communication.

6

Initial Crack Growth Tearing Resistance in Transformation Toughened Ceramics

J. W. HUTCHINSON

Division of Applied Sciences, Harvard University, Cambridge, Massachusetts 02138, USA

ABSTRACT

Transformation toughened ceramics display a resistance to crack advance requiring an increasing level of applied stress intensity to advance the crack tip. In this paper the initial slope of the resistance curve is determined. Two material models are considered. In each, only a net dilatational transformation is considered. In one, the transformation is assumed to be triggered at a critical value of the mean stress. In the other, the transformation is assumed to take place when the maximum shear stress reaches a critical value.

1. INTRODUCTION

Tearing resistance in transformation toughened ceramics was predicted in Refs. 1 and 2 and *R*-curves have been experimentally measured in Ref. 3 for monoclinic zirconia containing a second phase of partially stabilized tetragonal zirconia. The high stresses at the tip of a macroscopic crack cause tetragonal particles of zirconia to transform to a monoclinic form producing an irreversible transformation strain in the particles. Transformation of an unconstrained particle involves a dilatation of approximately 4% and a shear strain of about 16%. A particle embedded in an untransforming matrix transforms into a number of parallel twins with alternating signed shears so that the net shear in the particle is a small fraction of 16%. As discussed in Ref. 4, preliminary modeling which includes both shear and dilatational transformations indicates that the dilatational component is the more important in transformation tough-

ening of ceramics. This will be the assumption made here. The dilatational transformation strain in the particle is denoted by θ_p^T and the
dilatational transformation strain of the matrix-particle combination is
given by $\theta^T = c\theta_p^T$, where c is the volume concentration of the particles
which is typically in the range 20–40%.

Conditions for nucleation of the transformation in the matrix constrained particles are not well established [4]. Several candidates have
been put forth and used in the various investigations of the mechanics of
transformation toughening. Two of these will be used in this paper. The
simplest from a mathematical point of view is the assumption that
transformation will occur when the mean stress reaches a critical value,
i.e. θ^T occurs in the particle–matrix combination when

$$\sigma_m \equiv \sigma_{kk} = \sigma_m^c \qquad \text{(Case A)} \qquad (1)$$

The second is based on attaining a critical value of the maximum shear
stress, i.e. θ^T occurs in the particle–matrix combination when

$$\tau_{max} = \tau_c \qquad \text{(Case B)} \qquad (2)$$

This paper focusses on the behavior of a macroscopic crack immediately following initiation of crack growth. In particular, a theoretical
calculation of the initial slope of the resistance curve is made for each of
the two nucleation cases, A and B. The calculations are natural exten-
sions of the work in Refs. 1, 2 and 4, and it is assumed that the reader is
familiar with these papers. The present calculations invoke the following:

(I)   All particles in a volume element transform, resulting in the full
      dilatational transformation θ^T when the critical stress condition,
      either A or B, is reached. In the nomenclature of Ref. 2, supercriti-
      cal transformation is assumed.

(II) To determine the zone of transformation, the stress field at the tip
      of the crack without transformation will be used to locate the
      boundary where the critical stress condition is met. The perturb-
      ing influence of the transformation on the zone size and shape is
      ignored.

The limitations of these and other assumptions will be discussed at the
end of the paper.

The analysis is carried out within the framework of plane strain. The
transformation zone is assumed to be small compared to all in-plane
geometric lengths so that small scale transformation may be assumed.
Thus, the field surrounding the transformation zone is the classical

singular field for an elastic mode I crack of the form

$$\sigma_{ij} = \frac{K}{\sqrt{2\pi r}} \tilde{\sigma}_{ij}(\phi) \tag{3}$$

where K will be called the applied stress intensity factor and where r and ϕ are planar polar coordinates centered at the crack tip as indicated in Fig. 1. In particular, the distributions of the mean stress and the maximum shear stress are given by

$$\sigma_m = \frac{K(1+v)}{3} \left(\frac{\pi r}{2} \right)^{-1/2} \cos(\phi/2) \tag{4}$$

and

$$\tau_{max} = K(8\pi r)^{-1/2} \sin \phi \tag{5}$$

where v is Poisson's ratio.

A singularity of the form (3) exists at the tip of the crack within the transformation zone except that its amplitude, K_{tip}, is altered by the existence of the transformation itself. The solution to the small scale transformation problem supplies the relation between K_{tip} and K, and it is this relation which is used to predict the effects of transformation on the crack growth behavior. In Refs. 1, 2 and 4 it is assumed that a constant, critical value of crack tip intensity, K_{tip}^c, is required to initiate and sustain crack growth, and this assumption will be invoked here. The main aim of this paper is the prediction of dK/da just after initiation of crack growth under the assumption that

$$K_{tip} = K_{tip}^c \tag{6}$$

is maintained. It will be argued at the end of the paper that the full resistance curve should be computed under assumptions which are less

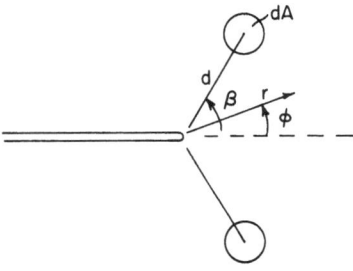

FIG. 1. Conventions at the crack tip.

restrictive than (I) and (II) invoked above. Such calculations will almost certainly require reasonably heavy numerical work. The present results, which are simple and exact under the stated assumptions, bring out essential trends.

The main result needed to carry out the present analysis is that for the effect on K_{tip} of spots of transformation. As in Fig. 1, suppose the material in the two symmetrically disposed cylinders undergoes an unconstrained transformation dilatation θ^T. With dA denoting the element of area of each spot, the change in K_{tip} due to the transformation in the two spots is [2]

$$dK_{tip} = \Gamma dA\, d^{-3/2} \cos\left(\frac{3}{2}\beta\right) \tag{7}$$

where

$$\Gamma = E\theta^T/[3(1-v)\sqrt{2\pi}] \tag{8}$$

and E is Young's modulus.

2. dK/da FOR CASE A

Assuming the mean stress is given by (4) and assuming that transformation occurs when (1) is met, the transformation zone associated with any history of crack advance is the union of all material points for which (1) is attained at any time in the history. For a stationary crack experiencing a monotonically increasing K the boundary of transformation zone is given by (see Fig. 2)

$$R(\phi) = C_A \left(\frac{K}{\sigma_m^c}\right)^2 \cos^2(\phi/2) \tag{9}$$

where

$$C_A = 2(1+v)^2/(9\pi) \tag{10}$$

Case A   Case B

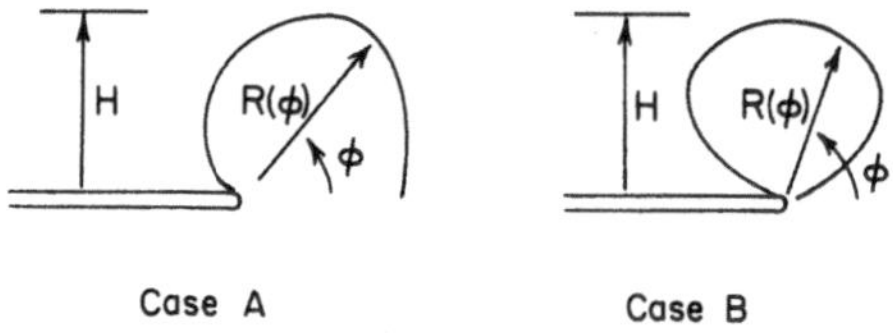

FIG. 2. Notation and sketches of boundary of transformation zones.

From (7), the near-tip intensity factor is given by

$$K_{tip} = K + \int_A \Gamma r^{-3/2} \cos(3\phi/2) dA \tag{11}$$

The integral in (11) is taken over the area A of the transformation zone in the upper half plane. For Case A, this integral is exactly zero giving for the stationary crack $K_{tip} = K$. Thus, by (6), initiation of crack growth in Case A occurs when

$$K = K_c \equiv K_{tip}^c \tag{12}$$

Now consider a small (infinitesimal) increment of crack growth Δa with K increased so as to maintain (6). That is, Δa and ΔK must be such that $\Delta K_{tip} = 0$. Refer to Fig. 3 which shows the initial transformation

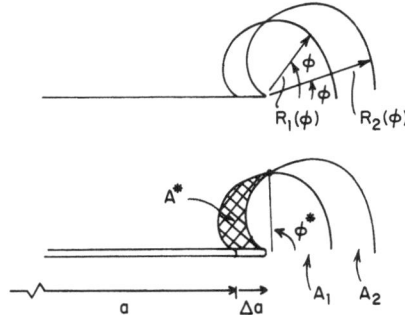

FIG. 3. Zone of stationary crack at initiation of crack growth at K, and region for which mean stress exceeds σ_m^c following crack advance Δa with applied stress intensity increased to $K + \Delta K$. Sketches are for Case A with similar sketches for Case B. A_1: transformation zone of initial stationary crack; A_2: region for which $\sigma_m \geq \sigma_m^c$ for crack of length $a + \Delta a$ at $K + \Delta K$.

zone and the region for which $\sigma_m \geq \sigma_m^c$ for the crack of length $a + \Delta a$ loaded at $K + \Delta K$. Since the transformed zone is the union of all points which have experienced a mean stress greater or equal to σ_m^c, it follows that K_{tip} is given by (11) where A is the union of A_1 and A_2 in Fig. 3. Furthermore, the integral over A_2 in (11) is identically zero for Case A, as is readily verified. Thus, for the initial increment of crack growth we require

$$\Delta K_{tip} = \Delta K + \int_{A^*} \Gamma r^{-3/2} \cos(3\phi/2) dA = 0 \tag{13}$$

where A^* is that part of A_1 not contained in A_2, i.e. the cross-hatched region in Fig. 3.

With the tip of the crack of length $a + \Delta a$ as origin for the planar polar coordinates,

$$\int_{A^*} r^{-3/2} \cos(3\phi/2) \, \mathrm{d}A = \int_{\phi^*}^{\pi} \mathrm{d}\phi \int_{R_2(\phi)}^{R_1(\phi)} r^{-1/2} \cos(3\phi/2) \, \mathrm{d}r$$

$$= 2 \int_{\phi^*}^{\pi} [R_1(\phi)^{1/2} - R_2(\phi)^{1/2}] \cos \frac{3}{2} \phi \, \mathrm{d}\phi \tag{14}$$

By (9),

$$R_2(\phi) = C_A \left(\frac{K + \Delta K}{\sigma_m^c} \right)^2 \cos^2(\phi/2) \tag{15}$$

while a direct calculation for small Δa gives

$$R_1(\phi) = C_A \left(\frac{K}{\sigma_m^c} \right)^2 \cos^2(\phi/2) + g(\phi) \Delta a \tag{16}$$

where

$$g(\phi) = \sin \alpha / \sin(\phi - \alpha) \tag{17}$$

and

$$\tan \alpha = (R' \sin \phi + R \cos \phi) / (R' \cos \phi - R \sin \phi) \tag{18}$$

with $(\;)' \equiv \mathrm{d}(\;)/\mathrm{d}\phi$. The term $g(\phi)\Delta a$ simply represents the change in the radial coordinate due to a small shift of the origin to the tip at $a + \Delta a$.

For Case A, a direct reduction of (18) gives $\alpha = 3\phi/2 - \pi/2$ and (17) gives

$$g(\phi) = -\cos(3\phi/2)/\cos(\phi/2) \tag{19}$$

The angle ϕ^* satisfies $R_1(\phi^*) = R_2(\phi^*)$. Equating the expressions in (15) and (16), dividing by Δa, and taking the limit as $\Delta a \to 0$, one finds

$$\lambda \equiv 2C_A \frac{K_c}{(\sigma_m^c)^2} \frac{\mathrm{d}K}{\mathrm{d}a} = -\frac{\cos(3\phi^*/2)}{\cos^3(\phi^*/2)} \tag{20}$$

Another direct calculation using (16) and (17) gives

$$R_1^{1/2} - R_2^{1/2} = \frac{1}{2C_A^{1/2}} \frac{\sigma_m^c}{K_c} \cos(\phi/2) \left[-\frac{\cos(3\phi/2)}{\cos^3(\phi/2)} - \lambda \right] \Delta a \tag{21}$$

to lowest order in Δa. Lastly, (13) with (14) and (21) require

$$\lambda = \zeta \int_{\phi^*}^{\pi} [\cos (3\phi/2)(\cos (\phi/2))^{-3} + \lambda]\cos (\phi/2)\cos (3\phi/2)\,d\phi \quad (22)$$

where

$$\zeta = \frac{2}{9\pi}\left(\frac{1+v}{1-v}\right)\frac{E\theta^{T}}{\sigma_{m}^{c}} \quad (23)$$

Equations (20) and (22) provide the relation between the initial slope of the resistance curve, as measured by the nondimensional combination λ, to the nondimensional parameter ζ. The integration in (22) must be performed numerically. The simplest way to generate the relation between λ and ζ is to treat ϕ^* as an independent variable and evaluate λ from (20) and ζ from (22); ϕ^* must be in the range $(\pi/3, \pi)$. The results will be presented in a different nondimensional form which may be more convenient for comparison with experiment. The half-height of the transformation zone *at initiation* is given by

$$H = \frac{\sqrt{3}(1+v)^2}{12\pi}\left(\frac{K_c}{\sigma_m^c}\right)^2 \quad (24)$$

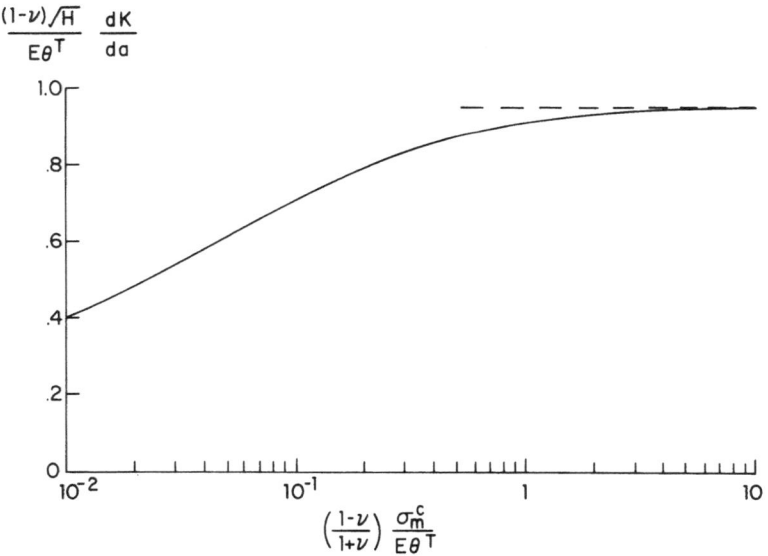

FIG. 4. Nondimensional initial slope of resistance curve for Case A where H is the transformation zone half-height at initiation of growth.

By eliminating K_c in favor of H, it is readily shown that

$$\frac{(1-v)\sqrt{H}}{E\theta^T}\frac{dK}{da} = \frac{1}{4(\sqrt{3\pi})^{1/2}}\frac{\lambda}{\zeta} \tag{25}$$

and this relation has been used to generate the plot in Fig. 4.

Before turning to Case B, we relate the present calculation for the initial slope of the resistance curve to an earlier calculation by McMeeking and Evans [1] which was performed to indicate the character of the resistance curve of a transformation toughened material.

3. McMEEKING AND EVANS' CALCULATION OF ΔK_{tip} VERSUS Δa FOR CASE A

From (7) it can be seen that transformed material lying behind the wedge specified by $|\phi| \le \pi/3$ reduces the near-tip stress intensity. Figure 5 shows the reduction in crack tip intensity $-\Delta K_{tip}$ as a function of finite amounts of crack advance Δa when the crack is advanced with K held

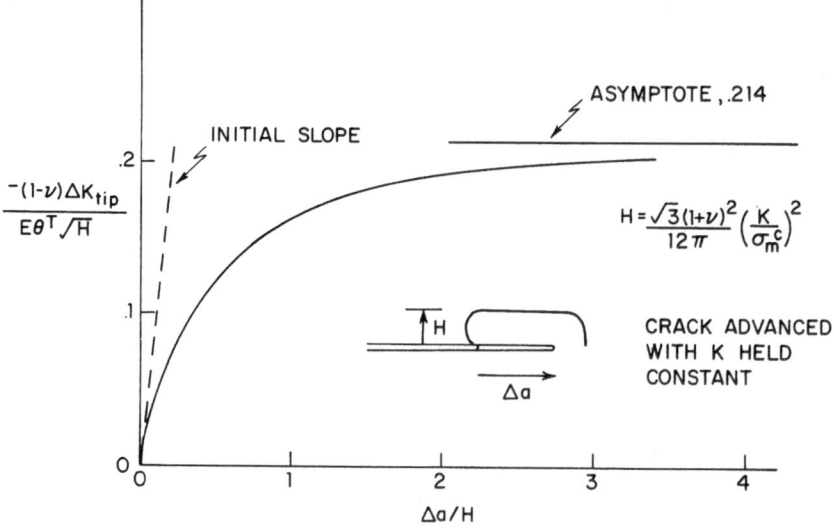

FIG. 5. Reduction in near-tip intensity as a function of crack advance when applied stress intensity factor is held constant [1].

constant. Using (11), one can show that this result is obtained from

$$-\frac{(1-v)\Delta K_{tip}}{E\theta^{T}\sqrt{H}}=\frac{16}{9\sqrt{6\pi}}\text{Real}\left\{\int_{\pi/3}^{\pi}([z(\phi)-\Delta a/H]^{-1/2}\right.$$

$$\left.-[z(\phi)]^{-1/2})\cos(\phi/2)\cos(3\phi/2)d\phi\right\} \tag{26}$$

where H is given by (24) with K replacing K_c and

$$z(\phi)=\frac{8}{3\sqrt{3}}\cos^{2}(\phi/2)[\cos\phi+i\sin\phi]$$

with $i=\sqrt{-1}$. The curve in Fig. 5 is the result of McMeeking and Evans [1] which was also plotted in their Fig. 5. This curve is not a resistance curve since it is computed with K held fixed so that K_{tip} diminishes as Δa increases. Nevertheless, it does clearly reveal the source of tearing resistance.

The full curve of $-\Delta K_{tip}$ versus Δa in Fig. 5 is also useful for present purposes in that it enables us to see how rapidly the curve departs from its initial slope. In the limit as $\Delta a \rightarrow 0$, (26) gives

$$-\frac{(1-v)\sqrt{H}}{E\theta^{T}}\frac{dK_{tip}}{da}=0.9518 \tag{27}$$

and this initial slope is shown in Fig. 5. Note that the full curve departs from the initial slope after very small amounts of crack advance. Similarly, we should expect that the initial slope of the resistance curve determined in this paper governs behavior only for a very small amount of crack advance.

4. dK/da FOR CASE B

The calculation of the initial slope of the resistance curve for Case B with the transformation condition in (2) follows closely that of Case A so less detail need be given. The main difference between the two cases is that K_{tip} and K are not equal for monotonic loading of the stationary crack in Case B, and this must be taken into account.

The zone boundary for the *stationary crack* (see sketch in Fig. 2, Case

B) is given by

$$R(\phi) = \frac{1}{8\pi} \left(\frac{K}{\tau_c}\right)^2 \sin^2 \phi \tag{28}$$

and, from (11), one can show by direct integration that

$$K_{\text{tip}} = K - \frac{2}{15\pi} \frac{E\theta^T K}{(1-v)\tau_c}$$

$$= K - \frac{8}{15\sqrt{2\pi}} \frac{E\theta^T}{(1-v)} \sqrt{H} \tag{29}$$

where, now,

$$H = \frac{1}{8\pi} \left(\frac{K}{\tau_c}\right)^2 \tag{30}$$

(We mention in passing that the result for the *steady-state problem* where the boundary is given by (28) for $|\phi| \leq \pi/2$ and by $R \sin \phi = \pm H$ for $|\phi| > \pi/2$ can also be obtained analytically as

$$K_{\text{tip}} = K - \frac{4(1+\sqrt{2})}{15\sqrt{\pi}} \frac{E\theta^T \sqrt{H}}{(1-v)}$$

$$= K - \frac{0 \cdot 3632 \, E\theta^T \sqrt{H}}{(1-v)} \tag{31}$$

An approximation to this result for Case B was obtained and discussed in Ref. 4.)

Imposition of the condition for crack advance (6) on (29) gives the value of K associated with initiation of crack advance, i.e.

$$K = K_c \equiv K_{\text{tip}}^c [1 - \mu]^{-1} \tag{32}$$

where

$$\mu = \frac{2}{15\pi} \frac{E\theta^T}{(1-v)\tau_c} \tag{33}$$

Note that K_c becomes unbounded for $\mu \to 1$. The present analysis, which neglects the perturbing influence of the transformation on the zone shape, is only accurate for relatively small values of μ, as will be discussed in more detail at the end of the paper. Nevertheless, there is an effect of transformation on initiation of growth from the stationary crack in Case B whose ramifications have not been fully considered.

The equation governing the first increment of growth in (13) is replaced by

$$\Delta K_{tip} = (1-\mu)\Delta K + \int_{A^*} \Gamma r^{-3/2} \cos(3\phi/2)\, dA = 0 \tag{34}$$

where A^* is again defined in Fig. 3, except that now regions A_1 and A_2 are defined by the transformation condition (2). Equation (14) continues to hold where now

$$R_2(\phi) = \frac{1}{8\pi}\left(\frac{K+\Delta K}{\tau_c}\right)^2 \sin^2\phi \tag{35}$$

and

$$R_1(\phi) = \frac{1}{8\pi}\left(\frac{K}{\tau_c}\right)^2 \sin^2\phi + g(\phi)\Delta a \tag{36}$$

In addition, after some algebraic manipulation one finds that $g(\phi) = -3\cos\phi$. The angle ϕ^* for which $R_1(\phi^*) = R_2(\phi^*)$ satisfies

$$\lambda_B \equiv \frac{1}{4\pi}\frac{K_c}{\tau_c^2}\frac{dK}{da} = -\frac{3\cos\phi^*}{\sin^2\phi^*} \tag{37}$$

Finally, in the limit $\Delta a \to 0$, (34) gives

$$(1-\mu)\lambda_B = \frac{5}{4}\mu \int_{\phi^*}^{\pi} [3\cos\phi\,(\sin\phi)^{-2} + \lambda_B]\sin\phi\cos(\tfrac{3}{2}\phi)\,d\phi \tag{38}$$

Equations (37) and (38) provide the relation between λ_B and the parameter μ. With

$$H = \frac{1}{8\pi}\left(\frac{K_c}{\tau_c}\right)^2 \tag{39}$$

as the half-height of the zone at *initiation of growth*, one also finds

$$\frac{(1-\nu)\sqrt{H}}{E\theta^T}\frac{dK}{da} = \frac{4}{15\sqrt{2\pi}}\frac{\lambda_B}{\mu} \tag{40}$$

This is the equation used to plot the curve in Fig. 6. The curve becomes unbounded as $\mu \to 1$, but the results are not trustworthy at values of μ this large as already mentioned.

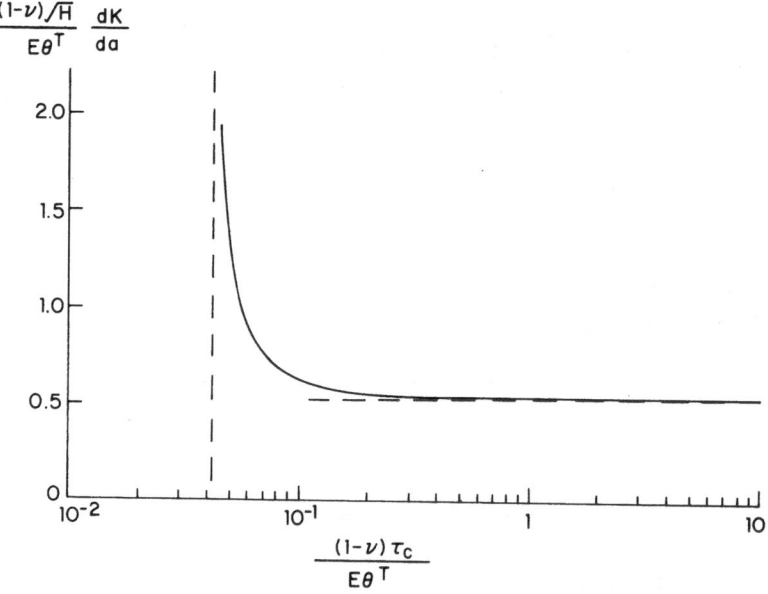

FIG. 6. Nondimensional initial slope of resistance curve for Case B where H is the transformation zone half-height at initiation of growth.

5. LIMITATIONS OF PRESENT ANALYSIS AND SUGGESTIONS FOR FURTHER WORK

As it becomes available, resistance curve data should supply valuable additional information to which the mechanics models can be compared and calibrated. The present calculations are just a first step in predicting resistance curves from the mechanics models. Hopefully, a comparison of theoretical predictions with experimental resistance curve data should shed further light on which, if either, of the two transformation conditions, A or B, is appropriate. Given either A or B as a starting point, there are three main limitations to the present results.

Firstly, as indicated by the example in Fig. 5, the initial slope of the resistance curve which has been computed here is expected to replicate the resistance curve for only small amounts of crack growth. For comparison with experimental data it is essential that the full resistance curves be calculated.

Secondly, the results in this paper, which were computed using the unperturbed stress field to locate the boundary of the transformation zone, are valid only for small ζ or μ. Analyses which account for the perturbation of the zone were carried out in Refs. 2 and 5 for the steady-state, small scale transformation problem for Case A. Those analyses indicate that the results of the unperturbed analysis are accurate for values of ζ as large as about $1/2$, and we expect the present results to be similarly limited. Comparable results for Case B do not exist. The unboundedness of K_c and of the nondimensional parameter involving the initial slope of the resistance curve in Fig. 6 should also emerge from a full analysis of Case B, but the value of μ at which the unboundedness occurs will not necessarily be unity.

Lastly, there is increasing evidence [4] that the supercritical transformation in which the material element undergoes complete transformation to θ^T may be the exception rather than the rule. Observations indicate that the density of transformed particles falls off smoothly, and not abruptly, with distance from the crack tip. This suggests that models based on subcritical transformation should also be analyzed.

ACKNOWLEDGEMENTS

This work was supported in part by the National Science Foundation under Grant MSM-84-16392, and by the Division of Applied Sciences, Harvard University. The author acknowledges helpful discussions with A. G. Evans and R. M. McMeeking.

REFERENCES

1. MCMEEKING, R. M. and EVANS, A. G., *J. Am. Ceram. Soc.*, **65** (1982), 242–6.
2. BUDIANSKY, B., HUTCHINSON, J. W. and LAMBROPOULOS, J. C., *Int. J. Solids Struct.* **19** (1983), 337–55.
3. SWAIN, M. V. and HANNINK, R. H. J., *Advances in Ceramics*, Vol. 12, American Ceramics Society, 1984, p. 225.
4. EVANS, A. G. and CANNON, R. M., *Acta Met.*, **34** (1986), 761–800.
5. ROSE, L. R. F., The size of the transformed zone during steady-state cracking in transformation-toughened materials, Aero. Research Report, Melbourne, Australia, 1986.

7

Some Aspects of the High Temperature Performance of Ceramics and Ceramic Composites

A. G. Evans and B. J. Dalgleish

Materials Program, College of Engineering, University of California, Santa Barbara, CA 93106, USA

ABSTRACT

Ceramics and ceramic composites are subject to creep rupture at elevated temperatures. The rupture strain in such materials has been shown to exhibit a major transition, from creep brittleness to creep ductility. The emphasis of the present article is on the definition of microstructures that provide ductility. For this purpose, the fundamental principles involved in high temperature flow and fracture are reviewed, and physical models of the ductile-to-brittle transition are presented. The mechancial phenomena involved in these considerations include: creep crack growth, crack blunting, flaw nucleation and stress corrosion.

1. INTRODUCTION

Ceramics are typically capable of withstanding higher temperatures than other materials. Hence, the substantial interest in such materials for heat engines [1, 2] bearings [3] etc. However, high temperature degradation phenomena exist that influence performance and reliability. The important degradation processes include: creep [4], creep rupture [5, 6], flaw generation [7], diminished toughness [8] and microstructural instability [9]. The fundamental principles associated with some of these degradation phenomena are reviewed, and prospects for counteracting the prevalent mechanisms are discussed.

The strength of a ceramic typically diminishes at elevated temperatures (Fig. 1), initially owing to the diminished potency of toughening mech-

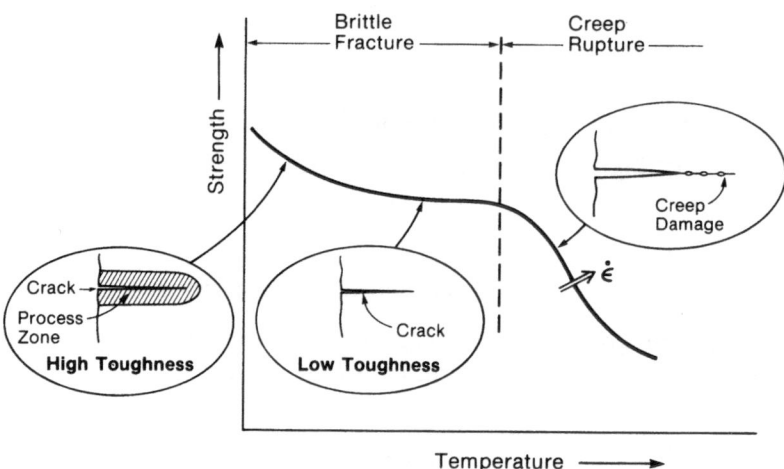

F<small>IG</small>. 1. A schematic illustrating trends in strength with temperature: the trends at lower temperature, in the brittle range, reflect the temperature dependence of the toughness: the trends at high temperature involve creep and creep rupture.

anisms [8, 10] and subsequently, following the onset of creep [11]. The degradation mechanisms that operate at the highest temperatures—in the creep regime—are emphasized in this article. A dominant microstructural consideration with regard to elevated temperature behavior is the existence of a *grain boundary phase* [9]. Such phases typically remain after liquid phase sintering and, frequently, are amorphous and silicate-based. The second phase constitutes a vehicle for rapid mass transport and dominates the creep [12], creep rupture [13] and oxidation [9] properties, as well as the microstructural stability. The *grain size* constitutes another important microstructural parameter, by virtue of its influence on the diffusion length and on the path density. Amorphous phase and grain size effects are thus emphasized in subsequent discussions of microstructural influences on high temperature properties.

The high temperature phenomenon that, in the broadest sense, has overwhelming practical significance especially for applications that allow only limited dimensional changes during operation, such as engine components, is the existence of a *transition* between creep brittleness and creep *ductility* [5, 6] (Fig. 2a, b). Fracture in the creep ductile regime occurs at large strains ($\varepsilon \gtrsim 0.1$, Fig. 2c), in excess of allowable strains in typical components. Consequently, when creep ductile behavior obtains, creep rupture is not normally a limiting material property. The current article thus emphasizes the material parameters that govern the brittle-to-ductile transition. However, it is recognized that this transition may not occur within a practical range in materials having undesirable microstructures. The emphases regarding microstructural design would thus differ from those presented in this article. Finally, some preliminary remarks and speculations regarding the influence of reinforcements, such as whiskers and fibers, on the high temperature performance are presented.

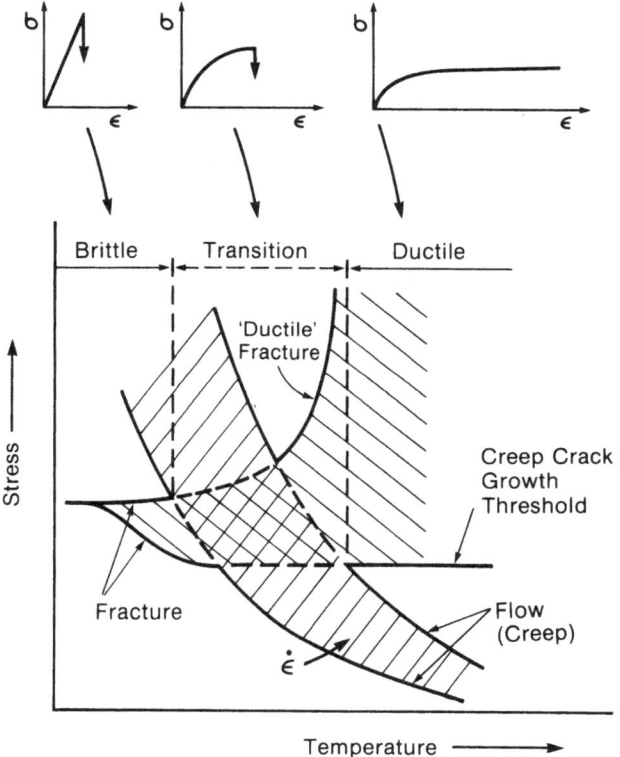

FIG. 2(a) Temperature dependent trends in flow and fracture revealing regions of brittle and ductile behavior.

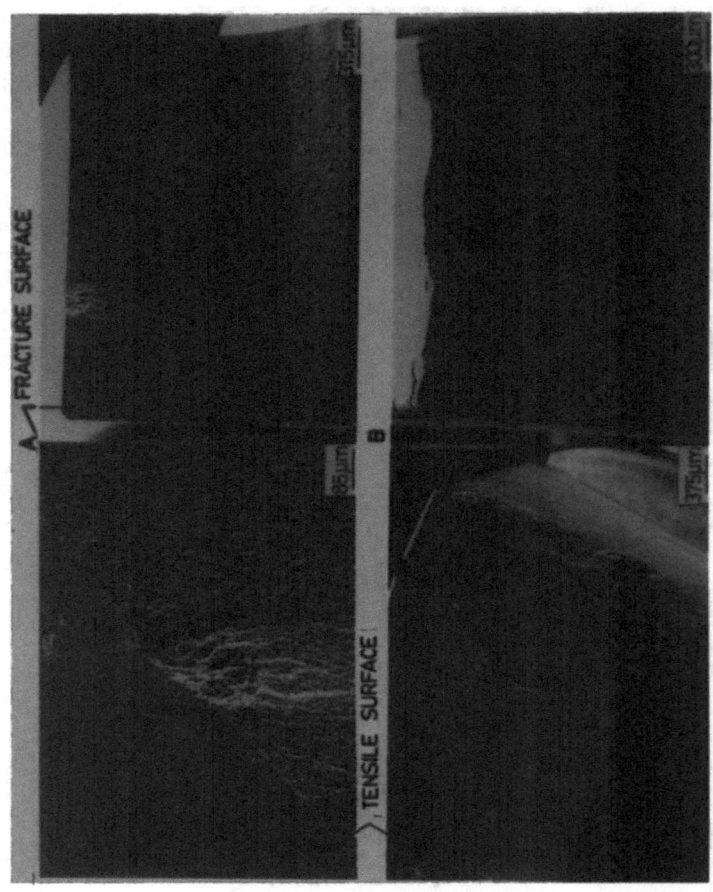

FIG. 2(b)   The differing fracture behaviors in the creep brittle (A) and creep ductile (B) regimes.

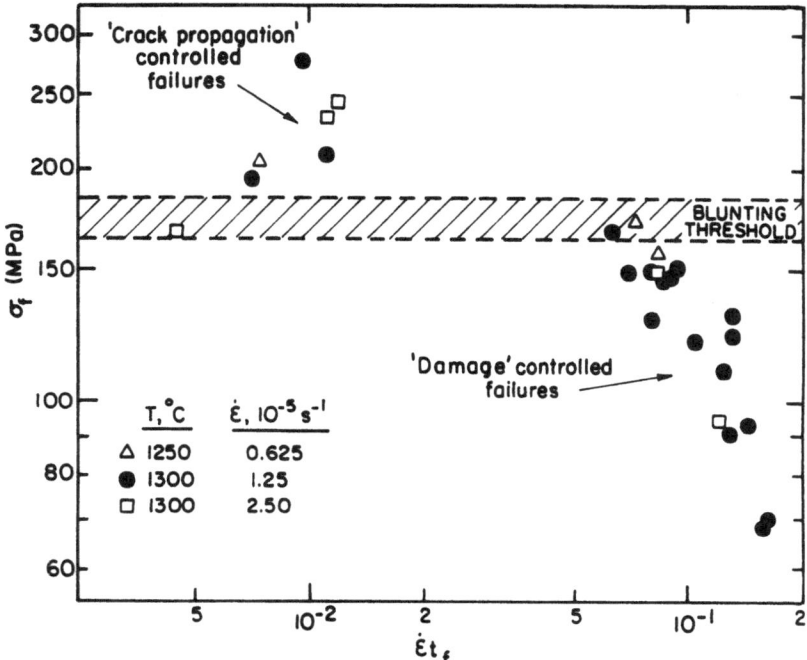

FIG. (2c) The increase in rupture strain that accompanies the transition between creep
brittleness and creep ductility. Data for Al_2O_3.

2. CREEP DUCTILITY

The transition to creep ductility represents, at the simplest level, a
competition between flow and fracture and thus occurs when the flow
stress becomes smaller than the stress needed to induce the unstable
extension of cracks (Fig. 2a). At a more sophisticated level, it is necessary
to specify the flow and fracture characteristics, subject to the imposed
loading. The *flow* in fine grained materials is supposedly governed by
diffusional creep and can usually be represented by a viscosity [4]

$$\eta = \frac{kTl^3}{D\delta\,[1 + D_v l/D\delta]\Omega} \tag{1}$$

where l is the grain size, D_v is the lattice diffusivity, Ω the atomic volume
and $D\delta$ is the diffusion parameter pertinent to either the grain boundary,
$D_b\delta_b$, or the grain boundary phase, $D_l\delta_o$. Some complicating effects occur
in very fine grained materials, involving nonlinearity at low stresses [14].

Such effects are not understood, but are presumed to relate to stress dependent interface limited phenomena (such as grain boundary sliding). Nonlinearities are also encountered in liquid phase sintered systems [15, 16], again for reasons not yet apparent.

The pertinent *fracture* processes are more complex. The fracture parameter seemingly having the greatest relevance to the brittle-to-ductile transition is the threshold stress intensity, K_{th}, that dictates the onset of crack blunting [6] (Fig. 3). Specifically, at stress intensities below K_{th}, crack growth is prohibited, whereupon creep ductility is assured (Fig. 2a). A conservative criterion for creep ductility is thus obtained by applying the inequality

$$K_{th}/\sqrt{a} \gtrsim 2\sigma_d/\sqrt{\pi} \tag{2}$$

where a is the radius of the largest crack that either pre-exists or may be

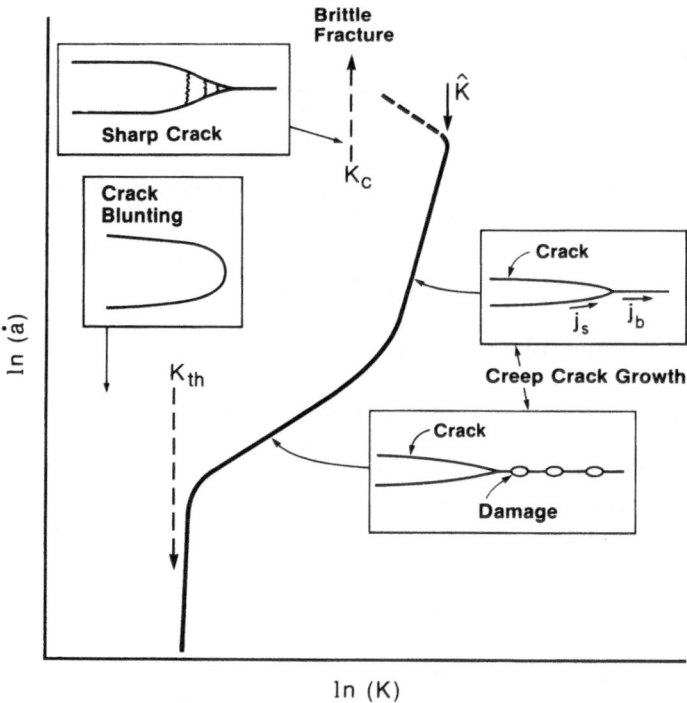

FIG. 3(a) A schematic illustrating the generalized dependence of high temperature crack growth rate, $\dot{a}$, on stress intensity, K, showing the differing regimes of crack growth.

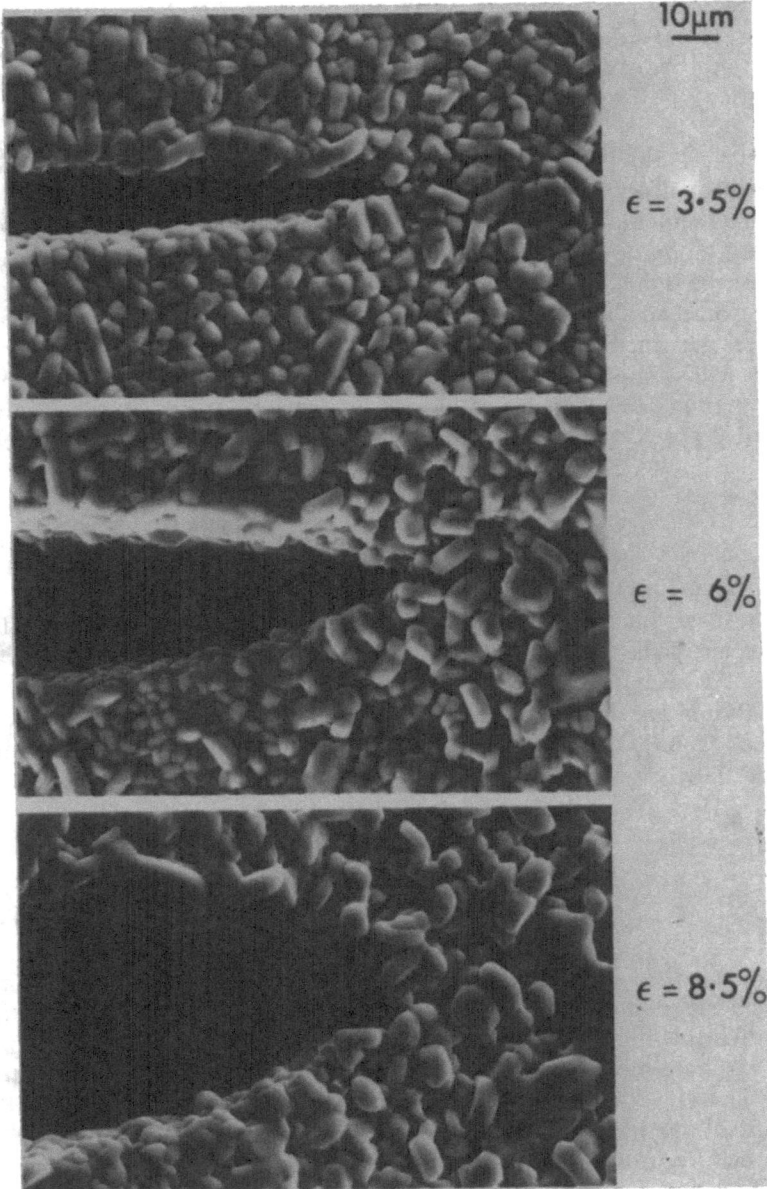

FIG. 3(b) A crack blunting sequence in SiC.

nucleated by heterogeneous creep, oxidation, etc., and σ_d is the design stress. However, it is also recognized that the permissible creep strain ε^* must not be exceeded within the lifetime, t^*, resulting in a second criterion,

$$\eta \gtrsim \sigma_d t^*/\varepsilon^* \tag{3}$$

The inequalities of eqns (2) and (3) must both be satisfied in order to assure adequate creep performance. Further progress thus requires appreciation of the creep crack growth threshold, as well as an understanding of the dominant high temperature flaws.

In some materials, significant creep crack growth is not encountered before the ductility transition. For such materials, the critical stress intensity, K_c is presumed to be the relevant fracture parameter, replacing K_{th} in eqn. (2). Consequently, K_c at elevated temperatures is also afforded consideration.

3. CREEP CRACK GROWTH

3.1. Creep Crack Growth Mechanisms

The basis for comprehending creep crack growth mechanisms is the character of the crack tip when diffusion operates, at elevated temperatures. At such temperatures, chemical potential continuity and force equilibrium are demanded at the crack tip [17]. Hence, since cracks are typically intergranular at high temperatures [5, 6, 15], the crack tip must be partially blunt (Fig. 4) in order to satisfy the equilibrium relations [17],

$$\gamma_b = 2\gamma_s \cos \psi$$
$$\gamma_s \kappa_o = \sigma_o \tag{4}$$

where ψ is the dihedral angle, γ_b and γ_s are the grain boundary and surface energies, respectively, κ_o is the surface curvature at the crack tip and σ_o is the normal stress on the grain boundary at the tip intersection. The resultant tip configuration, as well as the corresponding crack tip field are very different from those associated with the sharp cracks involved in brittle fracture. Consequently, the conditions for extension of the crack cannot be readily related to the ambient fracture toughness. Instead, the crack growth mechanisms involve the removal of material from the crack tip region (by diffusion or viscous flow), resulting in the creation of a new crack surface. Two categories of such mechanisms

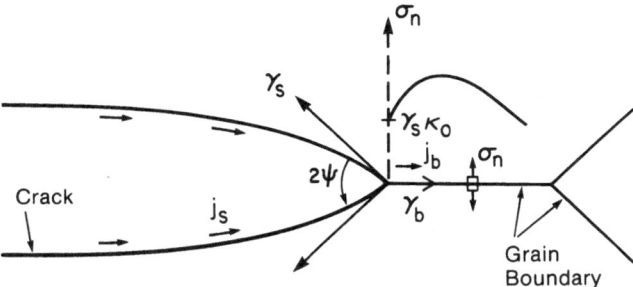

a) Diffusive Crack Growth

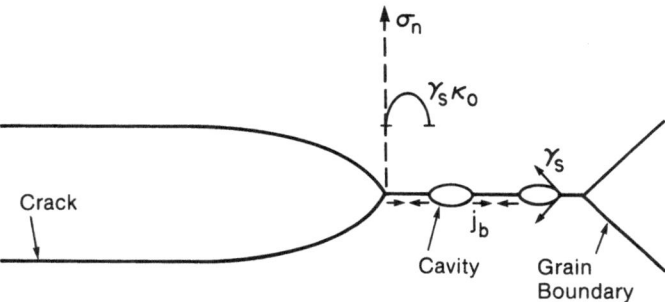

b) Damage Mechanism of Crack Growth

FIG. 4. Schematic illustrating the crack morphology and the mass fluxes accompanying intergranular high temperature creep crack growth, (a) direct extension mechanism, (b) damage mechanism.

typically dominate: *direct extension* mechanisms that entail matter transport over relatively large distances [17, 18] (Fig. 4a), and *damage mechanisms* that involve small scale mass transport within a zone directly ahead of the crack tip [19, 20] (Fig. 4b). However, the mechanistic details are sensitive to various aspects of the microstructure.

Creep crack growth rates in ceramics that exhibit Newtonian behavior typically satisfy the non-dimensional form:

$$K/\sigma_o\sqrt{L} = F(M) \qquad (5)$$

where L is a characteristic length for grain boundary diffusion, and F is a function of various microstructural features, such as grain size and cavity spacing. Typically, both σ_o and L depend on crack velocity, resulting in nonlinear crack growth rates

$$\dot{a} = \dot{a}_o(K/K_c)^n \qquad (6)$$

A. G. Evans and B. J. Dalgleish

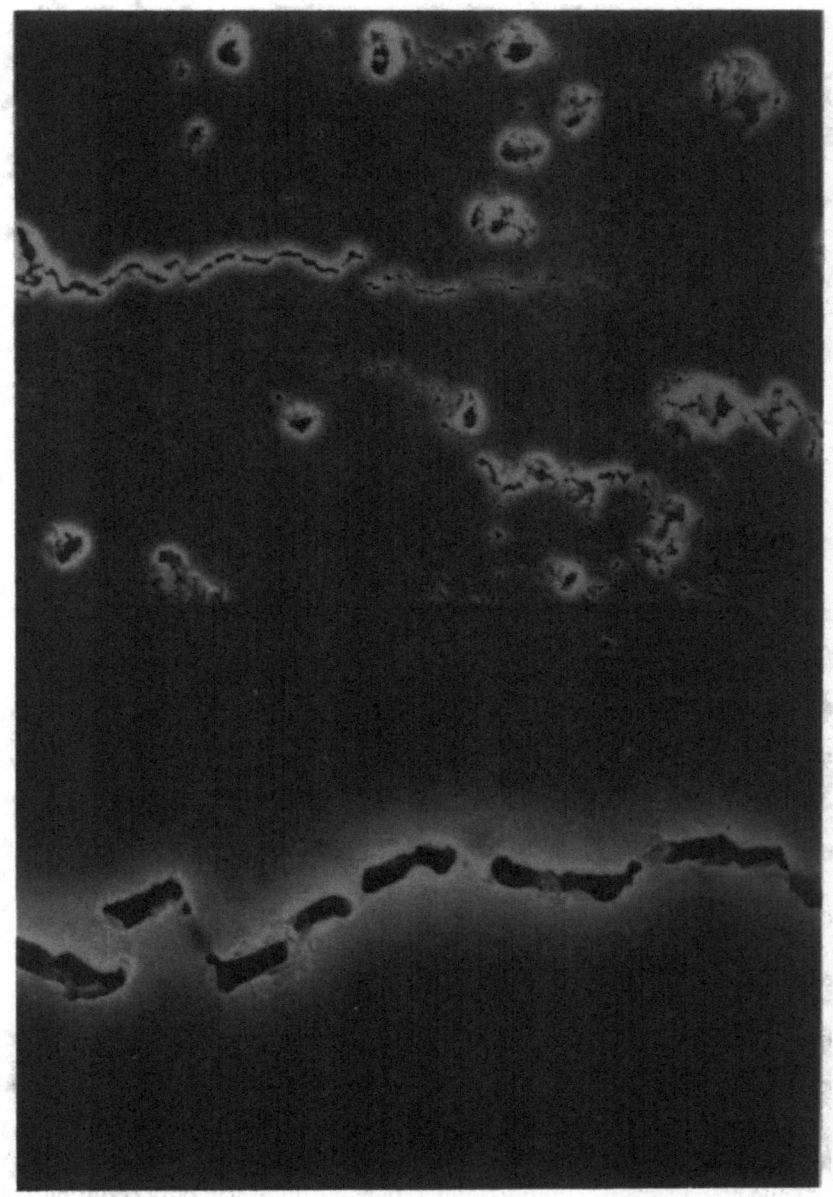

FIG. 5. Amorphous ligaments behind the crack tip: (a) observations for Al_2O_3/SiO_2.

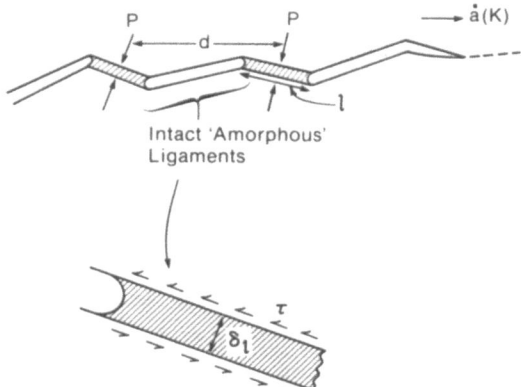

F IG. 5—*contd.* (b) schematic of crack surface tractions.

where $\dot{a}_0$ and n are material sensitive coefficients. In particular, the magnitude of n depends sensitively on the dominant mechanism and the choice of boundary conditions. Selection of conditions that pertain to the actual crack growth problem of interest is thus a crucial aspect of comparing crack growth measurements with predictions.

In some materials, especially those containing amorphous phases, intact ligaments of amorphous material remain behind the crack tip [16] (Fig. 5). These ligaments enforce crack surface tractions that reduce the tip K and thus impede crack growth. Such wake effects need to be incorporated into generalized models of creep crack growth. Some of the relevant models and the associated conditions are described below.

3.1.1. Direct extension mechanisms

Direct crack extension involves the mass flow depicted in Fig. 4a. The flux within the crack is directed toward the tip, while the local grain boundary flux occurs away from the tip, causing net removal of matter from the crack [17, 18]. The deposition of matter onto the grain boundaries is accommodated by grain displacements normal to the crack plane, resulting in *work* done on the system. The work done compensates for the increase in both surface energy and strain energy, thereby allowing crack extension to proceed with a net reduction in free energy. Crack growth rate predictions have been performed for cracks located at a bicrystal boundary, wherein matter deposition is accommodated elastically. The corresponding viscoelastic behavior pertinent to fine-grained polycrystals has yet to be evaluated. The importance of grain size is thus, presently, unknown. Nevertheless, the elastic results provide useful insights. The nondimensional crack growth rate when matter transport involves surface diffusion along the crack has the form [17],

$$\frac{\dot{a}\Omega^{2/3}}{D_b\delta_b} \approx 10 \left[\frac{K^2}{E\gamma_s(2-\gamma_b/\gamma_s)} \right]^6 \left(\frac{D_s\delta_s}{D_b\delta_b} \right)^4 \left(\frac{E^3\Omega^{5/3}}{kT\gamma_s^2} \right) \tag{7}$$

where E is Young's modulus and the subscript s refers to the surface. The corresponding relation when the crack contains an amorphous fluid phase that 'wets' the crack surfaces is [18]

$$\frac{\dot{a}\Omega^{2/3}}{D_b\delta_b} \approx 0.01 \left[\frac{K^2}{E\gamma_l(2-\gamma_b/\gamma_l)} \right]^2 (c_o\Omega) \frac{EkT\Omega}{(\eta_l D_b\delta_b)^2} \tag{8}$$

where c_o is the equilibrium concentration of solid dissolved in the liquid. These results clearly indicate the relative role of the mass flow parameters, D_b and η_l, as well as important effects of the dihedral angle (i.e. of γ_l/γ_b). Furthermore, it is noted that the crack growth rate is predicted to vary as a nonlinear function of K, due to the nonlinear relation between crack velocity and the predominant diffusion lengths (e.g. L in eqn. (5)).

Operation of the above mechanism in polycrystals is restricted by the ability of cracks to circumvent grain junctions. Specifically, when the crack does not contain a wetting fluid, the dihedral angle, ψ, is large and substantial mass flow is needed to achieve crack extension across a grain junction. Consequently, only the relatively narrow cracks that obtain at higher velocities extend by this mechanism. However, when a wetting fluid is located in the crack ($\psi \to 0$ or $\gamma_l \to \gamma_b/2$), the crack can remain as a narrow entity [18], even at low velocities, and extend beyond the grain junction. For this reason, a wetting fluid may be regarded as a prime source of high temperature stress corrosion.

Materials that contain a continuous amorphous phase may be subject to an alternative direct crack advance mechanism [13]. In this instance, an amorphous phase meniscus at the crack tip (Fig. 6) simply extends along the grain boundary, causing the crack to grow, and leaving amorphous material on the crack surface. Analysis of this process has been conducted subject to the conditions: the amorphous phase is thin, the grain displacements are discretized by the sliding of grain boundaries ahead of the crack and such displacements are accommodated by viscous creep of the surrounding solid. Then, crack growth is highly constrained

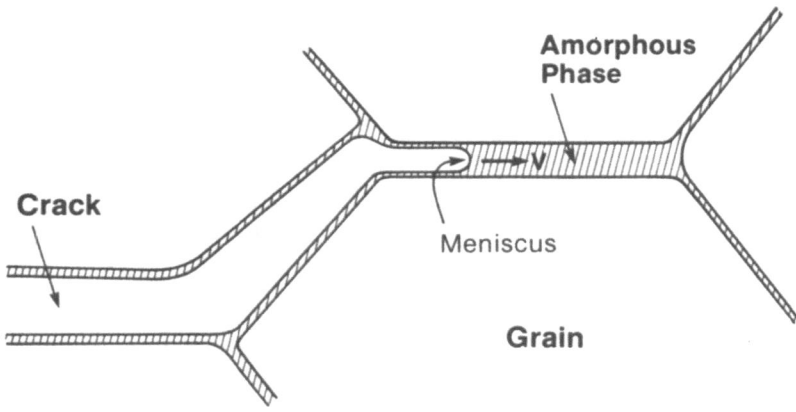

F<small>IG</small>. 6. A mechanism of creep crack growth in materials that contain a thin amorphous grain boundary phase.

and the crack growth rate has the form [13],

$$\dot{a} \approx \frac{K D_1 \Omega}{k T l^{3/2} [\delta_c / \delta_o - 1]} \tag{9}$$

where δ is now the amorphous phase thickness (the subscripts o and c refer, respectively to the initial value and the value when the grains at the crack tip separate). Unfortunately, it is not possible to compare eqn. (9) with eqn. (8) because of the very different material responses used to derive the results. Nevertheless, it is noteworthy that the crack velocity in eqn. (9) is insensitive to the thickness of the second phase, δ_o, but strongly dependent on grain size.

3.1.2. Damage mechanisms

The prevalent mechanism of damage enhanced crack growth involves the nucleation and growth of cavities on grain boundaries in a damage zone ahead of the crack [19, 20] (Fig. 7). The stress on the damage zone motivates growth of the cavities, once nucleated. Consequently, the crack progresses when the damage coalesces on those grain facets contiguous with the crack. The growth of the cavities in the damage zone generally causes displacements that modify the stress field ahead of the crack[20] (cf. Fig. 4a). Determination of the crack growth rates thus requires solution of simultaneous relations for the cavity growth rate (as determined by

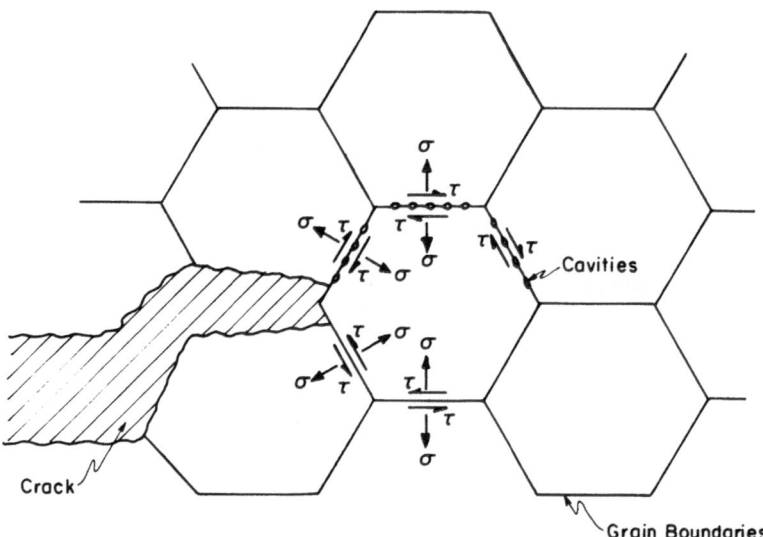

FIG. 7. A schematic illustrating the damage mechanism of crack growth.

the resultant normal stress) and the stresses (as dictated by the displacements induced by cavity growth). Such calculations have been conducted for a viscous solid [13, 20]. Then, when the damage zone is large (such that damage growth is relatively unconstrained) the steady-state crack growth rate has the form

$$\dot{a} \approx \frac{K\Omega D_b \delta_b}{kT \, l^{5/2} \, (\lambda/l)^3} \tag{10}$$

where λ is the spacing between cavities in the damage zone. Nonlinear behavior would obtain if λ/l were dependent on crack velocity. Zone size effects also emerge, and affect the linearity, when the zone size becomes small [13].

Comparison of the above crack growth rate predictions with data has been achieved by using independent measurements of λ and of the damage zone size obtained, on failed specimens [21] (Fig. 8). However, a full predictive capability does not exist, because there is no fundamental understanding of the effects of microstructure on λ. Nevertheless, certain important trends are apparent. In particular, the importance of the grain size, diffusivity and cavity spacing appear explicitly and have the expected influence on crack growth rates. When an amorphous phase is

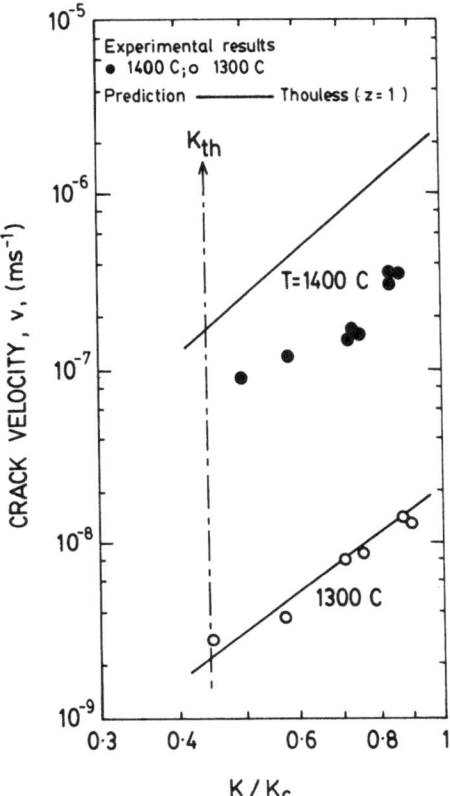

FIG. 8. A comparison of creep crack growth rates measured for Al_2O_3, with values predicted by the damage mechanism.

present [20], the velocity increases by l/δ, as well as by the increase in diffusivity (D_1/D_b).

3.1.3. Mechanism regimes

Various observations and predictions suggest that the direct extension and damage mechanism have differing realms of dominance. *Observations* of failed specimens [21] have revealed that cavitation damage exists on the fracture surface in the region of slow crack growth (Fig. 9A). By contrast, rapid propagation is accompanied by a faceted fracture surface (Fig. 9B). Such observations clearly suggest the prevalence of damage mechanisms at the lower crack velocities. Crack

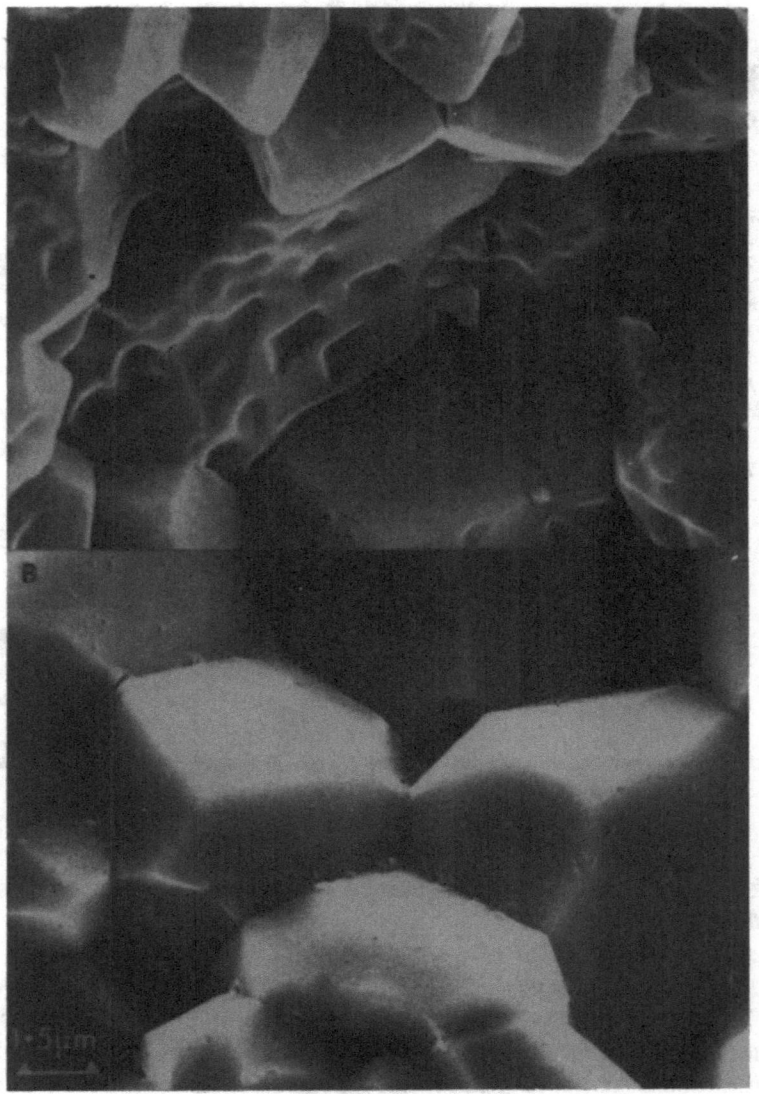

FIG. 9. The fracture surface morphology accompanying creep crack growth in Al_2O_3.
(A) Slow growth rates ($\dot{a} \gtrsim 10^{-6}\,ms^{-1}$) indicating cavitation damage; (B) rapid propagation
($\dot{a} \gtrsim 10^{-4}\,ms^{-1}$) revealing featureless grain boundary facets.

growth *models* predict similar features (Fig. 3), because the direct extension mechanisms have a larger n (eqn. 6), owing to additional velocity dependent parameters (notably, the crack width). This separation of the regimes of relevance has significant implications for two features of the fracture process: the crack growth threshold K_{th}, and the critical stress intensity factor, K_c, as discussed in the subsequent sections.

3.1.4. Effect of ligaments

When intact ligaments remain behind the crack tip, they exert forces on the crack surface that tend to reduce the tip K and thus diminish the creep crack growth rate. The general trends can be conceived from a simplified analysis, depicted in Fig. 5, based on observations by Wiederhorn *et al.* [15, 16]. The intact regions exert tractions that depend on the size, l, and viscosity, η_1, of the ligament material. The corresponding opening rate of the crack surface is governed by the viscosity η of the body and the resultant tip K. Hence, by utilizing a Dugdale analysis, it can be readily demonstrated that the change in K provided by the intact ligaments has the form,

$$\Delta K = -\chi(l/d)^2 (\eta_1/\eta)(\eta l)^{2/3} K^{1/3} \dot{a}^{2/3} \delta_1^{-1} \tag{11}$$

where d is the spacing between ligaments and χ is a constant ≈ 0.1. Then, the crack growth rate may be related to the applied K, by combining eqns (6) and (11) with

$$K_\infty = K + \Delta K \tag{12}$$

to give the relation

$$K_\infty = \dot{a}^{1/n} \frac{K_c}{\dot{a}_o^{1/n}} + \dot{a}^{(1+2n)/3n} \frac{K_c^{1/3}}{\dot{a}_o^{1/3n}} \chi(l/d)^2 (\eta_1/\eta)(\eta l)^{2/3} \delta_1^{-1} \tag{13}$$

The ligaments thus introduce a complex dependence between crack growth rate and stress intensity. Furthermore, strong effects on crack growth rate of the viscosity of the ligament material and ligament size and spacing are apparent. Ligament effects may be of considerable importance in the near threshold region and thus, some understanding of how ligaments form is regarded as an important topic for future research.

3.2. The Threshold Stress Intensity

The considerations of the preceding sections reveal that the threshold represents a process that intervenes while crack growth is occurring by a damage mechanism (Fig. 3). It thus seems appropriate to regard the

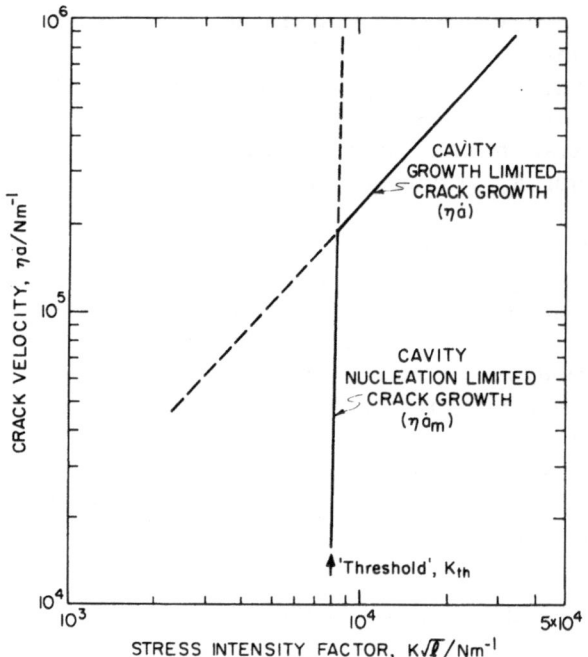

FIG. 10. Predicted crack growth rates when a damage nucleation step is incorporated, revealing an abrupt change in behavior, at a 'threshold' stress intensity, K_{th}.

threshold as a stress intensity level that inhibits the nucleation of damage in the crack tip region [22]. For a viscoelastic solid, typical of most ceramics, damage inhibition would require that the elastic stress on the first grain boundary facet (as modified by grain boundary sliding, at the crack tip) be less than a 'critical' stress for cavity nucleation. Indeed, considerations of cavity nucleation rates [22] indicate that crack growth can be nucleation limited (Fig. 10), resulting in a relatively abrupt decrease in the crack growth rate. A nucleation limited threshold thus seems plausible, with the threshold occurring at a stress intensity

$$K_{th} \approx \frac{\gamma_s \sqrt{l}}{\Omega^{1/3}} F(\psi) \tag{14a}$$

where, $F(\psi) = \sqrt{\pi} (8\pi/3)^{1/3} [2 - 3\cos\phi + \cos^3\psi]^{1/3}$, or in the presence of an amorphous phase,

$$K_{th} \approx \gamma_1 \sqrt{l/\delta_0} \qquad (14b)$$

This predicted threshold is larger than values observed experimentally (probably because of additional stresses induced by grain boundary sliding transients) [22]. Nevertheless, general trends in K_{th} with grain size and surface energy appear to be in accordance with the limited threshold data available in the literature. Specifically, the threshold is apparently lower in materials having a fine grain size (however, it is cautioned that the effect of grain size on viscosity introduces some subjectivity into the interpretation of grain size trends) and in the presence of an amorphous phase that both reduces the surface energy pertinent to damage nucleation, and allows an increase in the characteristic nucleation dimension (δ_0 replaces $\Omega^{1/3}$).

Comparison of eqn. (14) with eqn. (2) reveals the explicit influence on the ductile-to-brittle transition of such parameters as the grain size, diffusivity, surface energy, dihedral angle and amorphous phase content. In particular, amorphous phases substantially reduce K_{th} and thus encourage creep brittleness [13, 23]. The major remaining uncertainty is the flaw size, a. High temperature flaws are discussed in the following sections.

3.3. The Critical Stress Intensity

The preceding discussion of mechanism regimes suggests that unstable crack growth by bond rupture is most likely to intervene while creep crack growth is proceeding by a direct extension mechanism. However, the criterion that dictates the transition is unknown. Furthermore, in most ceramic materials, high temperature stable, slow crack growth may occur at stress intensities substantially in excess of the ambient K_c, as illustrated in Fig. 3 [16, 24]. This phenomenon reflects the 'blunt' character of the crack tip, during creep crack growth, as elucidated in Section 3.1.

Recognition that direct creep crack extension processes are accompanied by a peak tensile stress, $\hat{\sigma}$ at a distance $\hat{x}$, ahead of the crack tip (Fig. 4a) suggests two plausible criteria for the transition to brittle propagation. Either $\hat{\sigma}$ exceeds the stress needed to nucleate a brittle crack at $\hat{x}$, or $\hat{x}$ diminishes to the atomic dimension. Both criteria give a peak stress intensity, $\hat{K}$, in excess of the ambient K_c (Fig. 3), in qualitative accordance with the previously stated measurements of creep crack growth. The quantity $\hat{K}$ would represent the 'critical stress intensity factor' measured using the usual fracture mechanics techniques.

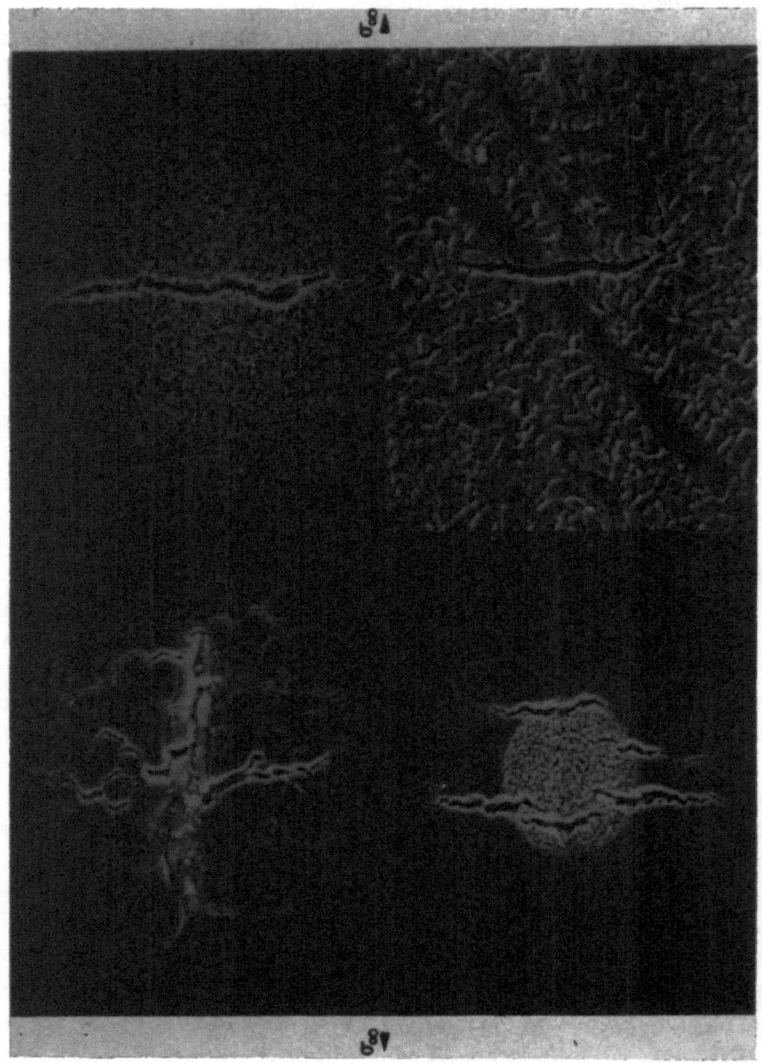

FIG. 11. Scanning electron micrographs of typical high temperature flaws.

4. HIGH TEMPERATURE FLAWS

Observations of fracture origins and of flaw initiation sites at high temperature are less prevalent than those available at lower temperatures. Nevertheless, present evidence [5, 6] strongly infers that the predominant high temperature flaws are generally different from the flaws that dominate the ambient mechanical strength. In particular, flaws are frequently found to originate at various chemical and microstructural heterogeneities (Fig. 11), as summarized in Table 1. Such regions evolve into flaws, either because local strain concentrations result from viscosity differentials, oxidation strains, etc. or because phases are formed that locally degrade the creep crack growth resistance. In either case, the zone of influence is typically of the order of the heterogeneity size, resulting in flaws that scale with the heterogeneity diameter [6].

TABLE 1

High temperature flaw	Material
Large grained region	Al_2O_3/MgO SiC/B
Amorphous zone	Al_2O_3/MgO SiC/B
Machining flaw	Si_3N_4 (all alloys) SiC (all alloys) Al_2O_3 (all alloys)
Oxidation pit	Si_3N_4 (all alloys)
'Blocky' heterogeneity	Al_2O_3/SiO_2 Si_3N_4/MgO
Chemical heterogeneity	$Al_2O_3/MgO/NiO$ SiC/Al_2O_3

While the quantitative understanding of high temperature flaws is lacking, it is deemed useful to present some results that have relevance to flaw formation and initial growth. In particular, it is noted that stress concentration effects can be estimated from elastic solutions, by replacing the shear moduli with the equivalent viscosities. Furthermore it is noted that the important flaw problems usually involve two stress intensities: a localized value, K_R, associated with the concentrated stress around the heterogeneity, and an applied value K_∞ (cf. indentation fracture) [25]. Typically, these stress intensities have opposing trends with crack

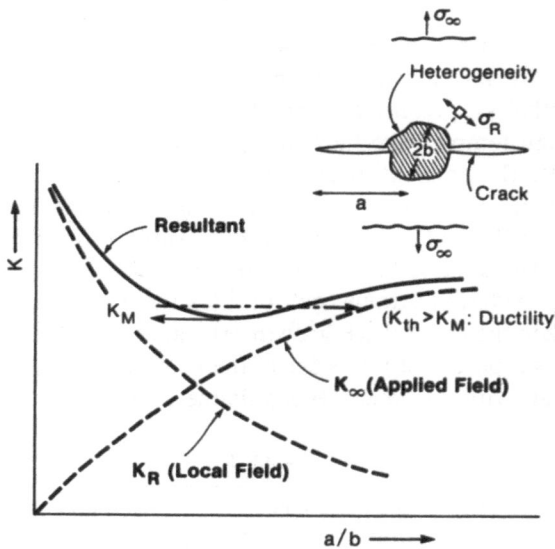

FIG. 12. Trends in stress intensity with crack length, incorporating the local (residual) term, K_R, and the applied term, K_x, revealing the existence of a minimum K_m. Creep ductility is assured when $K_m < K_{th}$.

length, resulting in a minimum, K_m (Fig. 12). When this situation obtains, creep ductility can be assured, by requiring that $K_m < K_{th}$. Explicit expressions for creep ductility can thereby be derived.

Of particular interest are planar, low viscosity faults [16]. Such faults, when inclined to the imposed tension, cause the sliding of relatively large 'blocks' of material resulting in values of K_m of order,

$$K_m \approx \tau_d \sqrt{d} \tag{15}$$

where τ_d is the shear stress along the fault and $2d$ is the length of the fault. Hence, creep ductility is assured when,

$$\frac{\gamma_s \sqrt{l/d}}{\Omega^{1/3}} > \frac{\tau_d}{F(\psi)} \tag{16}$$

This inequality constitutes a conservative ductility criterion, because stress relaxation by local mass transport reduces the stress at the fault tip and eliminates the singularity (cf. Fig. 4a). The maximum stress then varies with time t, after the sliding event, as [23]:

$$\sigma_m = \tau_d \sqrt{d} \left[\frac{4(1-v^2)kT}{ED_b\delta_b\Omega t} \right]^{1/6} \tag{17}$$

Consequently, large values of the diffusivity and slow sliding rates can reduce the local stress and may result in peak stresses less than the critical level needed to nucleate flaws. Such effects may be used, advantageously, to encourage creep ductility.

Oxidation induced flaws have various manifestations, depending on the nature of the heterogeneity having the greatest susceptibility to oxide formation. The flaws may either form externally, as perturbations on the surface oxide [26], or internally. Such oxidation sites usually evolve into high temperature flaws because of the residual stresses associated with the oxidation strain-rate ($\dot{\varepsilon}^T$). Crack formation at sites of local dilatation in a viscous solid is accompanied by a residual stress intensity,

$$K_R \approx 3\sqrt{\pi}\,\eta\,b^2\,\dot{\varepsilon}^T/a^{3/2} \tag{18}$$

where b is the radius of the oxidation zone. Hence, by superimposing the stress intensity associated with the design stress,

$$K_\infty \approx (2/\sqrt{\pi})\sigma_d\sqrt{a} \tag{19}$$

K_m may be evaluated. Then, by setting $K_{th} > K_m$, the following creep ductility criterion results,

$$\frac{\gamma_s\sqrt{l/b}}{\Omega^{1/3}(\eta\dot{\varepsilon}^T)^{1/4}} > \xi\frac{\sigma_d^{3/4}}{F(\psi)} \tag{20}$$

where $\xi \approx 3$. The trends associated with the important material parameters ($\gamma_s, \dot{\varepsilon}_T, \eta, \psi$) are clearly prescribed by this result. In particular, a critical size of oxidation prone defect can be defined, such that, ductility is assured if,

$$b^2 < \frac{\gamma_s\sqrt{l}\,F(\psi)}{\xi\Omega^{1/3}(\eta\dot{\varepsilon}^T\sigma_d^3)^{1/4}} \tag{21}$$

5. CERAMIC COMPOSITES

5.1. Creep Rates

Ceramic composites typically consist of a creep susceptible matrix and creep resistant reinforcements [27]. For this case, the creep characteristics depend on the relative dimensions of the whiskers and the grains. When the whiskers are relatively large and have a width, $w \gtrsim l$, the matrix behaves as a *continuum*. Then, the steady-state creep-rate of the composite has the *same stress dependence* as the matrix, but deviates from the matrix creep rate by a fixed multiple ω, that depends on the creep resistance, volume fraction, and shape of the reinforcement, as well as the shear resistance of the interface. For a linearly viscous matrix, the magnitude of ω can be obtained from composite elastic modulus solutions, by replacing the shear modulus with the viscosity. Typical trends are illustrated in Fig. 13 for randomly oriented, rigid whiskers [28] having a *shear resistant interface*. Similar values of ω would obtain for $n \gtrsim 2$, typical of most ceramics.

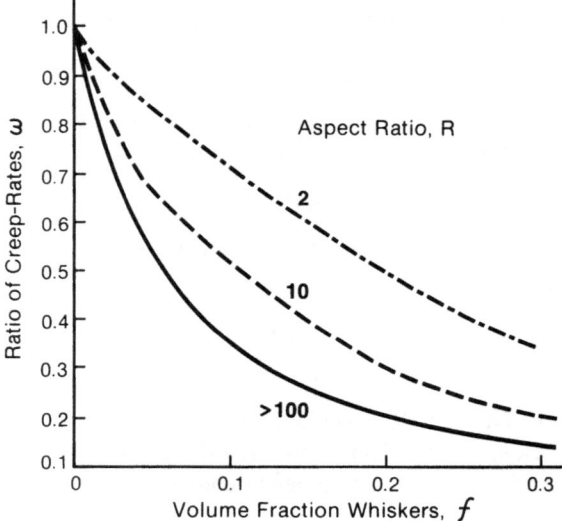

FIG. 13. Predicted trends in creep rate with volume fraction of long aspect ratio whiskers, according to a continuum model.

When the interface has a relatively low viscosity compared with that of the matrix, the magnitude of ω diminishes. Such behavior is expected to be typical of many reinforced ceramics, due to the tendency to form thin amorphous phases at the interface [29]. Sliding at the interface clearly enhances the creep rate, by means of a change in ω. However, sliding may also induce stress concentrations that result in creep damage and a

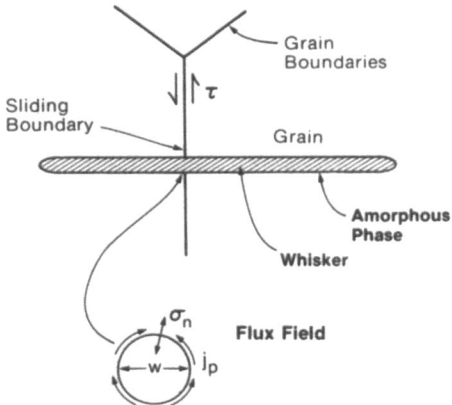

FIG. 14. Inhibition of grain boundary sliding by a small whisker, indicating the flux field through the grain boundary phase.

consequent acceleration of the creep rate. In particular, the component of the stress resolved along the major axis causes stress concentrations at the tip of the reinforcement, that may nucleate cavities. The reinforcement then becomes partially ineffectual as a creep inhibitor.

When the fiber width is *small*, $w < l$, a continuum description is inappropriate. Then, the role of the whisker is to inhibit grain boundary sliding, as sketched in Fig. 14. Sliding occurs at a rate dictated by the transport of matter from one side of the whisker to the other, through the amorphous interphase. Simple analysis indicates that this process can be characterized by a viscosity

$$\eta \approx \frac{kT w^2 l}{D_1 \delta_1 \Omega} \tag{22}$$

Comparison of eqn. (22) with eqn. (1) reveals that, since $w < l$ and $D\delta$ for the amorphous phase is expected to be larger than that for the grain boundary, small whiskers should not exert a significant influence on the creep rate.

A comparison of the preceding predictions with creep data obtained for Al_2O_3 reinforced with SiC whiskers reveals several features of interest (Fig. 15). In particular, the composite creep rate data have a different slope to the matrix data and hence, the results deviate from the continuum prediction for a composite containing stiff, bonded whiskers. Another disparity between experiment and theory is the relatively low

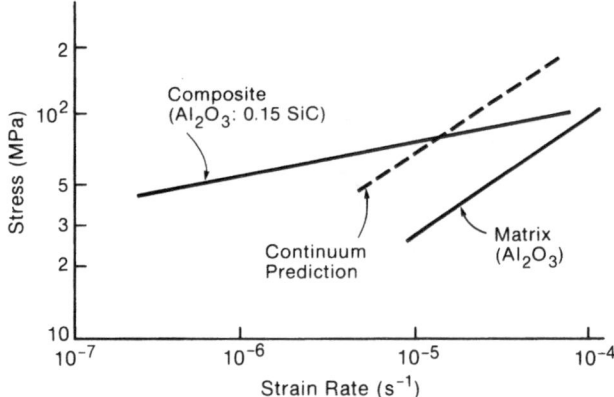

F IG . 15. A comparison of creep rates measured on an Al_2O_3/SiC composite [27] with values predicted by the continuum analysis (utilizing Fig. 13).

creep rate achieved by the composite, at low stresses. Such low creep rates (small ω) are not predicted from composite theory. One explanation of the disparity is that the diffusivity and grain boundary sliding rate are affected by the chemical changes that result from incorporation of the whiskers. Clearly, however, further research is needed to understand the creep behavior of reinforced ceramics.

5.2. Creep Crack Growth

Fibers and whiskers have well-known effects on the ambient fracture resistance, due to tractions imposed on the crack surfaces by intact fibers in the wake. Intact whiskers would exert a similar influence on the high temperature crack growth resistance. The phenomenon should be generally similar to the effect of intact ligaments discussed in Section 3, with l becoming the whisker length.

6. CONCLUDING REMARKS

This article has summarized present understanding of the reliability of ceramics at elevated temperatures. Particular emphasis has been placed on the transition to creep ductility, based on the notion that creep rupture would not normally be performance limiting when creep ductility conditions prevail. A material parameter having major importance, in this context, is the stress intensity threshold at which cracks seemingly blunt, rather than propagate. Analysis of available models of creep crack growth suggests that the threshold coincides with a transition to crack growth limited by the rate of nucleation of damage in a crack tip damage zone. However, the details are not well understood. Further understanding of this phenomenon should thus be a priority for future research on high temperature reliability.

Models of creep crack growth have limited applicability because, in most cases, the material response considered in the models does not coincide with the behavior of typical ceramic polycrystals. A substantial need thus exists for the development of models that incorporate both the viscoelastic character of the ceramic and specific microstructural events (such as localized grain boundary sliding).

High temperature flaws in ceramics, in many instances, differ from the flaws that control brittle failure at ambient temperatures. Some understanding of these flaws is beginning to emerge. However, a systematic attempt should be made to locate and analyze the flaws having the major influence on creep rupture, in the creep brittle range.

Finally, it is noted that certain ceramic composites have interesting high temperature characteristics, such as creep and creep rupture resistance. Little is known about these materials, suggesting the need for systematic investigation.

REFERENCES

1. KATZ, R. N., *Mater. Sci. Engng*, 7 (1985), 227.
2. SUZUKI, H. *Mater, Sci. Engng*, 7 (1985), 211.
3. KATZ, R. N. and HANOOSH, J. G., *Int. J. High Tech. Ceramics*, 1 (1985), 69.
4. RAJ, R. and ASHBY, M. F., *Met. Trans.*, 2 (1971), 1113.
5. WIEDEHORN, S. M. and FULLER, E. R., *Mater. Sci. Engng*, 7 (1985), 169.
6. DALGLEISH, B. J., SLAMOVITCH, E. and EVANS, A. G., *J. Am. Ceram. Soc.*, 68 (1985), 575.
7. DALGLEISH, B. J., JOHNSON, S. M. and EVANS, A. G., *J. Am. Ceram. Soc.*, 67 (1984), 741.
8. EVANS, A. G. and CANNON, R. M., *Mechanical Properties and Phase Transformations in Engineering Materials* (S. D. Antolovich, R. D. Ritchie and W. W. Gerberich eds), AIME Publications, 1986, p. 409.
9. CLARKE, D. R. and LANGE, F. F., *J. Am. Ceram. Soc.*, 63 (1980), 586.
10. LUH, E. Y. and EVANS, A. G., *J. Am. Ceram. Soc.*, in press.
11. EVANS, A. G., *J. Am. Ceram. Soc.*, 65 (1982), 127.
12. TSAI, R. L. and RAJ, R., *Acta Met.*, 30 (1982), 1043.
13. THOULESS, M. D. and EVANS, A. G., *Acta Met.*, 34 (1986), 23.

14. HEUER, A. H., TIGHE, N. J. and CANNON, R. M., *J. Am. Ceram. Soc.*, **63** (1980), 53.
15. WIEDERHORN, S. M., HOCKEY, B. J., KRAUSE, R. F. and JAKUS, K., *J. Mater. Sci.*, **21** (1986), 810.
16. WIEDERHORN, S. M., CHUCK, L., FULLER, E. R. and TIGHE, N. J., *J. Am. Ceram. Soc.*, in press.
17. CHUANG, T. J., *J. Am. Ceram. Soc.*, **65** (1982), 93.
18. HSUEH, C. H., CAO, H. C. and EVANS, A. G., *J. Am. Ceram. Soc.*, in press.
19. RAJ, R. and BAIK, S., *Metal Sci.*, **14** (1980), 385.
20. THOULESS, M. D. and EVANS, A. G., *Acta Met.*, **31** (1983), 1675.
21. BLUMENTHAL, W. and EVANS, A. G., *J. Am. Ceram. Soc.*, **67** (1984), 751.
22. THOULESS, M. D. and EVANS, A. G., *Scripta Met.*, **18** (1984), 1175.
23. EVANS, A. G., RICE, J. R. and HIRTH, J. P., *J. Am. Ceram. Soc.*, **63** (1980), 358.
24. EVANS, A. G. and WIEDERHORN, S. M., *J. Mater. Sci.*, **9** (1974), 270.
25. CHANTIKUL, P., ANSTIS, G. R., LAWN, B. R. and MARSHALL, D. B., *J. Am. Ceram. Soc.*, **64** (1981), 539.
26. TIGHE, N. J., WIEDERHORN, S. M., CHUANG, T. J. and McDANIEL, C. L., *Deformation of Ceramic Materials*, Plenum, New York, 1984, p. 587.
27. CHOKSI, A. and PORTER, J. R., *J. Am. Ceram. Soc.*, **68** (1985), C144.
28. CHOU, T. W. and KELLY, A., *Ann. Rev. Mater. Sci.*, **10** (1980), 229.
29. CLAUSSEN, N., WEISSKOPF, K. L. and RUHLE, M., *J. Am. Ceram. Soc.*, **69** (1986), 288.

8

Fracture and Elevated-temperature Static-fatigue of Ceramics Containing Small Flaws

S. Usami, I. Takahashi, H. Kimoto, T. Machida and H. Miyata

3rd Department, Mechanical Engineering Research Laboratory, Hitachi, Ltd, Kandatsu-machi 502, Tsuchiura, Ibaraki, Japan.

ABSTRACT

Studies concerning effects of flaw size on brittle fracture and elevated-temperature static-fatigue strengths are reviewed. Although ceramic materials fracture elastically, fracture stress has a nonlinear relation to flaw size, and the critical stress intensity factor for a small crack is lower than that for a large crack. The relationship is explained well by a fracture model, which takes into account the interaction between a flaw and the microstructure of the ceramic. A crack can be arrested below a certain stress level, i.e. below the static-fatigue limit. The static-fatigue limit also shows a nonlinear relationship to flaw size similar to that for brittle fracture. The effects of temperature, environment and material on static-fatigue limit are also discussed.

1. INTRODUCTION

In severe environments, ceramic materials show many characteristics which are superior to metals. However, when they fail, it is in a brittle manner, and their strength values vary widely. The conventional strength evaluation method for ceramic components is based on Weibull statistics. Although small flaws, which are frequently observed at fracture origins, are considered to be the cause of the strength scattering, the relationship between crack size and strength is not sufficiently clear. On the other hand, because the cracks in ceramics may grow under a sustained static load at elevated temperatures, component lives are estimated using the power law relationship for large cracks, namely, $dc/dt = AK_I^n$. However, small-crack behavior during static-fatigue is not thoroughly understood.

This paper reviews the effects of flaw and grain sizes on brittle fracture strength [1–12], as well as the static-fatigue behavior of small cracks [13–16] in ceramics.

2. BRITTLE FRACTURE

2.1. Effect of Flaw Size

Relationships between equivalent crack length, a_e, and net section fracture stress, σ_c, for soda-lime glass [2,8,11] and ceramics (i.e. sialon, silicon nitride, silicon carbide and alumina) [11] are shown in Fig. 1a. The relationships for silicon nitrides [3–5, 9–11] are shown in Fig. 1b. When the other flaws were not observed by SEM in the fracture surfaces, the artificial small flaws were considered to be the cause of the fractures. The equivalent crack length, a_e, which is half the length of an equivalent through crack in an infinitely wide plate, represents a flaw with a shape and a size. This is shown by the following equation:

$$a_e = \frac{1}{\pi}\left(\frac{K_I}{\sigma_n}\right)^2 \tag{1}$$

where K_I is the stress intensity factor and σ_n is net section stress. If a is the crack depth, a_e is $1.258a$ for any edge crack in a semi-infinite plate. If D is the diameter of a surface crack, a_e is $0.25D$ for any semicircular surface crack in a semi-infinite body. The edge crack and an internal circular crack ($a_e = 0.20D$) have the same a_e when they have the following relationship:

$$a = \frac{D}{6.2} \tag{2}$$

Therefore, one should be careful to accurately detect the depth of scratch cracks caused by grinding, which are frequently more than ten times the depth of ordinary measured surface irregularities.

The relationships between a_e and σ_c for surface scratches, large crystals, pores, controlled surface flaws (CSF) introduced by Vickers- or Knoop-indenting, and for acute notches are almost identical for all the materials shown in Fig. 1b. The inherent strength scattering seems to be very small. The strengths of specimens with surface scratches or a

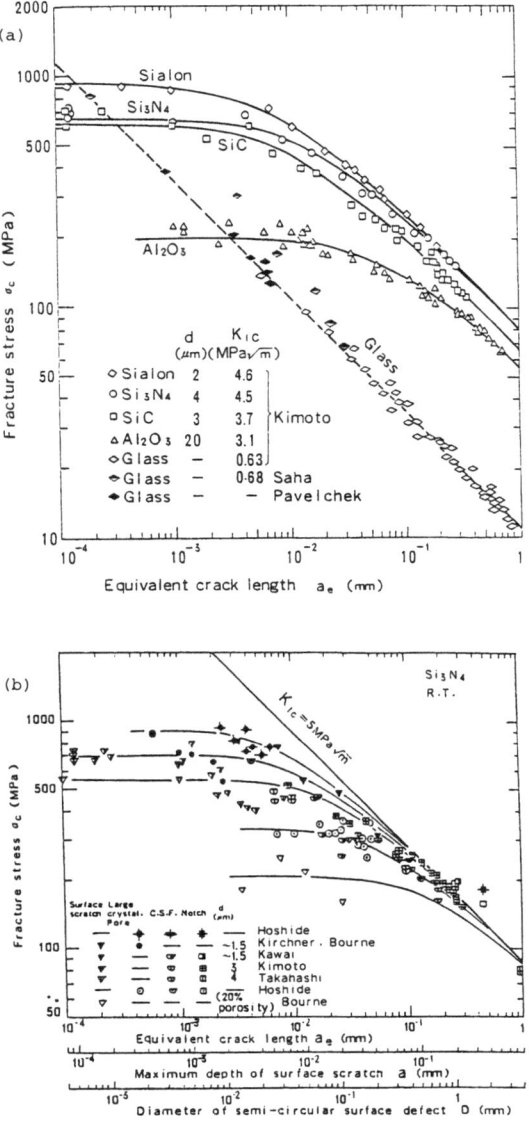

FIG. 1. Relationships between equivalent crack length and fracture stress. (a) Glass and ceramic materials; (b) silicon nitrides.

controlled surface flaw, which are thought to sustain crack-mouth opening residual forces due to machining[17], do not differ greatly from the strengths of specimens with large crystals or pores which do not sustain crack-mouth opening forces. Therefore, the supposed crack-mouth opening force is not a dominant factor in strength reduction for small flaws. The force is reduced by tensile load.

The fracture stress in amorphous glass of very small structural size maintains the following LEFM relationship, even for small flaws:

$$K_I = K_{IC} \qquad (3)$$

Although large flaws abide by the linear elastic fracture mechanics (LEFM) relationship, small flaw strengths deviate from the relationship and level off for polycrystalline ceramics, as shown in Figs. 1a and 1b. The fracture stress of materials with larger grains levels off for larger cracks, which results in lower small-crack strength. The many flaws which control component strengths, are outside the applicability of the LEFM. In other words, the LEFM having a large-crack toughness value, K_{IC}, may give unsafe side estimations of component strength.

The K_{IC} is highest for sialon and silicon nitride, followed by silicon

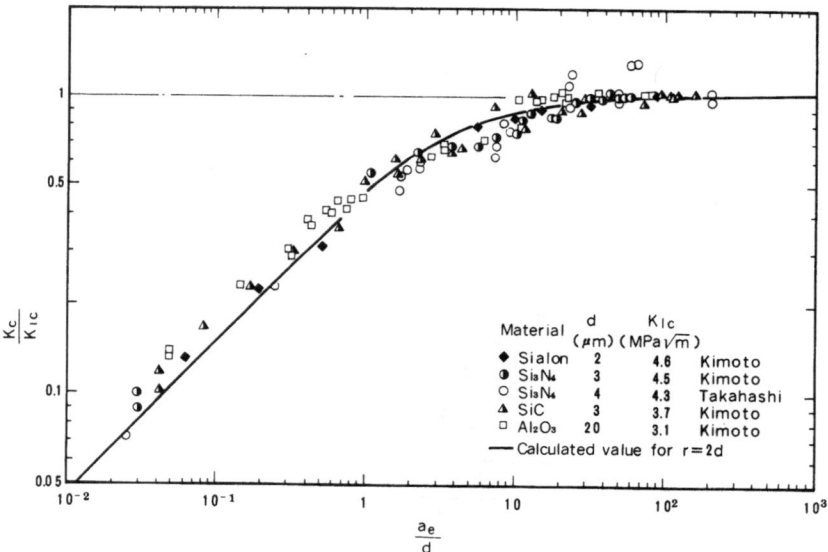

FIG. 2. Relationships between the ratio of equivalent crack length and mean grain diameter, a_e/d, and the ratio of critical stress-intensity factor values for short and long cracks, K_C/K_{IC}.

carbide, alumina and glass. On the other hand, the critical stress-intensity factor is not influenced by the kind of silicon nitride, including the sintering processes, the amount of difference in chemical composition and the grain size.

2.2. A Fracture Model

The test results [10,11] for polycrystalline engineering ceramics are rearranged according to the relationship between the ratio of equivalent crack length and mean grain diameter of the material, a_e/d, and the ratio of critical stress intensity factors for short and long cracks, $K_C K_{IC}$. The test results for different materials and grain sizes all fall in a relatively narrow scatter band, as shown in Fig. 2. Similar results can be seen in an earlier study [18]. The ratio, K_C/K_{IC}, decreases from unity when a_e/d is lower than about 30. The figure suggests that the reduction in small-crack toughness is due to the interaction between a crack and the grains in the ceramic.

A crack is forced to change direction at a grain boundary, according to either an intergranular or transgranular cleavage fracture. At the grain boundary it needs a higher stress intensity for consecutive propagation [19]. The degree of premature small fractures is detected by acoustic emission monitoring at stress well below the specimen failure. Certain relationships can be obtained from a grain-fracture model, in which a semicircular fracture of radius r_0 occurs prior to an original crack tip in a large, weak grain. The condition of the crack's consecutive propagation at the deepest point, A, shown in Fig. 3, governs the

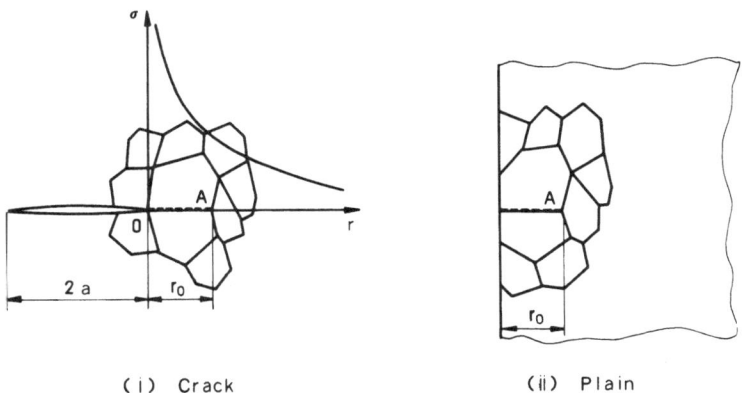

(I) Crack   (ii) Plain

FIG. 3. A grain-fracture model for a polycrystalline ceramic.

specimen strength. These relationships are:

$$\frac{K_C}{K_{IC}} = \frac{(1 + r_0/2a_e)^{1/2}}{1 + r_0/a_e} \quad \text{for cracked specimens} \tag{4a}$$

and

$$\sigma_{0.c} = \frac{0.943 \, K_{IC}}{(2\pi r_0)^{1/2}} \quad \text{for plain specimens} \tag{4b}$$

Equation (4b) has a form similar to that introduced by Evans[20]. The calculated values, assuming the weakest grain radius (r_0) is twice the mean grain diameter (d), are in good agreement with the test results, as shown in Fig. 2.

2.3. Plain Specimen Strength

The relationships between mean grain diameter and fracture stress of plain specimens for structural ceramic materials [1, 6, 7, 10] are shown in

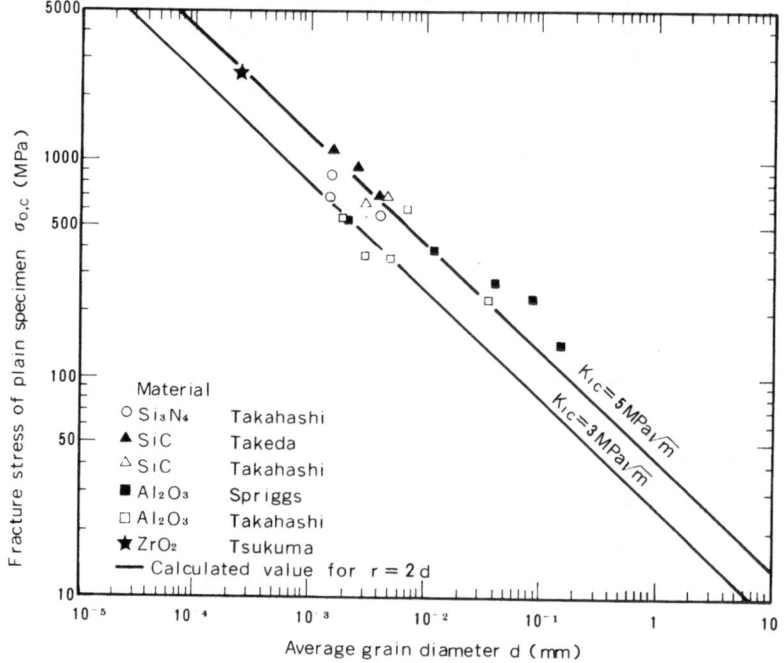

FIG. 4. Relationships between mean grain diameter and plain-specimen fracture stress for ceramic materials.

Fig. 4. They fall in a relatively narrow scatter band and the strengths are essentially inversely proportional to the square root of the mean grain diameter. The values calculated by eqn. (4b), utilizing typical values of K_{IC} for the ceramics of 3 and 5 MPa$\sqrt{m}$ and the relationship $r_0 = 2d$, are also shown in Fig. 4. They are close to the experimental values.

The above considerations suggest guidelines for microstructural strengthening of ceramic materials, namely, smaller flaw size, smaller mean grain diameter and higher toughness for a large crack, K_{IC}.

2.4. Crack Healing

The room temperature strengths after 1 h of no-load holding in elevated-temperature air are shown in Fig. 5 [12]. The heat treatment increased the strength of both CSF samples about 20% at 800°C. It also improved the strength up to that for plain specimens at 1200°C for silicon nitride and 1400°C for silicon carbide. The cracks were almost eliminated due to evaporation of the material or due to silicon oxide formation in the cracks. Excessive oxidation at higher temperature, however, caused pits on specimen surfaces and resulted in strength reduction for plain specimens. The heat treatments both in air and in vacuum environments yielded similar strength improvement for silicon nitride [22], but that in air resulted in a larger improvement for silicon carbide [23] than in a vacuum. This may have been caused by an abundance of silicon oxide in the grain boundaries of the silicon nitride.

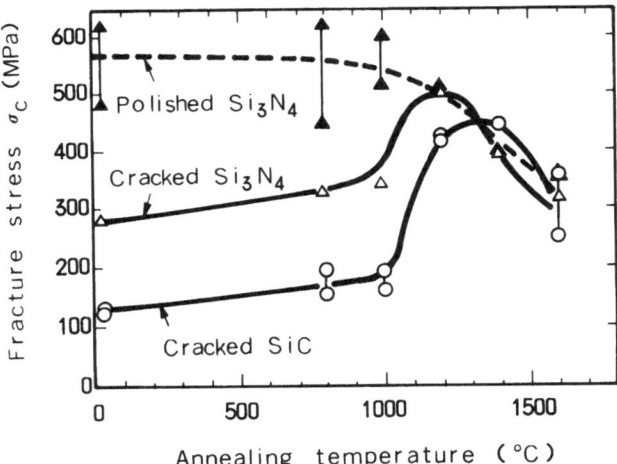

FIG. 5. Relationship between temperature of in-air heat treatment and room temperature strength for silicon nitride and silicon carbide.

3. ELEVATED-TEMPERATURE STATIC-FATIGUE

3.1. Static-fatigue Limit

The following power law relationship [24] between stress intensity, K_I, and static-fatigue (delayed fracture) crack growth rate, dc/dt, is used for estimating ceramic component lives:

$$\frac{dc}{dt} = AK_I^n \qquad (5)$$

Some test results [25,26] suggested that a crack does not grow below a threshold stress intensity factor. This indicates that ceramic components can be designed for elevated temperature operation on the basis of static-fatigue limit.

The relationships between the initial stress intensity [27] and static-fatigue life for silicon nitride with a CSF of about 700 μm in diameter are shown in Fig. 6 [14,16]. The diameter is large enough to apply LEFM to

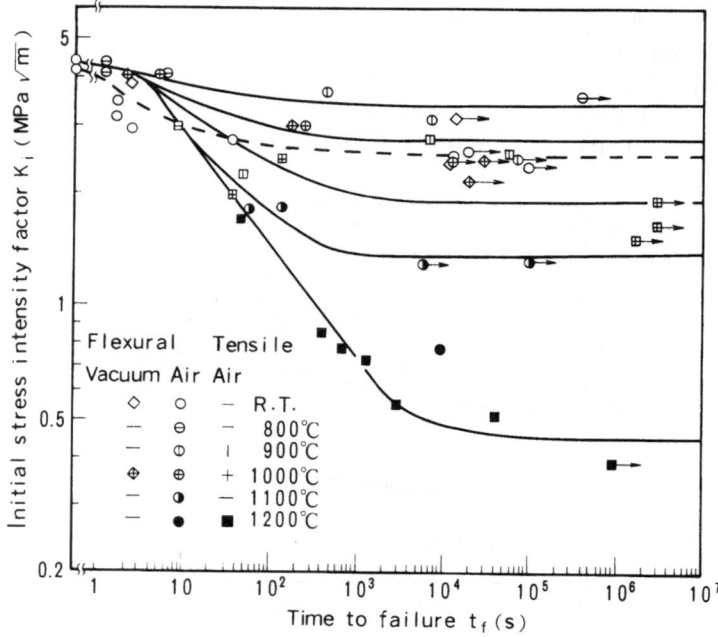

FIG. 6. Relationship between initial stress-intensity factor and static-fatigue life for large CSF (of 700 μm in diameter) in silicon nitride.

the brittle fracture, as was shown in Fig. 1b. Although the short time strengths were almost identical for different temperatures, the long term static-fatigue strength decreased significantly at higher temperatures. Furthermore, the fatigue strength showed threshold, i.e. a fatigue limit, for longer than 10^4 s. The time it took to reach the static-fatigue limit was longer for higher temperatures. The static-fatigue crack threshold values were about 80, 50 and 7% of the short time fracture toughness values at 800, 1000 and 1200°C, respectively.

Even in an air environment (15–25°C, 60–80% relative humidity), static-fatigue occurred. The fatigue limit value was about 70% of the short time strength. On the other hand, strength reduction did not significantly occur in a room temperature vacuum. Therefore, the humidity in air is thought to cause the degradation of the glass phase in the grain boundaries of silicon nitride. This degradation results in delayed fractures as in soda lime glass [28]. The static-fatigue strengths at 1000°C were almost identical in air and in a vacuum.

3.2. Crack Growth and Arrest

The static-fatigue crack behaviors from CSFs of 540–610 μm in diameter in sialon at 1000°C are shown in Fig. 7 [16]. The growth rate of each crack initially decreased despite an increase in stress intensity. The growth rates increased again and accelerated to fast fracture in the fractured specimen. In the nonfractured specimen, they decreased monotonically and the crack became nonpropagating. The cracks in the silicon nitride grew a little and were arrested at 1000°C even when the initial stress intensity was as low as one-quarter of the threshold stress intensity. The cracks did not shrink after being arrested. The crack arrest behavior was not observed at room temperature either in air or in a vacuum. The crack growth behavior and the sizes of nonpropagating cracks were almost identical in air and in vacuum at 1000°C. The crack arrest behavior confirms the existence of the static-fatigue limit. The cracks parallel to the tensile stress did not grow and the cracks in the specimens annealed for 1 h at 1200°C showed growth behavior similar to that in specimens which were not annealed. Therefore, the effect of the supposed residual crack-mouth opening forces caused by CSF indenting is thought to be negligible. Residual stress is relaxed in a short time [29].

The static-fatigue crack growth rates do not have any clear relationships with the stress intensity factors, even for large cracks to which LEFM is applicable in brittle fracture. Although other data for CSF [13]

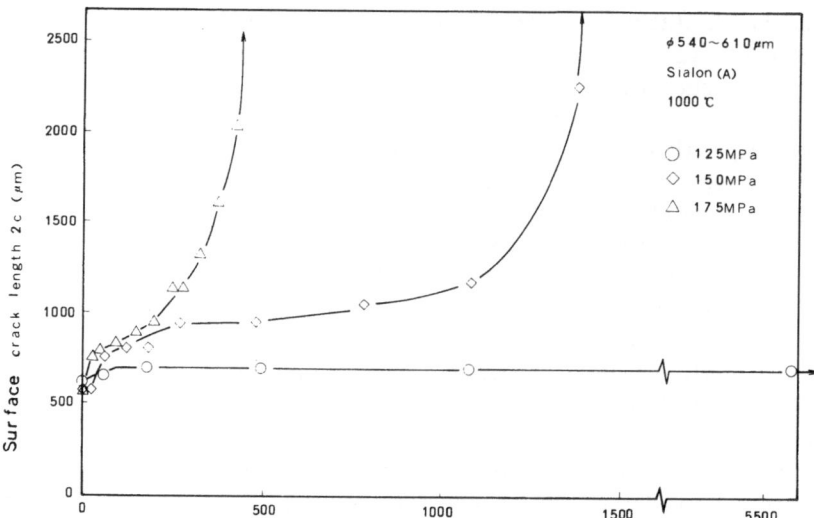

FIG. 7. Static-fatigue crack growth and arrest behavior in sialon (A) at 1000°C.

are close to those from this experiment, the data for double-torsion specimens with a long crack [24] yielded a significantly lower growth rate and higher threshold values for silicon nitride. When the initial stress intensity factors were less than about $2.5 \, MPa\sqrt{m}$, the growth rates for the CSFs in the silicon nitride decreased monotonically and resulted in nonpropagation at 1000°C.

These peculiar growth behaviors may be caused by the combined effect of crack acceleration factors (e.g. stress intensity, grain boundary sliding and grain boundary creep-void formation) and crack deceleration factors (e.g. evaporation of the material at newly-formed activated crack surfaces and narrow crack openings). The chemical activity of the material may function in both factors, while grain boundary sliding also contributes to crack tip stress relaxation.

Above 900°C, the fracture surfaces were essentially transgranular for brittle fracture and intergranular for the static-fatigue fracture. Although silicon oxide as deep as 25 μm was formed on the specimen surface, crack tip sticking was not clearly observed for the arrested crack. This may suggest that the interaction which occurred was not strong enough to sinter the crack tip, or that stress relaxation ahead of the crack tip contributes to the crack arrest phenomenon under the elevated temperature static-fatigue.

3.3. Effect of Flaw Size

The relationship between the initial equivalent crack length, a_e, and the flexural static-fatigue limit for the silicon nitride at 1000°C, together with the short time strengths at room temperature and at 1000°C, is shown in Fig. 8 [15,16]. The fatigue limit has a unique relationship with the equivalent crack length for different flaws, i.e. for the plain surface, the surface scratch, the acute notch (SENB) and the CSF.

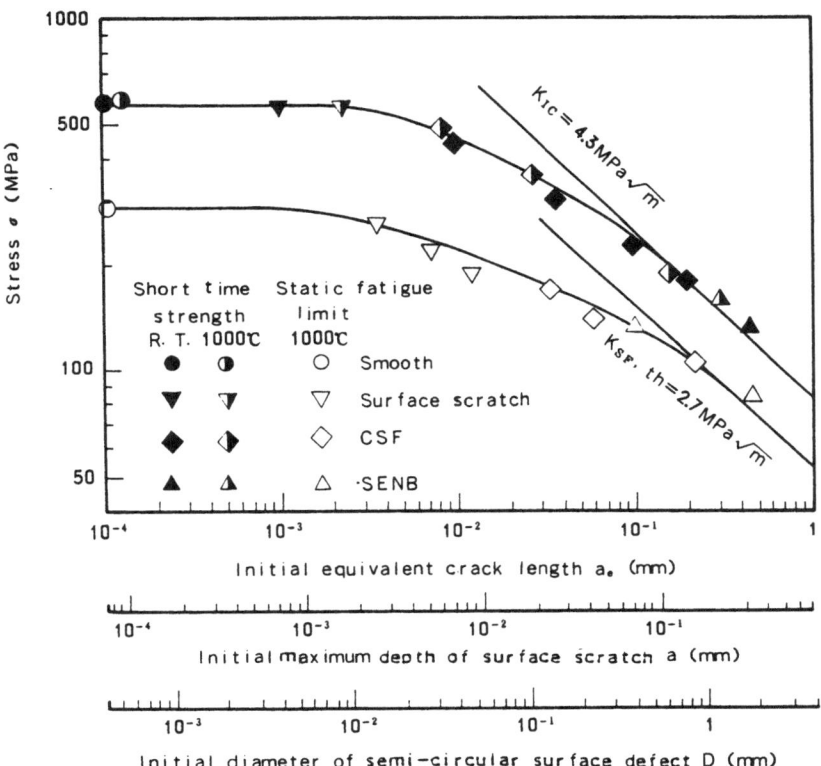

FIG. 8. Effect of initial equivalent crack length on static-fatigue limit at 1000°C and on fracture stresses at room temperature and 1000°C in silicon nitride.

The short-time fracture stresses are identical at room temperature and at 1000°C. The static-fatigue limit is 300 MPa for the plain specimen, which is 50% of the short time strength (600 MPa), whereas the threshold stress intensity factor for the large cracks, $K_{SF,th}$ is 2·7 MPa$\sqrt{m}$, which is 63% of the short time fracture toughness (4·3 MPa$\sqrt{m}$). Although the

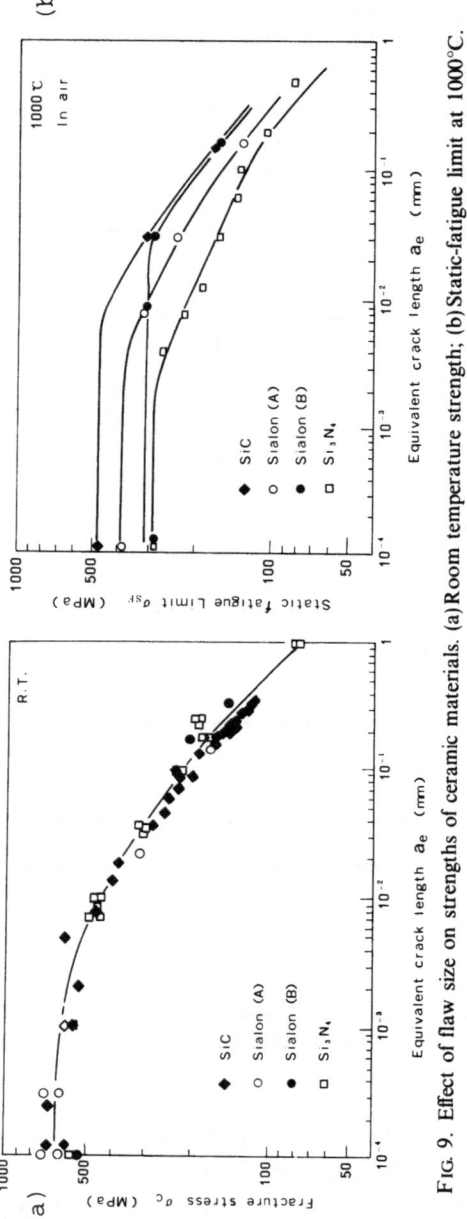

FIG. 9. Effect of flaw size on strengths of ceramic materials. (a) Room temperature strength; (b) Static-fatigue limit at 1000°C.

reduction is more significant for smaller flaws, the flaw size dependency of the static-fatigue limit is similar to that of the short time strength. In other words, the threshold stress intensity factor decreased for the initial equivalent crack length below about 200 μm, and the fatigue limit value levelled off for small defects. Hence, the LEFM employing a long crack threshold stress intensity yields higher fatigue limit values for small flaws in ceramic components.

The smaller cracks show a higher growth rate than the larger cracks for the same stress intensities. When the cracks are above the fatigue limit, the cracks extend to fracture, where their lengths coincide with the critical crack lengths. On the other hand, they were arrested before reaching the critical length when they were below the fatigue limit. The average stress over a certain length ahead of a crack may determine the fatigue limit. This criterion allows for lower threshold stress intensities for small flaws [12].

3.4. Effects of the Material

The effects of equivalent crack length on room temperature strength and on static-fatigue limit at 1000°C (for silicon carbide, two kinds of sialon and the silicon nitride) are shown in Figs. 9a and 9b [15], respectively. Although the room temperature strengths do not much differ for each material, the static-fatigue limits do. The difference between the static-fatigue limit values and the short time strength values increased as the flaw size decreased. The silicon carbide, which has the least glass phase in the grain boundaries, shows the highest fatigue limit values, followed by the sialon (A) and the silicon nitride. The sialon (B) shows the fatigue limit values close to those of the silicon nitride for the plain specimen and those of the silicon carbide for large cracks. This may be due to larger glass phase in the sialon (B) than in the sialon (A). The large glass phase reduces the plain specimen fatigue strength and yields a healing effect in the tips of the large cracks.

A proof test load can be applied at the level of the short time fracture stress, which corresponds to the fatigue limit defect size for the operating load.

4. CONCLUDING REMARKS

Crack size has unique relationships both with the brittle fracture and the static-fatigue strength of ceramic materials. Therefore, methodologies for the design, manufacture and operation of ceramic components based on small crack properties should lead to greater reliability of components.

REFERENCES

1. SPRIGGS, R. M., MITCHELL, J. B. and VASILOS, T., *J. Am. Ceram. Soc.*, **47** (1964), 323.
2. PAVELCHEK, E. K. and DOREMUS, R. H., *J. Mater. Sci.*, **9** (1974), 1803.
3. KIRCHNER, H. P., GRUVER, R. M. and SOTTER, W. A., *Mater. Sci. Engng*, **22** (1976), 147.
4. BOURNE, W. G. and TRESSLER, R. E., *Int. Symp. Fract. Mech. Ceram.*, Vol. 3, 1978, p. 113.
5. KAWAI, M., ABE, H. and NAKAYAMA, J., *Proc. Int. Symp. Factors Densification Sinter. Oxide Nonoxide Ceram.*, 1978, p. 545.
6. TAKEDA, Y., KOSUGI, T., IIJIMA, S. and NAKAMURA, K., *Proc. Int. Symp. Ceram. Components for Engines*, 1984, p. 529.
7. TSUKUMA, K., UEDA, K. and TSUKIDATE, T., *Ann. Mtg Ceram. Soc., Japan*, 1984, p. 119.
8. SAHA, C. K. and COOPER, A. R. Jr, *J. Am. Ceram. Soc.*, **67** (1984), C-158.
9. HOSHIDE, T., FURUYA, H., NAGASE, Y. and YAMADA, T., *Int. J. Fract.*, **26** (1984), 229.
10. TAKAHASHI, I., USAMI, S., NAKAKADO, K., MIYATA, H. and SHIDA, S., *J. Ceram. Soc. Japan*, **93** (1985), 186.
11. KIMOTO, H., USAMI, S. and MIYATA, H., *Trans. Japan Soc. Mech. Engng*, **51** (1985), 2482.
12. USAMI, S., KIMOTO, H., TAKAHASHI, I. and SHIDA, S., *Engng. Fract. Mech.*, **23** (1986), 745.
13. KAWAI, M., ABE, H. and NAKAYAMA, J., *Int. Symp. Fract. Mech. Ceram.*, Vol. **6**, 1983, p. 587.
14. TAKAHASHI, I., USAMI, S. and MACHIDA, T., *Trans. Japan Soc. Mech. Engrs*, **52** (1986), 2378.
15. MACHIDA, T., USAMI, S. and TAKAHASHI, I., to appear in *Trans. Japan Soc. Mech. Engrs*, **53** (1987).
16. USAMI, S., TAKAHASHI, I. and MACHIDA, T., *Engng, Fract. Mech.*, **25** (1986), 483.
17. KIRCHNER, H. P. and ISAACSON, E. D., *Int. Symp. Fract. Mech. Ceram.*, Vol. **5**, 1983, p. 57.
18. RICE, R. W., FREIMAN, S. W. and MECHOLSKY, J. J. Jr, *J. Am. Ceram. Soc.*, **63** (1980), 129.
19. GELL, M. and SMITH, E., *Acta Met.*, **15** (1967), 253.
20. EVANS, A. G., *Proc. 3rd Int. Conf. Mech. Behav. Mater.*, Vol. 1, 1979, p. 279.
21. TAKAHASHI, I., NAKAKADO, K. and MIYATA, H., Paper presented at *61st Ann. Mtg of Japan Soc. Mech. Engng*, No. 830–10, 1983, p. 31.
22. PETROVIC, J. J. and JACOBSON, L. A., *Ceramics for High Performance Applications*, 1974, p. 397.

23. PETROVIC, J. J. and MENDIRATTA, M. G., *Fracture Mechanics Applied to Brittle Materials*, ASTM STP 678, 1979, p. 83.
24. EVANS, A. G. and WIEDERHORN, S. M., *J. Mater, Sci.*, **9** (1974), 270.
25. EVANS, A. G., *J. Mater. Sci.*, **7** (1972), 1137.
26. BLUMENTHAL, W. and EVANS, A. G., *J. Am. Ceram. Soc.*, **67** (1984), 751.
27. NEWMAN, J. C. and RAJU, I. S., *Engng. Fract. Mech.*, **15** (1981), 185.
28. WIEDERHORN, S. M., *J. Am. Ceram. Soc.*, **50** (1967), 407.
29. MENDIRATTA, M. G. and PETROVIC, J. J., *J. Am. Ceram. Soc.*, **61** (1978), 226.

9

Protection of Alloys from Erosion and Corrosion at Elevated Temperatures

F. S. Pettit

Department of Metallurgy and Materials Engineering, 848 Benedum Hall, University of Pittsburgh, Pittsburgh, PA, 15261, USA.

ABSTRACT

At elevated temperatures degradation of the alloys can occur due to reaction with the gas environment. Such gaseous-induced degradation can be affected by the erosive action of particles in the gas stream. The principal forms of elevated temperature, gaseous-induced corrosion are briefly described and the effects of erosive conditions on these corrosion processes are examined. Procedures that can be used to attempt to inhibit such degradation processes are presented and discussed.

1. INTRODUCTION

In order to consider procedures by which alloys may be made resistant to severe environments some of the important forms of high temperature corrosion will be discussed and then techniques to inhibit this degradation will be examined. A great variety of alloys could be considered but the alloys of significance are those that can be made resistant to high temperature corrosion. When alloys are subjected to corrosive environments at elevated temperatures, the most effective technique to develop resistance to attack by this environment is to have a reaction product barrier formed upon the surface of the alloy which separates the alloy from the environment. In order for this approach to be successful the reaction product that is to be formed must possess the following properties:

— It must be the most thermodynamically stable phase of all the phases that could form by reaction of the alloy with the environment.

— Since the rate of attack of the alloy will be controlled by diffusion of the reactants through this reaction product, it must be formed as a continuous layer over the surface of the alloy and diffusion through it must occur as slowly as possible.
— It must be adherent to the alloy and resistant to thermally induced stresses and the erosive action of particulate matter.

In the following sections of this paper the principal mechanisms by which selectively formed reaction product barriers are developed and degraded will be described. The conditions causing the degradation will first involve oxygen (oxidation) and then more complex gas environments containing more than one reactant will be considered (mixed gas corrosion). More severe conditions will then be discussed by examining the effects of deposits which may accumulate on the surfaces of alloys and affect the gas-induced corrosion (hot corrosion), or damage the reaction product barrier upon impact (erosion). Finally, techniques to be followed to attempt to obtain alloys with improved resistance to such forms of degradation will be discussed.

It is not the intention of the present paper to describe the degradation of alloys by various environments in great detail. The approach to be followed will give a brief overview accompanied by references for those interested in more detailed discussions.

2. OXIDATION

When alloys are exposed at elevated temperatures to oxygen, depending upon the composition of the alloys and the pressure of oxygen in the gas, a large variety of oxide phases initially can be formed upon the surface of the alloys. However the different elements in the alloy do compete with each other for oxygen and it is possible for certain alloy compositions to have only one oxide phase develop as a continuous layer over the surface of the alloy. Such a process is called selective oxidation and has been described in textbooks [1] and papers [2–4]. It is sufficient in this paper to indicate that the only oxides that have properties appropriate for use as selectively formed reaction product barriers are Al_2O_3, Cr_2O_3 and SiO_2 [5]. In Fig. 1 the selective oxidation of chromium and aluminium in Ni–Cr–Al alloys is illustrated schematically. It can be seen that by appropriate choice of the chromium and aluminium concentrations in these types of alloys, continuous, external layers of Al_2O_3 or Cr_2O_3 can be formed on the alloys upon exposure to oxygen.

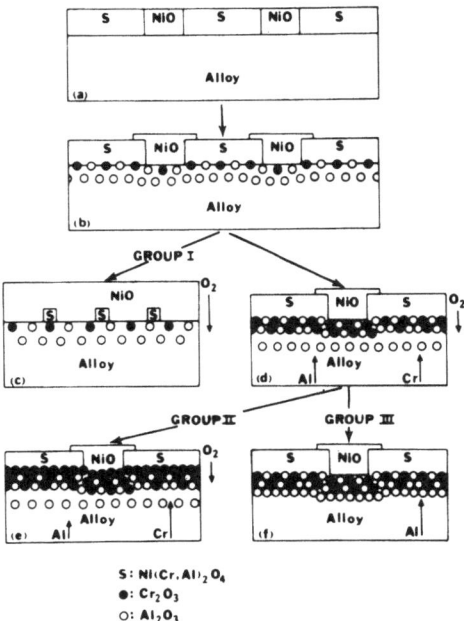

S : Ni(Cr,Al)$_2$O$_4$
●: Cr$_2$O$_3$
O: Al$_2$O$_3$

FIG. 1. Schematic diagram to illustrate oxide scale development on Ni–Cr–Al alloys for alloys with different compositions.

Even when continuous and protective reaction product barriers are formed on alloys, these barriers crack and spall when used. The selective oxidation process must then be capable of having the scale form again in those areas where metal has been exposed due to cracking and spalling. Eventually less protective reaction products must be formed and the alloys will then be degraded at more rapid rates. Typical results obtained for the degradation of alloys due to cracking and spalling of selectively formed barriers of oxide are presented in Fig. 2. When the most protective barriers are no longer developed upon alloys the alloys must be removed from service. The times over which an alloy will maintain protective oxide barriers depend upon service conditions and the composition of the alloy. Higher temperatures and more thermal excursions will shorten this time. Higher concentrations of chromium and aluminium will enhance selective oxidation in the case of Cr$_2$O$_3$ or Al$_2$O$_3$ formers, but limits are reached in regard to the levels to which certain elements can be increased due to mechanical property requirements.

F. S. Pettit

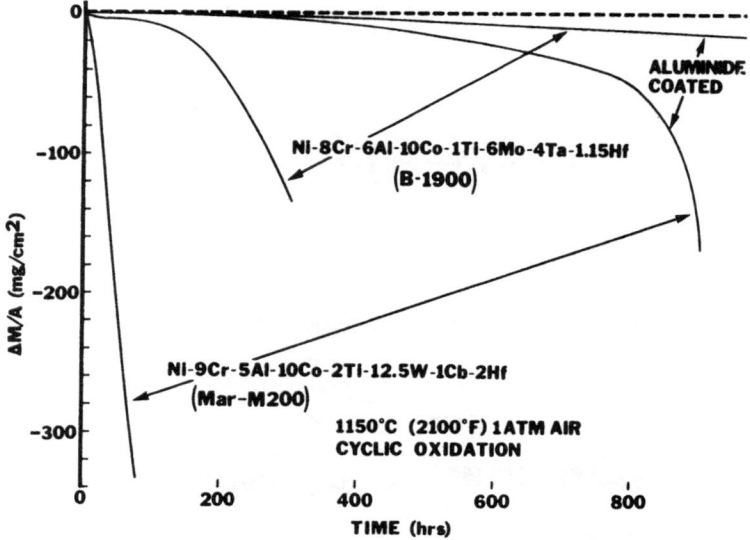

F$_{\text{IG}}$. 2. Cyclic oxidation data for two nickel-base superalloys and these same alloys with diffusion aluminide coatings. Initially Al$_2$O$_3$ scales are formed on all of these alloys via selective oxidation.

Oxygen active elements can be used to improve the adherence of Cr$_2$O$_3$ and Al$_2$O$_3$ scales [6–8] to metallic substrates.

3. MIXED GAS ATTACK

When alloys are exposed to gases containing other reactants in addition to, or in place of, oxygen, the approach is the same as that discussed previously for oxygen. To be resistant to the gas, it is necessary to selectively form the most protective reaction product barrier that can be developed in the particular gas under consideration. Transport of re-actants through oxides is usually slower than through other phases that could be formed upon the surface of the alloys, and therefore the most resistant systems are those upon which scales of Al$_2$O$_3$, Cr$_2$O$_3$ or SiO$_2$ can be formed. Since most metals have greater affinities for oxygen than the other reactants commonly encountered in practical environments, selective oxide barrier development is usually used to obtain resistance to attack by mixed gases. Selective oxide barrier formation is often possible in gases containing oxygen at pressures as low as about 0·1 Pa (10^{-6} atm).

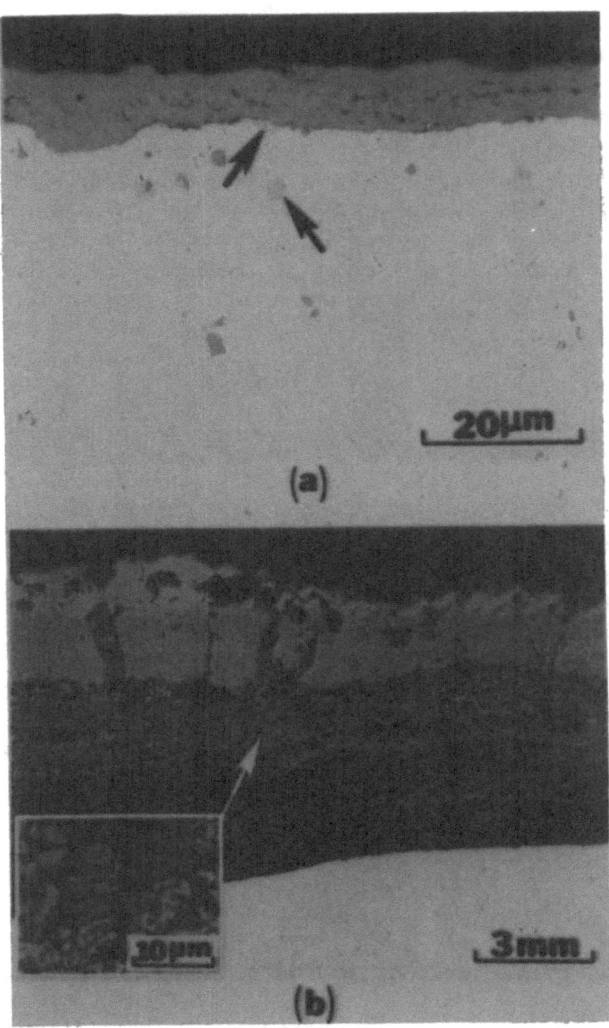

FIG. 3. Photomicrographs showing typical microstructures developed during cyclic oxidation of Fe–15Cr at 900°C in flowing argon–5% SO_2. (a) During the earlier stages, an external Cr_2O_3 scale covers the alloy surface and chromium sulfide particles (arrows) are formed within the alloy beneath the scale. (b) As more severe attack occurs the reaction product is composed of an oxide and sulfide mixture beneath an outer zone of iron oxides.

When mixed gases do not contain enough oxygen for selective oxide barrier formation, the alloy composition must be selected such that the next most protective reaction product barrier can be formed.

As discussed for oxygen environments, since alloys are used in mixed gases eventually the selective barrier can no longer be formed and the alloy must be removed from service. Cracking and spalling of the protective barriers play important roles in the degradation of the alloys, but in addition, products resulting from reaction of elements in the alloy with other reactants in the gas are formed just below the alloy surface and their subsequent attack by the gas (e.g. sulfide particle in alloy attacked by oxygen in the gas) results in more rapid degradation of the alloy (Fig. 3) [9–11].

4. HOT CORROSION

In some environments deposits can condense from the gas environment and accumulate upon the surfaces of alloys. These deposits separate the alloy from the gas and, as indicated in Fig. 4, significantly alter the

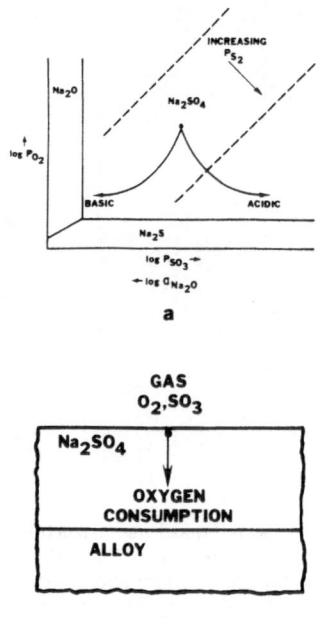

FIG. 4. Schematic stability diagram for Na–S–O system. (a) depicting the types of compositional variations that can be developed across a layer of Na_2SO_4 on an alloy (b).

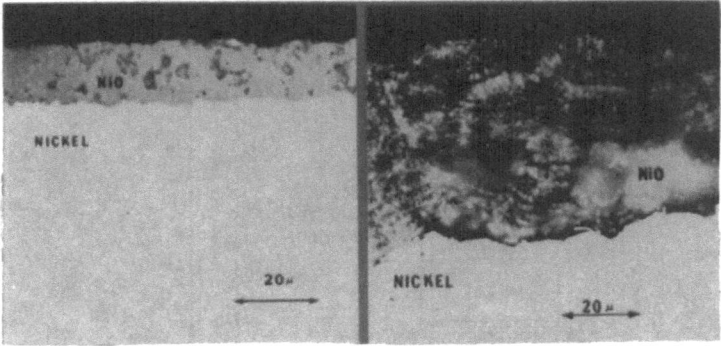

FIG. 5. Protective oxide formed on nickel after 3 h of oxidation at 1000°C in air and nonprotective scale that forms after 1 min under the same conditions but with a deposit of Na_2SO_4 on the specimen.

environment in which the reaction product barrier must be formed. It is sufficient in this paper to state that some deposits (e.g. $(Na,K)_2 SO_4$, NaCl) are especially effective in rendering reaction product barriers nonprotective (Fig. 5). A number of papers have described these deposit-enhanced corrosion processes which are called hot corrosion [12–14]. The problem is difficult to overcome and the approach that is to be used to control hot corrosion depends upon the conditions that cause it; in particular, there are several mechanisms by which hot corrosion can proceed. In the case of hot corrosion referred to as Type II, which occurs at temperatures between about 600 and 800°C and in gases containing SO_3, alloys with 30–35% Cr are more resistant to attack but there are some cases where the addition of chromium can produce adverse effects [15]. Type I hot corrosion occurs at temperatures between 800 and 1000°C. This form of hot corrosion can be minimized by using alloys with high chromium concentrations (e.g. > 20%Cr) (all compositions are given in weight percent) and controlling the refractory metal concentrations.

5. EROSION

When particulate matter hits the surfaces of alloys at elevated temperatures, the high temperature corrosion processes can be significantly

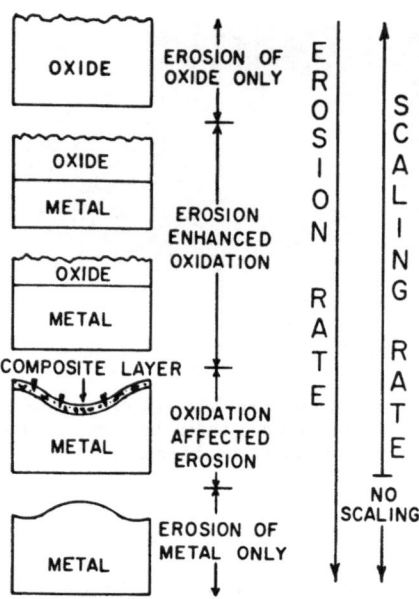

FIG. 6. Relationship between type of degradation observed under conjoint erosion–oxidation conditions and the relative severity of the erosion and oxidation components.

modified. A number of interactions between erosion and corrosion processes have been identified [16]. The nature of these interactions are dependent upon a number of parameters which include: the size, shape and velocity of the erosive particles, the growth rate of the reaction product barrier, and its physical characteristics (e.g. adherence to the metallic substrate, porosity). It has been found that when the growth rate of the reaction product is large, the erosive particles can reduce its thickness which causes the degradation rate to be increased. This condition is called erosion-enhanced oxidation. If the growth rate of the reaction product barrier is small, the erosive action of the particles can be increased due to the oxidation processes that are occurring concomitantly on the alloy surface. This condition is called oxidation-affected erosion. In Fig. 6 a few of the oxidation–erosion interactions are indicated schematically. It is sufficient in the current paper to state that most erosive interactions produce adverse effects on the degradation rates of alloys. Moreover, it is extremely difficult to develop new systems that possess resistance to this form of degradation short of removing the erosive particles from the system.

6. PROCEDURES TO DEVELOP RESISTANCE TO CORROSION AND EROSION AT ELEVATED TEMPERATURES

In order to develop resistance to severe environments at elevated temperatures conditions must be established such that a protective reaction product barrier is maintained upon the surfaces of alloys. In many instances it is not possible to alter the alloy composition to the extent required for protective barrier development while still having the alloy possess the necessary mechanical properties. Hence, the approach usually followed is to coat the alloy with a material that provides corrosion resistance to the structural alloy. A variety of coating fabrication techniques are available [17–19]. The most simple procedure is to enrich the surface of the structural alloy with the element that is necessary to form the desired reaction product barrier. Such diffusion coatings are usually obtained by aluminizing, chromizing, or siliconizing. Some typical aluminide coatings are shown in Fig. 7. In the case of diffusion coatings, their compositions are dependent to some extent upon the compositions of the structural alloys upon which they are formed. This limitation can be removed by utilizing overlay coatings which may be deposited by

FIG. 7. Photomicrographs of two diffusion aluminide coatings formed upon a nickel-base superalloy. The aluminum concentration at the surface of the alloy has been increased to about 31–33% and a β-NiAl phase has been formed.

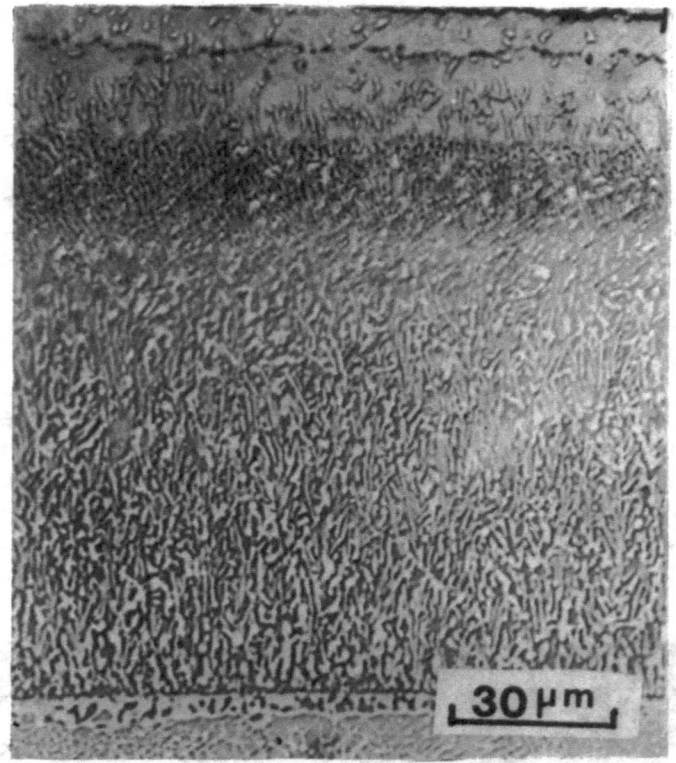

Fig. 8. Overlay coating deposited on a superalloy by using electron beam melting and vapor deposition. This coating was subsequently siliconized to enrich its surface in silicon.

electron beam vapor deposition or plasma spray processes. A typical overlay coating is shown in Fig. 8.

The compositions of coatings that should be used for different applications depend upon the conditions in which they are to provide protection. In the case of high temperature oxidation at temperatures about 1000°C, Al_2O_3 and SiO_2 are more effective barriers than Cr_2O_3 since gaseous CrO_3 becomes significant. Coating lives can be extended by adding elements to improve oxide scale adherence such as yttrium or hafnium. Other elements (e.g. Si, Pt) can be used to make the coatings more resistant to oxidation although the mechanisms by which some of these additions improve the lives of the coatings are not understood.

In mixed gas environments coatings can be used rather effectively to provide resistance to attack. Diffusion aluminide coatings appear to provide resistance to sulfur–oxygen gas mixtures as well as many other gas mixtures. The specific composition depends upon the gas composition in which the coating is to provide protection. In the case of gases with only small concentrations of oxygen the concentration of aluminum must be increased to inhibit the effects of sulfur. Overlay coatings can also be used but diffusion coatings when available are preferred due to cost, however many of the plasma spray techniques can be used to deposit overlay coatings at costs competitive with that for diffusion coatings.

There are a great number of coatings that can be used to develop resistance to hot corrosion attack. Aluminia and Cr_2O_3 scales have been used to obtain resistance to this form of degradation. Silica scales are extremely resistant to hot corrosion attack induced by acidic deposits, but very susceptible to attack caused by basic deposits. Consequently, silica barriers have not been used extensively to develop resistance to hot corrosion because in many applications the deposits causing hot corrosion attack may alternate between basic and acidic conditions. In the case of selecting coatings for resistance to hot corrosion attack the conditions causing the attack are extremely important. Some coating compositions can be extremely resistant to one form of hot corrosion but extremely susceptible to other forms.

Coatings are available to provide resistance to erosion. Usually these coatings are hard and typically may be carbides or nitrides. Unfortunately, virtually no coatings are available for resistance to the combined effects of erosion and corrosion. Composite coatings consisting of an outer layer for erosion resistance and an inner layer for corrosion resistance are worth consideration. Thermal barrier coatings [20] have an outer porous oxide layer and an inner more oxidation resistant, metallic layer. The outer oxide can be used for protection from erosion but often the porous nature of the thermal barriers do not result in a sufficiently tough coating, and cracking is usually encountered in an erosive medium.

While coatings are a very effective means to develop resistance to severe environments, cyclic temperature fluctuations, especially at temperatures above 1000°C, can cause thermal fatigue cracks [21] to develop in the coatings. This condition is then aggravated by oxidation of the coating exposed due to cracking. When thermal fatigue of the coating is a problem, differences between the coefficients of expansion of the coatings and the substrates must be minimized. Thermal barrier coatings can also be used to decrease the magnitude of the cyclic stresses developed in coatings.

7. CONCLUDING REMARKS

In attempting to develop materials with resistance to severe environments the characteristics of the particular environment under consideration must be analyzed. Based upon this analysis an approach can be formulated. Often the approach will involve the use of coatings. Most metallic coatings will afford protection by reacting with the environment to form a protective barrier between itself and the environment. The useful lives of such coatings will be determined by the time over which the protective barrier can be maintained upon the coatings. Ceramic coatings should be considered, but those currently available must be porous to prevent cracking, and hence can be used only as thermal barriers.

REFERENCES

1. MEIER, G. H. and BIRKS, N., *Introduction to High Temperature Oxidation of Metals*, London, Edward Arnold, 1983.
2. WAGNER, C., *Corr. Sci.*, **5** (1965), 751.
3. WOOD, G. C. and CHATTOPADHYAY, B., *Coor. Sci.*, **10** (1970), 471.
4. GIGGINS, C. S. and PETTIT, F. S., *J. Electrochem. Soc.*, **118** (1971), 1782.
5. PETTIT, F. S. and GOWARD, G. W., in *Metal Treatises* (J. K. Tien and J. F. Elliott eds), Warrendale, PA, The Metal Society, 1979.
6. TIEN, J. K. and PETTIT, F. S., *Met. Trans.*, **3** (1972), 1587.
7. WHITTLE, D. P. and STRINGER, J., *Phil. Trans. R. Soc. Lond.*, **295** (1980), 309.
8. STOTT, F. H., WOOD, G. C. and FOUNTAIN, J. G., *Oxid. Metals*, **14** (1980), 135.
9. GIGGINS, C. S. and PETTIT, F. S., *Oxid. Metals*, **14** (1980), 363.
10. CHU, W. F. and RHMEL, A., *Oxid. Metals*, **16** (1980), 175.
11. MEIER, G. H., COONS, W. C. and PERKINS, R. A., *Oxid. Metals*, **17** (1982), 235.
12. BORNESTEIN, N. S. and DeCRESCENTE, M. A., *Met. Trans.*, **2** (1971), 2875.
13. GOEBEL, J. A., PETTIT, F. S. and GOWARD, G. W., *Met. Trans.*, **4**, (1973), 261.
14. LUTHRA, K. L. and SHORES, D. A., *J. Electrochem. Soc.*, **127** (1980), 2202.
15. LUTHRA, K. L., *J. Electrochem. Soc.*, **132** (1985). 1293.
16. KANG, C. T., CHANG, S. L., BIRKS, N. and PETTIT, F. S., *Japan Inst. Met.* **24** (1983), 87.

17. LANG, E., *Coatings for High Temperature Applications* (London, Elsevier Applied Science, 1983).
18. GOWARD, G. W. and BOONE, D. H., *Oxid. Metals*, **3** (1971), 475.
19. LEVINE, S. R. and CAVES, R. M., *J. Electrochem. Soc.*, **121** (1974), 1051.
20. SINGHAL, S. C. and BRATTON, R. J., *Trans. ASME*, **102** (1980), 770.
21. STRANGMAN, T., Doctoral Dissertation, The University of Connecticut, 1978.

10

Lifetime Analysis of First Wall Materials Exposed to High Temperature and High Energy Neutrons in a Fusion Reactor

K. MIYA, H. HASHIZUME, H. OOMURA* and M. AKIYAMA**

Faculty of Engineering, Nuclear Engineering Research Laboratory, University of Tokyo, Tohkai,-mura, Naka-gun, Ibaraki, Japan.

Ishikawajima-Harima Heavy Industries, Co., Ltd, Koto-ku, Tokyo, Japan.

**Faculty of Engineering, University of Tokyo, Bunkyo-ku, Tokyo, Japan.*

ABSTRACT

A first wall of a fusion power reactor will be subjected to neutrons, charged particles and radiation, leading to neutron irradiation damage, decrease of thickness by physical sputtering, and high heat flux, respectively. As a consequence, the present state of the art requires first wall replacement after only a few years of operation. On-going efforts to solve this problem include work on the mechanisms of such irradiation induced damages as swelling, void formation and relevant mechanical properties, plasma–wall interaction and thermal stresses. Because of the complexities a computer code is required to account for such factors. As an extreme case of abnormal operation, a plasma disruption scenario is considered leading to analysis of melting, resolidification and evaporation, including elastoplastic stress.

1. INTRODUCTION

The structural integrity of in-vessel components (first wall, limiter and divertor plate) has recently been a key issue concerning the technological feasibility of a fusion power reactor. Twenty percent of the energy from

D–T reactions is deposited on the surface of the in-vessel components resulting in surface heating during normal operation. The remaining energy is carried by high-energy neutrons and is deposited in the blanket structure.

Crucial problems with the in-vessel components include the three effects of radiation damage, plasma major disruption and sputtering erosion. Another important problem is the fatigue-related failure of the components that is a natural consequence of the cyclic operation of the power reactor. Therefore, the in-vessel components are exposed to a severe and complex environment and it is not easy to evaluate their lifetime and performance. A numerical scheme is required for the prediction of the lifetime and performance of the components and a design methodology.

The second effect is directly related to the plasma major disruption. A model treated in the present study considers three processes, melting, evaporation and resolidification. The melting can be handled with consideration of temperature rise to melting points and latent heat needed for further temperature rise. An application of the Marcal iteration proposed for elastoplastic stress analysis is made here to deal with the melting.

Daenner [1] showed numerical results on the lifetime of a tubular first wall based on five criteria. The criteria were based on tensile properties of the material, large deformation due to swelling and progressive deformation due to irradiation creep. Mattas [2] performed a one-dimensional analysis of temperature and stress distribution taking into account changes of material properties due to neutron irradiation and the reduction of thickness due to sputtering erosion. An important result from the study was a drastic change of stress in the first wall with operation. Watson *et al.* [3] applied fracture mechanics to the lifetime prediction of a thin-walled thickness with a mechanical membrane load. A crack growth due to the fatigue damage is predicted based on a stress intensity factor and failure is predicted due to either unstable fracture by a larger K_I over K_{IC} (material constant) or a full penetration. Ghoniem [4] investigated the lifetime of a tubular first wall made of ferritic stainless steel. Adegbulugbe [5] discussed the failure criterion based on cyclic damage at elevated temperature and propagation of flaws due to fatigue damage. In the present paper a numerical scheme is investigated for a one-dimensional analysis of temperature distribution and stress history under plasma-on and plasma-off conditions.

2. MODEL AND MATHEMATICAL FORMULATION FOR LIFETIME ANALYSIS

2.1. Modeling and Material Constants for Lifetime Evaluation

The first wall is assumed to be a flat plate and consists of base metal and coating material. A one-dimensional analysis is a reasonable approximation for the first wall with a large radius of curvature compared to the wall thickness. Stress analysis was carried out with the assumption that the first wall is free to expand but not to bend.

The first wall will be exposed to a high neutron flux and a high heat flux. Material constants, therefore, will vary depending on the temperature and irradiation conditions. In the code, material constants are stored as the temperature- and displacement per atom (dpa)-dependent tables in a file and every time the material constant is needed, two-dimensional interpolation will be performed to provide proper values for the analysis. Tabulated material constants are the thermal expansion coefficient, Young's modulus, yield strength, Poisson's ratio, thermal conductivity, accumulated swelling, yield strain, ultimate tensile strength and allowable stress.

2.2. Analysis of Temperature and Stress Distribution

The heat conducting equation with volumetric heating for the one-dimensional analysis is;

$$\frac{d}{dz_i}\left[k_i(T,D)\frac{dT}{dz_i}\right]+Q_{vi}=0 \tag{1}$$

where $i=1$ for the base metal; $i=2$ for the coating material; z is the coordinate; k is the thermal conductivity; T is the temperature; D is the dpa; Q_{vi} is the nuclear heating rate.

Equation (1) is solved numerically at the discrete nodes with boundary conditions. For the thin first wall, the normal stress is almost zero and only in-plane stresses $\sigma_x(z)$ and $\sigma_y(z)$ were analyzed by the following equation,

$$\sigma_x(z)=\frac{N_x}{t_b}-\frac{E}{(1-v^2)}(e_x+ve_y)+\frac{E}{(1-v^2)t_b}\int_{-t_b/2}^{t_b/2}(e_x+ve_y)\,dz \tag{2}$$

where e_x and e_y are the total inelastic strains and N_x is the membrane load in the x direction. t_b is thickness of the wall. e_x can be represented as follows,

$$e_x(z, D) = \alpha[T(z) - T_0] + \frac{1}{3}S(z, D) + \int_0^\tau \psi\left(\sigma_x - \frac{1}{2}\sigma_y\right)d\tau' \tag{3}$$

$\sigma_y(z)$ and $e_y(z, D)$ are obtained by exchanging x with y. Here α is a thermal expansion coefficient, S is swelling and ψ is creep compliance given by

$$\psi = \dot{\varepsilon}_{cr}/\sigma_{eq} \tag{4}$$

where $\dot{\varepsilon}_{cr}$ is the irradiation-creep rate given later, and σ_{eq} is the equivalent stress.

2.3. Analysis of Crack Growth and Failure

The following empirical equation for the stress intensity factor was adopted, which was developed by Newman and Raju (6).

$$K_1 = (\sigma_t + H\sigma_b)\sqrt{\pi a/Q}\, F \tag{5}$$

where σ_t and σ_b are remote uniform tensile and bending stresses respectively. H, Q and F are the correction factors given in Ref. 6.

For an analysis of the crack growth, contributions from fatigue (da/dN) and creep (da/dt) were evaluated by the following formulas;

$$\frac{da}{dN} = \frac{C_1 \lambda^m [f\Delta K - \Delta K_0]^{2\cdot95}}{K_{IC} - \lambda f\Delta K} \tag{6}$$

$$\frac{da}{dt} = C_2 \exp[-Q_c/RT](K_{max})P \tag{7}$$

where C_1, C_2, Q_c, R and P are constants, λ, m, ΔK_0 are functions of the stress intensity factor K_1 and f is $E(T^0)/E(T)$. Equations (6) and (7) are the same as employed by Watson et al. [3]. The dpa dependence of the mode I critical stress intensity factor is given by

$$K_{IC}(D) = 115\exp[-0\cdot25D] + 35\exp[-0\cdot0134D] \tag{8}$$

3. MODEL AND MATHEMATICAL FORMULATION FOR MELTING AND RESOLIDIFICATION AND STRESS

3.1. Thermomechanical Analysis for Melting and Resolidification

A starting point for the thermomechanical analysis is to solve the following equation of heat conduction,

$$\rho c \dot{T} = k \nabla T + Q \tag{9}$$

where ρ = density; c = specific heat per unit mass;
$\quad k$ = thermal conductivity; Q = heat generation rate;
$\quad T$ = temperature; $\cdot = \partial/\partial t$, derivative in time.

The conventional scheme of finite element formulation of eqn. (9) is applied to derive the following system equation,

$$[A]\{T\} + [B]\{\dot{T}\} = \{F\} \tag{10}$$

where

$$[B] = \int \rho c [N]^T [N] \, dv$$

$$[A] = \int k [[N]_{,x}^T [N]_{,x} + [N]_{,y}^T [N]_{,y} + [N]_{,z}^T [N]_{,z}] \, dv$$

$$+ \int \alpha_t [N]^T [N] \, dS + \int \alpha_y [N]^T [N] \, dS$$

$$\{F\} = \int Q [N]^T dv - \int q [N]^T dS + \int \alpha_t T_0 [N]^T dS + \int \alpha_y T_y [N]^T dS$$

where q = heat flux, α_t = heat transfer coefficient, T_0 = temperature of coolant, T_y = temperature of radiated source and α_y is given by

$$\alpha_y = \varepsilon \sigma F_0 (T + T_y)(T^2 + T_y^2) \tag{11}$$

where ε = emissivity, σ = Stefan–Boltzmann constant, F_0 = shape constant.

In order to apply eqn. (10) to the melting phenomenon, an artificial technique is adopted. Although the temperature of the melting zone is kept constant, T_m (= melting temperature), a very small increment of temperature, ΔT, is introduced in the melted zone. This corresponds to allowing the melted portion to be treated like a solid possessing a pseudo-specific heat capacity defined by $\Delta T/L$ (L is latent heat).

On the other hand, heat is consumed in vaporization of the first wall material. The heat of vaporization is large compared with the latent heat and must be considered in the calculation of melted zone size. An evaporation flux of atoms is then given by [4]

$$J(t) = \frac{P}{\sqrt{2\pi mkT}} \qquad\qquad t \leqq t_v \qquad (12)$$

$$J(t) = \frac{P}{\sqrt{2\pi mkT}} \left[0 \cdot 8 + 0 \cdot 2 \exp\left(\frac{t_v - t}{10\tau_c}\right) \right] \quad t > t_v \qquad (13)$$

where τ_c is the collision time of atoms, t_v is preheat time, m is atomic mass and k is the Boltzmann constant. From the evaporation flux, the heat of vaporization can be evaluated by multiplying the atomic volume and the vaporization heat per unit mass. This quantity is treated as a negative heat flux.

3.2. Elastoplastic Stress Analysis

Dependence of mechanical properties on temperature has to be considered especially in the stress analysis of the problem where temperature change is large enough to result in melting and then resolidification of metal. For example, Young's modulus of stainless steel at 800°C is about half of that at room temperature. Another important factor to be considered is the temperature derivative of elastic constants. Stress increment is given as follows for elastic and plastic stress states,

$$\{\Delta\sigma\} = [D^e]\{\Delta\varepsilon^t - \Delta\varepsilon^\theta\} - [D^e]\frac{\partial[D^e]^{-1}}{\partial T}\{\sigma\}\Delta T \quad \text{(for elastic)} \quad (14)$$

$$\{\Delta\sigma\} = [D^p]\{\Delta\varepsilon^t - \Delta\varepsilon^\theta\} - [D^p]\frac{\partial[D^e]^{-1}}{\partial T}\{\sigma\}\Delta T +$$

$$\sigma_Y[D^e]\{\sigma\}\frac{\partial\sigma_Y}{\partial T}\frac{2}{3}\frac{\Delta T}{S_0} \qquad\qquad \text{(for plastic)} \quad (15)$$

where
$$\{\Delta\sigma\} = \text{vector of stress increment;}$$
$$\{\Delta\varepsilon^t\} = \text{vector of total strain increment;}$$
$$\{\Delta\varepsilon^\theta\} = \text{vector of thermal strain increment;}$$
$$\{\sigma\} = \text{vector of elastoplastic stress;}$$
$$\{\sigma'\} = \text{vector of deviated stress;}$$
$$[D^e] = \text{elastic stiffness matrix;}$$
$$[D^p] = \text{plastic stiffness matrix;}$$

σ_Y = yield stress;

ΔT = temperature increment;

$S_0 = \frac{4}{9}H'\sigma_{eq}^2 + \{\sigma'\}^T[D^e]\{\sigma'\};$

H' = strain hardening rate.

4. ANALYSIS OF EXPERIMENT

In Fig. 1 the experimental and numerical results of temperature change are compared. The temperature was measured at two points, point 'A' on the top surface and point 'B' on the bottom surface as indicated in the figure. Since 'A' is very close to the heated region, the temperature at that

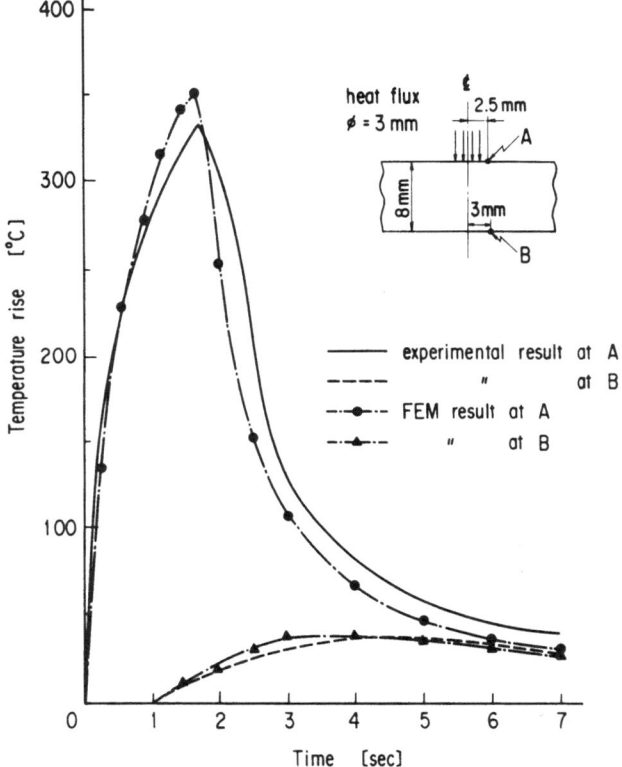

FIG. 1. Temperature change at the surface of plate subjected to heat flux.

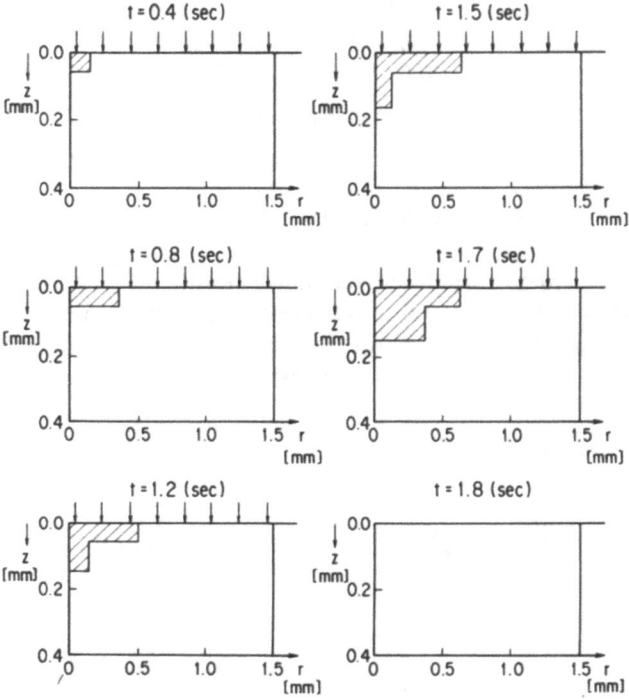

FIG. 2. Progress of melting zone size (numerical results).

point follows immediately a change of the beam power. The effect of evaporation is not considered in the numerical analysis of the data, which may be justified by the fact that the experiment was carried out in atmospheric environment and as such evaporation is suppressed. A plasma disruption, however, takes place in vacuum and the evaporation of first wall material is not suppressed. Therefore, the effect of evaporation must be included in analyses of the plasma disruptions, as will be shown later.

Figure 2 shows progress of a melting zone with time. The boundary between liquid and solid is not smooth, since the size of an element is too large compared with that of the melted zone. The size is a maximum at $t = 1.7$ s, and resolidification of it is completed in a very short time, 0.1 s. The maximum depth is about $180\,\mu m$.

A comparison of tangential strains is shown in Fig. 3. The strain is

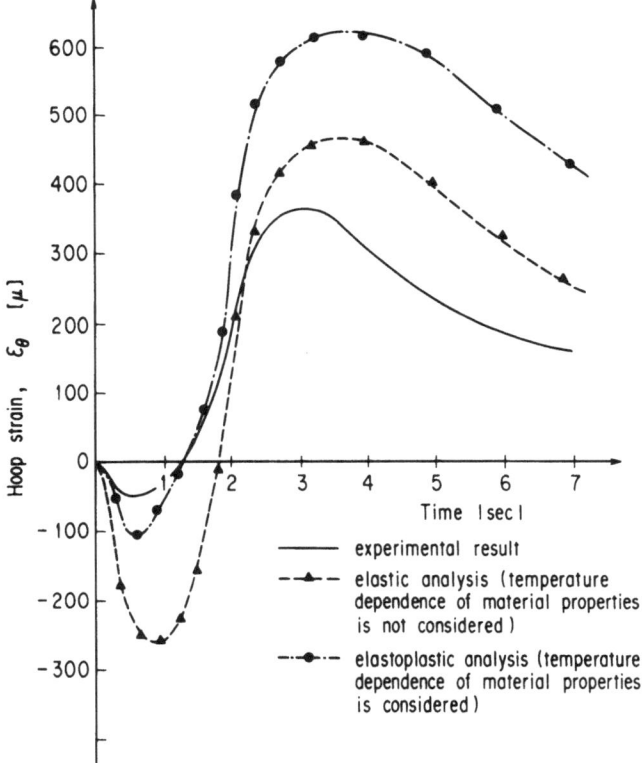

FIG. 3. Thermal strain change at the surface of plate subjected to heat flux.

compressive after the beginning of heat input and becomes tensile roughly in 2 s. Peaks of the tangential strain are observed between 3 and 4 s. It should be noted that the resolidification of the melted zone is completed in at least 0·1 s after completion of preheating. The temperature change lasts for 10 s.

Figure 4 shows distributions of radial stresses at various times. The stress is taken on the line A–B, 0·1 mm below the top surface. A location near the point 'A' is not melted 0·05 s after the start of heating, but does melt in 0·3 s resulting in zero stress. The stress distributions are almost the same for elapsed times of 0·3 and 1·7 s. However, when the resolidification is completed and the temperature decreases, the stress at the melted region changes to tensile due to the contraction of this

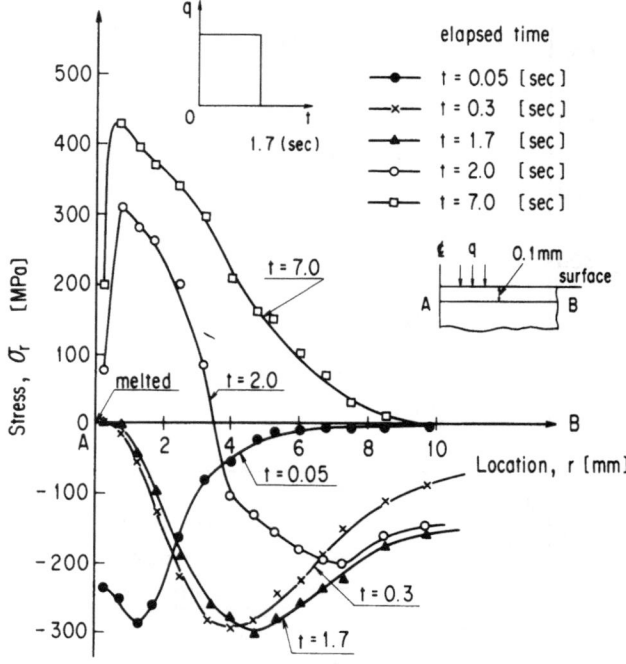

FIG. 4. Thermal stress distributions in stainless steel plate subjected to heat flux.

portion. The stress distribution at 7·0 s may be considered as a residual
stress distribution. Thus the maximum tensile stress of 42 kg/mm² remains. This high residual stress could cause microcracks around the
melted region. There were many cases where hair-line cracks were
observed in metallurgically polished specimens. Initiation of the hair-line
cracks during the process of resolidification is very hazardous to the
structural integrity of the first wall when subjected to large plasma
disruption. It is very important to be able to find some engineering
solution for this problem of cracking, but the problem is difficult.

5. NUMERICAL RESULTS AND DISCUSSION

5.1. Temperature and Stress Distributions

Figure 5 shows the time history of temperature distribution for the
5 mm thick first wall. The temperature of the plasma side will increase as

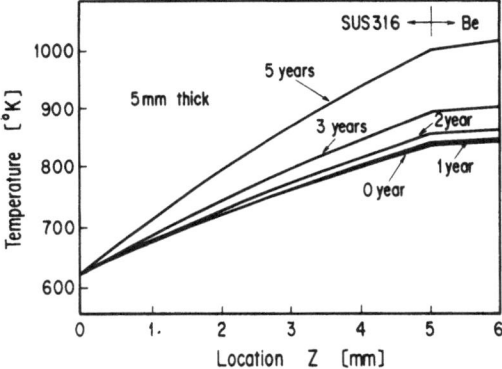

FIG. 5. Distribution of temperature in first wall.

time elapses. This is mainly because thermal conductivity decreases. The dpa-dependency of thermal conductivity, $\lambda(T)$, was taken into account by the following formula,

$$\lambda(T, D) = \lambda(T)[(1 - S(D)/(1 + 25D)]\tag{16}$$

where $S(D)$ is the swelling which is a function of the dpa D. It is seen that initial temperature differences between the plasma side and the coolant side are about 50% of those expected after 5 years' operation. This means that time-dependent temperature distribution must be considered when the stress distribution is analyzed.

In considering a time history of temperature distribution and irradiation effect for the stress distribution, Gittus's formulation [7] for the irradiation creep rate is used for 20% CW SUS 316, which is considered to be the base metal of the first wall. Therefore, under the typical TOKAMAK reactor conditions, the first term of the following equation is dominant because of the large swelling rate of SUS 316.

$$\dot{\varepsilon}_{CR} = \frac{\sigma}{\sigma_Y} \dot{S} + 0{\cdot}5\,\sigma D/E\tag{17}$$

Figure 6 shows the stress history of the plasma side surface. High thermal stress is induced at the coolant side at the beginning of the operation and is reduced rapidly after 1 year of operation. Adversely, tensile inelastic stress is growing larger at the plasma side of the first wall for the plasma-off condition. The change of stress is caused mainly by the irradiation creep.

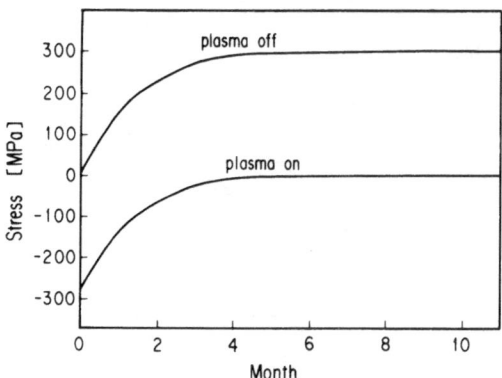

FIG. 6. History of stress at plasma side surface.

5.2. Crack Growth Behavior

Only the crack at the plasma side is important for lifetime evaluation of the first wall. This is mainly because Gittus's formula for the irradiation creep rate was used. Figure 7 shows the crack growth behavior for the 0·1 mm initial length crack at the plasma side surface. Values of swelling rate taken till the incubation dose are $2·66 \times 10^{-10}$, $1·08 \times 10^{-10}$, $2·66 \times 10^{-11}$ (per second) and the incubation dose taken is 40 dpa, 100 dpa and 100 dpa, respectively. It is seen that the calculated lifetime for these cases are almost the same. It turned out, therefore, that swelling rate does not play an important role in the determination of first wall lifetime.

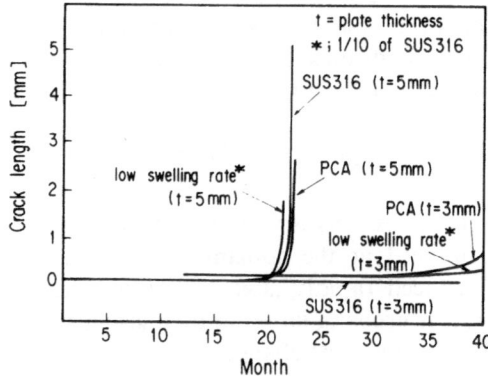

FIG. 7. Crack growth behavior (effect of swelling rate).

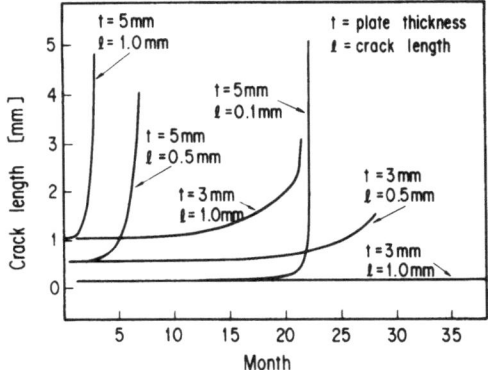

F IG. 8. Crack growth behavior (effect of initial crack length).

The effect of initial crack length on the lifetime of the first wall is shown in Fig. 8. After plasma disruption, many hair-line cracks of 0·1 mm length will appear at the plasma side surface. If the length is less than 0·1 mm, the lifetime of the first wall is about 2 years for the 5 mm thick case and about 4 years for the 3 mm thick first wall. As the length of the initial crack becomes larger, the lifetime decreases rapidly.

Qualitatively, then if the wall loading is reduced, the lifetime will be prolonged because the dpa rate, heat flux and volumetric heating will decrease. It means that the decrease of dpa rate causes a reduction of the irradiation damage of the first wall material. Figure 9 shows that when

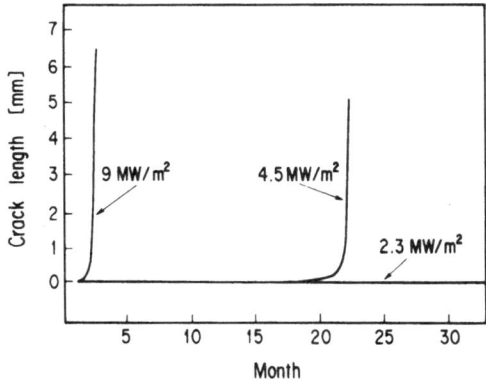

F IG. 9. Crack growth behavior (effect of wall loading).

the wall loading becomes twice that expected for the standard case, the lifetime of the first wall will then be reduced by about one order of magnitude. If the wall loading is one half, then the lifetime will increase to about twice the standard case for the 5 mm thick case.

Fracture toughness decreases exponentially if the dpa-dependency of eqn. (8) is assumed. Accordingly, fracture toughness is a most important limiting factor for the lifetime of the first wall. Recently it has been said that the fracture toughness tends to saturate after a neutron exposure of approximately 15 pda. Saturated value of K_{IC} is about 90 MPa$\sqrt{m}$ for SUS 316. If this is true, then the lifetime will be longer than that predicted by the present study.

5.3. Evaporation and Melting

Figure 10 shows the dependence of the evaporation thickness on the deposited energy. It is evident from the results that the evaporation thickness increases with the deposited energy and decreases with the disruption time. When the disruption time increases, temperature rise could be suppressed because of diffusion of heat. Evaporation thicknesses are 70 μm for a disruption time of 10 ms and 12 μm for 100 ms. A difference of 70 μm and 10 μm is influential in a design specification of the first wall from a viewpoint of wall thinning. For example, an

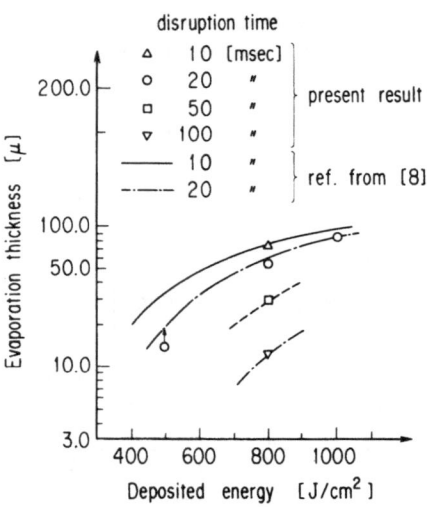

FIG. 10. Relation between evaporated thickness and deposited energy at plasma disruption.

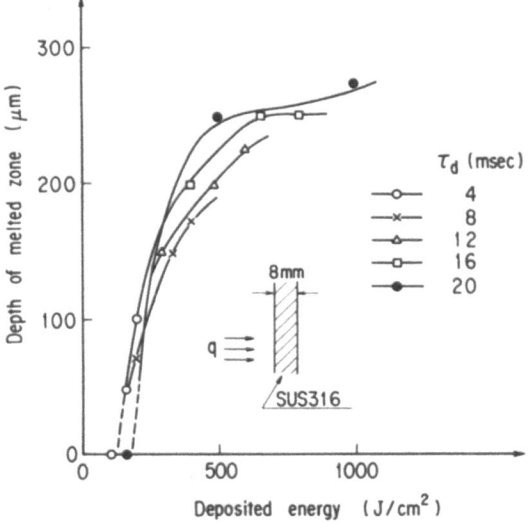

FIG. 11. Relation between melting depth and deposited energy with disruption time as the parameter.

assumption that the disruption of 10 ms and 800 J/cm^2 takes place 100 times during a lifetime of the first wall will erode the surface of the wall by 0·7 mm in the case of stainless steels. Thickness of the first wall of a fusion power reactor has to be less than 10 mm, or very large thermal stresses will be induced, resulting in a shorter lifetime at the wall due to fatigue and creep damage. The evaporation thickness of 1 μm per disruption could be allowed if there are 1000 or less. Therefore, it is very important to obtain a data base on the evaporation thickness as a function of the disruption time and the deposited energy. In the figure are shown numerical results from Hassanein [8] to compare with the present results. Solid and dotted lines show the evaporation thickness for the disruption times of 10 ms and 20 ms, respectively. Agreement is quite excellent. Figure 11 shows a relation between melting depth and deposited energy with a parameter of the disruption time. In the range of the disruption time between 4 and 20 ms, the melting depth does not always depend on the disruption time but strongly depends on the deposited energy. There is clearly a threshold energy for melting, which is about 200 J/cm^2 for stainless steel. This is a very important piece of information for the thermomechanical design of the first wall. Below this threshold value, many heat deposition cycles are allowed since fatigue damage due to thermal cycling can be neglected. Other important information in the figure is that the melting depth does not increase significantly for energy deposition values over $500 \, J/cm^2$. As stated previously, this is primarily due to the energy consumption in the evaporation. This result is remarkable for the shorter plasma disruption. This fact could be different for materials such as graphite that are easy to evaporate or sublimate.

6. CONCLUSIONS

Conclusions obtained from the present study are summarized as follows;

(1) Crack propagation from the plasma side is more important than that from the coolant side.
(2) Temperature and stress change with progress of operation.
(3) The amount of evaporated mass increases with a decrease of the disruption time.
(4) There exists a threshold energy for initiation of melting, which is about $200 \, J/cm^2$ for stainless steels.

REFERENCES

1. DAENNER, W., *J. Nucl. Mater.*, **103, 104** (1981).
2. MATTAS, R. F., *Nucl. Tech./Fusion*, **4** (1983), 1257.
3. WATSON, R. D., PETERSON, R. R. and WOLFER, N. G., *ASME J. Press. Vess. Technol.*, **105** (1983), 144.
4. GHONIEM, N. M., *Nucl. Technol.*, **4** (1983), 769.
5. ADEGBULUGBE, A. O., *Nucl. Eng. Des./Fusion*, **1** (1984), 301.
6. NEWMAN, J. C. and RAJU, I. S., *Engng Fract. Mech.*, **15** (1981), 185.
7. GITTUS, J., *Irradiation Effects in Crystalline Solid*, London, Elsevier Applied Science, p. 285.
8. HASSANEIN, A. M., KULCINSKI, G. L. and WOLFER, W. G., *Nucl. Engng. Des./Fusion*, **1** (1984).

11

Physical and Mechanical Approaches to Cyclic Constitutive Relationships and Life Evaluation of Structural Materials in Biaxial Low-cycle Fatigue at High Temperatures

M. OHNAMI, M. SAKANE, S. NISHINO* and T. ITSUMURA*

Faculty of Science and Engineering, Department of Mechanical Engineering, Ritsumeikan University, Tojiin-Kitamachi, Kita-ku, Kyoto, Japan, 603.

*Graduate School of Ritsumeikan University, Tojiin-Kitamachi, Kita-ku, Kyoto, Japan, 603.

ABSTRACT

This report intends to clarify, by means of TEM observation, the difference in the cyclic constitutive relationships between proportional and nonproportional loadings in high-temperature low-cycle fatigue. Another problem which will be discussed is the study of crack opening displacement as a parameter for correlating fatigue data in biaxial stress states at high temperatures.

1. INTRODUCTION

Recent progress in computer-aided numerical analysis calls for an accurate stress/strain equation for materials used at high temperatures. This is especially necessary for cyclic loading conditions, since any deviation in the equation from the actual material may result in a gross error in material evaluation. In this report, we will discuss the problem in cyclic constitutive equations which reflects the structure of materials. We will also discuss the difference in the cyclic constitutive relationships between proportional and nonproportional loadings in strain-controlled high-temperature low-cycle fatigue.

Although many mechanical parameters have been proposed for cor-

165

relating the fatigue failure life data in biaxial stress states [1–3], the studies on crack initiation and propagation at high temperatures are scarce. Also there is no parameter which can correlate the data well in a variety of stress states at high temperatures. The crack opening displacement (COD) basis can be used effectively in studying biaxial low-cycle fatigue at high temperature where there is a wide range in the strain ratio ϕ which is the ratio of the minimum principal strain, ε_3, to the maximum strain, ε_1, and which covers the range of -1 to 1. In this paper, a micro X-ray Laue method was used and the high-temperature biaxial low-cycle test using a cruciform test specimen was newly performed.

2. PROBLEMS IN CYCLIC STRESS–STRAIN RELATIONS AT HIGH TEMPERATURES, BIAXIAL STRESS STATES

We will address the following problems; (1) how can we construct an accurate cyclic constitutive relation in which substructural change during strain cycles in biaxial stress states is considered and (2) what is the difference between the cyclic constitutive relation under proportional loading and that under nonproportional loading.

The test material was a type 304 solution heat treated stainless steel with a chemical composition by wt.% ratio of 0·38Si, 1·13Mn, 0·008P, 0·021S, 8·74Ni, 18·52Cr, 0·06C and remainder Fe with ASTM No. 3·5 grain size. We used a smooth, hollow cylindrical specimen of 12 mm OD., 9 mm ID. and 20 mm gauge length. A unique high-temperature biaxial low-cycle fatigue apparatus (electro-hydraulic-servo type with microcomputer) [4] performed a von Mises' equivalent total strain range-controlled test in the range of $-1 \leqq \phi \leqq -0·5$ at 0·1 Hz and 823 K in air. The strain wave was a fully reversed triangular wave (strain ratio $R = -1$). JOEL JEM100C(100 kV) TEM and Rigaku Roterflex RU-200 X-ray diffraction instruments were used to analyze the substructures of the material.

Figures 1a and 1b [5] show the geometrical configurations of ladder and the maze type dislocation structures of the material in strain-controlled push-pull (abbreviated as 'P', Fig. 2a) and reversed torsional (abbreviated as 'T', Fig. 2b) low-cycle fatigue tests at cycles $N_3 = 0·5\,N_f$, respectively, where the equivalent strain range, $\Delta\bar{\varepsilon}$, is 1%, and N_f is the number of cycles to failure. Since the electron beam penetrates from the top to the bottom of the specimen shown schematically in the figure, it is clear that the dense parts of the ladder or maze structures are on $\{111\}$

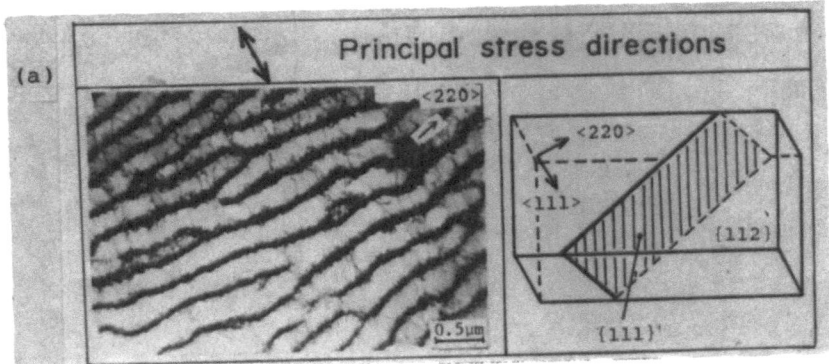

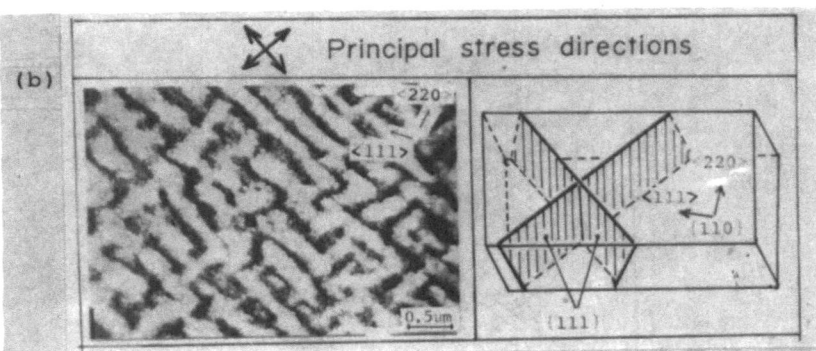

FIG. 1. Dislocation structures, primary slip planes ($\Delta \bar{\varepsilon} = 1\%$, $N/N_f = 0.5$). (a) Ladder type in P loading; (b) maze type in T loading.

planes. This was also confirmed by tests at cycles $N_1 = (0.05-0.1)\ N_f$ and $N_2 = (0.2-0.3)\ N_f$.

Regarding the relationship between cyclic stress response and dislocation structure, the findings are as follows. Generally, in uniaxial low-cycle fatigue tests at room temperature, a cell structure is formed. At elevated temperatures, however, ladder or maze structures form because of a thermally activated process which rearranges the dislocations into more ordered arrays having lower elastic strain energy. Mura *et al.* [6] reported that the maze and ladder structures have less strain energy than the cell structures according to elastic calculations. On the other hand, in the alternation test, the maximum principal shear strain changes direction by 45° in each cycle (abbreviated as 'APT', Fig. 2c), so that a larger

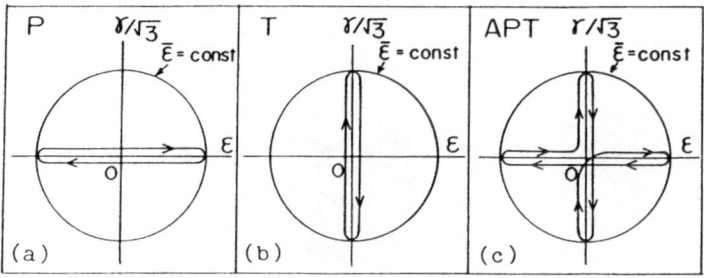

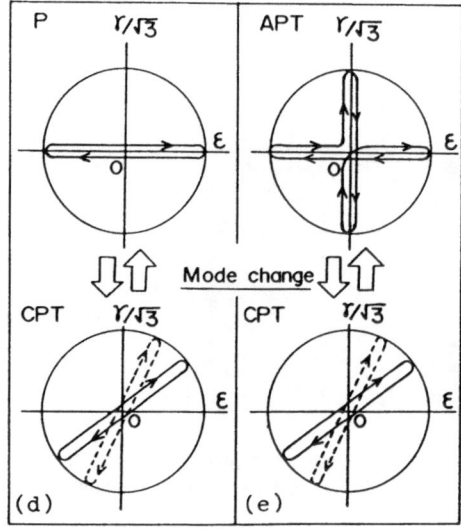

FIG. 2. Test programs.

interaction between the slip system occurs. In the reversed cycle tests, this larger interaction results in a cell structure, which prevents the rearrangement of the dislocations by the thermally activated process. The resistance to the dislocation glide is of course greater in the cell structure so that larger cyclic strain hardening results.

The total misorientation, β_t, obtained by micro beam X-ray diffraction patterns under the three loading modes at N_1 are: $\beta_t = 1.31 \times 10^{-2}$ rad, 2.35×10^{-2} rad and 11.71×10^{-2} rad for P, T and APT modes, respectively. As total misorientation β_t corresponds with the continuous change of crystal orientation over one or two grains, a much greater intensity of plastic deformation in grains occurs in the APT test and contributes to the increase in cyclic stress amplitude by about 40% [5].

Next, we examined how the prestrained material in the P and APT tests hardened, i.e. isotropically or anisotropically. Cyclic yield curves of the prestrained material were drawn by changing the loading direction at $\Delta\bar{\varepsilon} = 1.0\%$ and 0.7%. Figures 2d and 2e show the test program. In test (d), material prestrained for several cycles in push–pull mode was subsequently strained 3–5 cycles in a direction with a principal strain ratio $\varepsilon_3/\varepsilon_1$ for obtaining a data point of cyclic stress amplitude on the yield ellipse of that material. This loading was called 'CPT' loading. The change of mode from P to CPT was the mode at null stress and strain. The material was then reloaded in the push–pull mode to initiate it before the CPT test, and a test to obtain a different point on the curve was performed. This sequence completed the cyclic stress amplitude and yield curve of the prestrained material in the P mode. As shown in Fig. 2e, the same procedure was carried out for the prestrained material in the APT mode. Note that there was almost no effect of CPT loading on the material behavior after the initiation in the P mode. This was confirmed by the fact that after initiation the material yielded at almost the same cyclic stress amplitude as the virgin material. Also, the dislocation structure after the initiation, as well as in the P test, was of the simple ladder type.

Figures 3a and 3b [5] show an example of both the equivalent stress amplitude ellipses and the cyclic yield stress ellipses in P and APT modes with those of the virgin material at $\Delta\bar{\varepsilon} = 1.0\%$, respectively. The equivalent stress amplitude was taken as the tensile and compressive peak stresses of a hysteresis loop and the cyclic yield stress was defined here as the stress at a strain of 0.4% measured from the peak stresses. The figures show that the virgin material exhibits an almost isotropic stress response macroscopically because it is represented by the von Mises' ellipse. However, the cyclic stress response is not macroscopically isotropic for the prestrained material in the P mode. The shape of the ellipse of that material is being elongated along the shear stress axis. On the other hand, the cyclic response for the prestrained material in APT mode is macroscopically isotropic, but the ellipse expands in an isotropic manner from the virgin material. The equivalent stress amplitude in the APT mode is about 1.5 times larger than those of the virgin material. These results suggest that the cyclic constitutive equation such as the kinematic hardening rule, ORNL kinematic hardening rule and MARC type

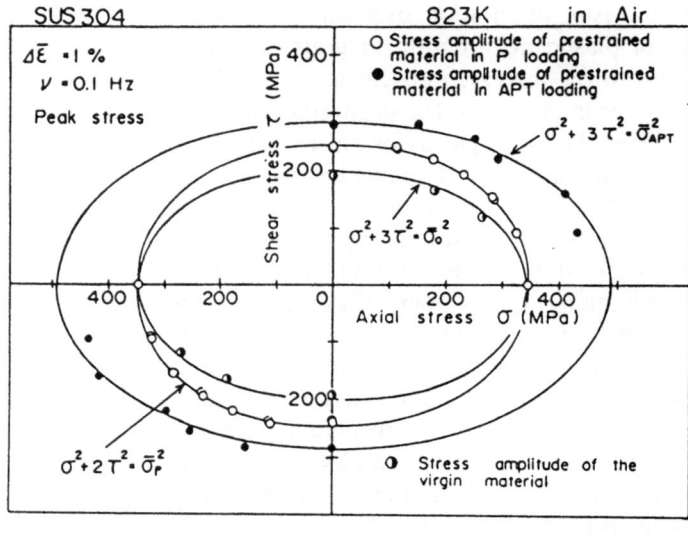

(a)

(b)

Fig. 3. (a) Equivalent stress amplitude ellipses and (b) equivalent cyclic yield stress ellipses
($\Delta\bar{\varepsilon} = 1\%$).

combined hardening rule can be applied to the cyclic hardened material with an isotropic dislocation structure such as cell formation. These results, however, cannot always be applied to cyclic hardening material with an anisotropic dislocation structure such as ladder and maze types.

3. PROBLEMS IN LIFE EVALUATION AND MECHANICAL PARAMETER OF BIAXIAL LOW-CYCLE FATIGUE AT HIGH TEMPERATURE

First, we describe a mechanical parameter of both crack initiation and propagation in strain-controlled push–pull low-cycle fatigue of the same test material with an average grain diameter of 0·9 mm at a frequency of 0·02 and 0·1 Hz and 873 K in air. A microscope with a vernier caliper was used to measure the crack length, l, and the crack opening displacement, COD, which was defined as the displacement at the notch root of a circular hole with a 1 mm diameter. Tests were interrupted about five times to perform X-ray diffraction on two parts of the specimen; one was the notch root or crack tip (cracked part) and the other was the opposite side of the notch hole (unnotched part). We confirmed that the crack had no effect on the strain intensity of the opposite side. Prior to X-ray diffraction, the 30 μm surface oxide film was removed by electropolishing. The position, where the X-ray diffraction was done, was 25 μm ahead of the notch root before crack initiation and also was 250 μm ahead of the crack tip after crack initiation.

Figure 4 [7] shows an example of the change in misorientation, β, with orientation function, Γ, before crack initiation. The orientation function, Γ, was calculated from Miller's indices. Before loading and at the 50th cycle, the misorientation does not depend on the diffraction plane and the misorientations of all the diffraction planes exhibit almost the same value. At the 140th cycle, however, the misorientations of special diffraction planes exhibit the larger value. This indicates that larger cumulative plastic deformation occurs on the planes. The planes having special larger misorientation are not the planes of the maximum shear stress or the primary slip plane of $\{111\}$ which is expected from Fig. 1a. The notably larger plastic deformations on those planes may be related to the micromechanism of crack initiation, but it is not clear whether the crack occurs at a critical value of maximum shear strain (γ^*_{max}) on the plane intersecting the free surfaces or the critical value of combined γ^*_{max} and the normal strain, ε^*_n [8]. We can say at least, that since the micro-

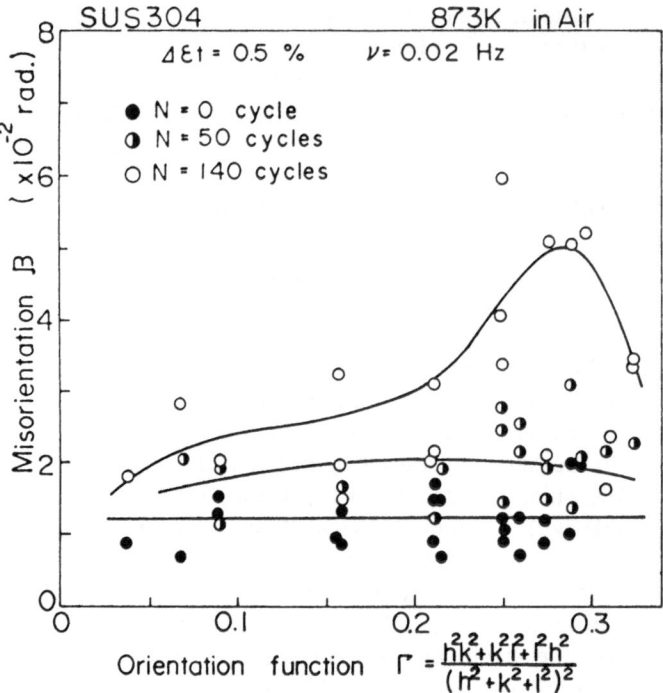

F IG. 4. Variation of misorientation against orientation function ($\Delta\bar{\varepsilon} = 0.5\%$).

mechanism of low-cycle fatigue-crack initiation of metal at high temperature is not clear, the misorientation obtained by the micro X-ray Laue method is a good method to estimate the crack initiation life. This finding is extended to the crack propagation behavior as follows.

Figures 5a and 5b [7] show the experimental relation between COD and the average misorientation, $\bar{\beta} - \bar{\beta}_0$, and that between the crack propagation rate, dl/dN, and the average misorientation, respectively, where the average misorientation is the arithmetic mean of β on various diffraction planes. Both COD and dl/dN are well correlated with the misorientation at the crack tip in all of the frequencies and the strain ranges tested. Since the average misorientation relates to the cumulative plastic strain range, $\Sigma\Delta\varepsilon_p$, examined in the unnotched part of the specimen, we can conclude that the average misorientation at the crack tip measured by the micro X-ray Laue method is a good parameter for estimating the crack propagation rate in low-cycle fatigue at high

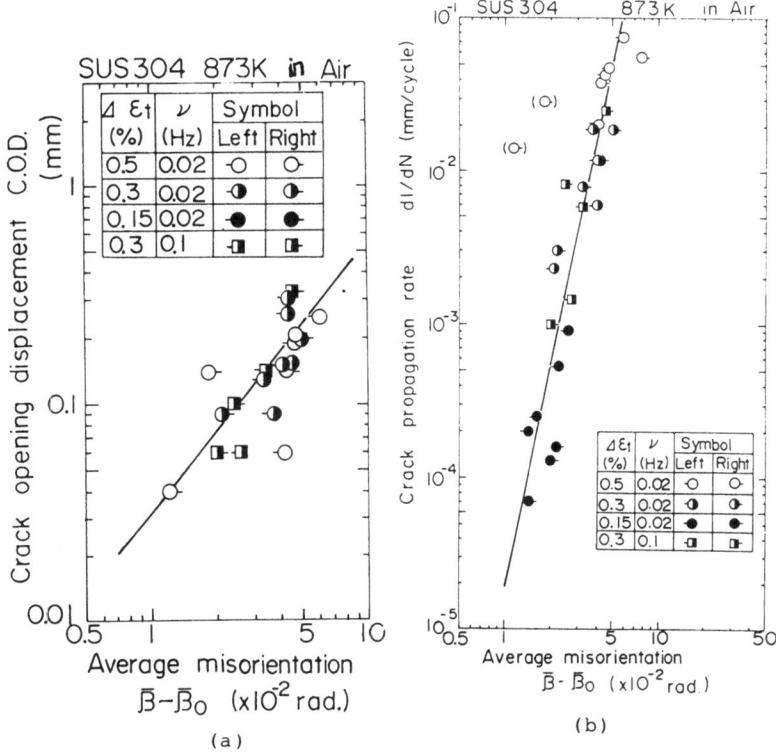

FIG. 5. (a) Experimental relation between COD and average misorientation and (b) that between the crack propagation rate and the average misorientation.

temperature. Therefore we can say that COD is a good parameter for evaluating the crack propagation rate and crack propagation life from both the physical and the mechanical viewpoints.

Figure 6 [9] shows the experimental relation between dl/dN and COD in stress-controlled low-cycle fatigue in P and T loading modes for the same notched hollow cylindrical specimen with a fine grain. The tests were performed at room and elevated temperatures in air. In the figure λ is the stress ratio of the minimum principal stress, σ_3, to the maximum stress, σ_1. The tests were carried out by the same biaxial low-cycle fatigue test machine. The figure shows good correlation of the data such that if we can formulate COD by a simple equation using the applied stress or strain, the equation will provide a good comparison parameter for correlating the biaxial fatigue data. In the following we derive the

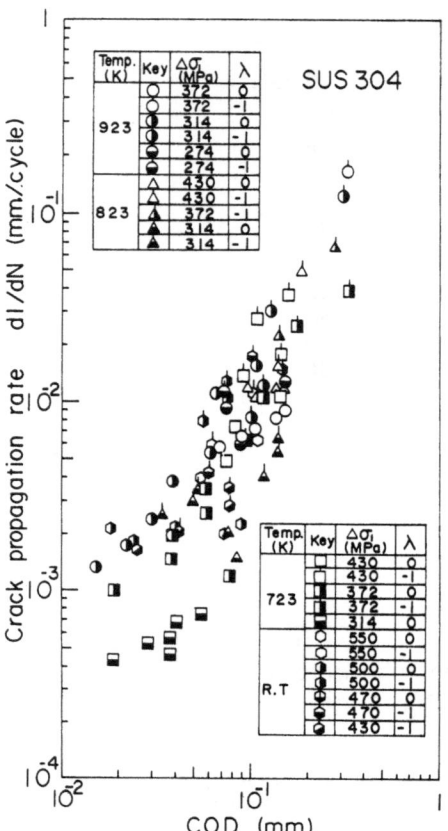

FIG. 6. Experimental relation between crack propagation rate and COD.

necessary relation by inelastic finite element method analysis.

Figures 7a and 7b [10] are the results of an FEM analysis of COD for a centrally cracked plate subjected to monotonic biaxial loading. These figures show the analytical relation between (COD/ε_1) and $(2-\phi)$ and that between (COD/σ_1) and $(2-\lambda)$. In these analyses, the cyclic constitutive relation of 1 Cr–1 Mo–0·25 V steel, commonly used for steam turbine rotors at 823 K in air, was used in the multilinear approximating representation and the constitutive equation was obtained by a step-up test method. If the data are approximated by a straight line at each COD, we can, for example, obtain

$$(COD/\varepsilon_1) = A(2-\phi)^m \qquad (1)$$

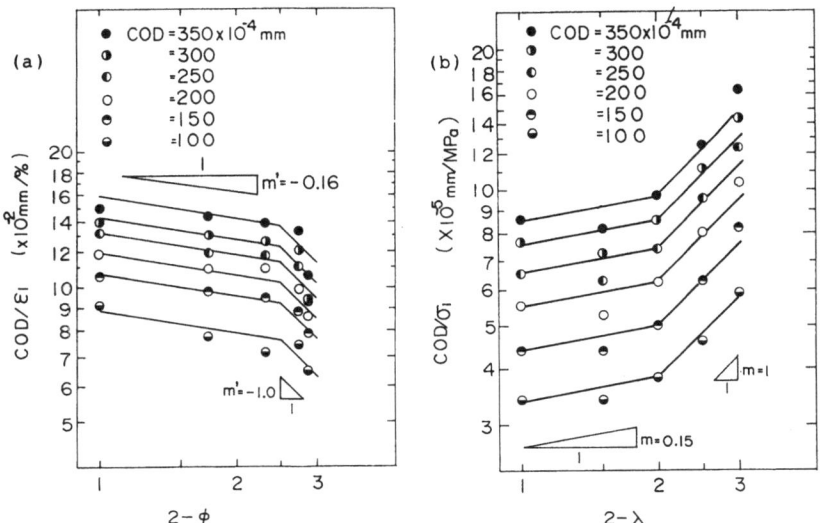

FIG. 7. (a) Analytical relation between COD/ε_1 and $(2-\phi)$ and (b) that between COD/σ_1 and $(2-\lambda)$ for the centrally cracked plate subjected to monotonic biaxial loading.

where m is -1 and -0.16 for $-1 \leqq \phi \leqq -0.5$ and $-0.5 \leqq \phi \leqq 1$, respectively, and A is a function of COD, i.e. a function of the applied strain.

In order to apply the equations to the low-cycle fatigue test, the principal strain, ε_1, is converted into the principal strain range, $\Delta\varepsilon_1$, and coefficient, A is modified as $\alpha A'$. Equation (1) is thus reduced to

$$(COD/A') = \alpha\Delta\varepsilon_1(2-\phi)^m \tag{2}$$

Equation (2) shows the crack opening displacement in biaxial stress states. Considering that the crack propagation rate in the biaxial stress state can be correlated by COD, as shown in Fig. 6, and the major part of the fatigue life is concerned with the crack propagation period. Equation (2) is an appropriate mechanical parameter to describe low-cycle fatigue life in the biaxial stress state.

The right hand side of eqn. (2) is the principal strain range modified by the principal strain ratio, ϕ. So, we define an equivalent strain range, as

$$\Delta\varepsilon^* = \alpha\Delta\varepsilon_1(2-\phi)^m \tag{3}$$

The nondimensional coefficient, α, is determined by satisfying the relation, $\Delta\varepsilon^* = \Delta\varepsilon_1$, for the uniaxial case, that is $\alpha = 2.5$ and 1.16 for

$-1 \leqq \phi \leqq -0.5$ and $-0.5 \leqq \phi \leqq 1$, respectively. Using the same procedure for deriving eqn. (3) the following equation of the stress basis is obtained;

$$\Delta\sigma^* = \beta\Delta\sigma_1(2-\lambda)^m \qquad (4)$$

where m' is 1 and 0.15 for $-1 \leq \lambda \leq 0$ and $0 \leq \lambda \leq 1$, respectively, and β is 0.5 and 0.9 for $-1 \leq \lambda \leq 0$ and $0 \leq \lambda \leq 1$, respectively. Thus, $\Delta\varepsilon^*$ and $\Delta\sigma^*$ are the equivalent stress and strain, respectively, and account for the effect of the stress parallel to the crack; $\Delta\varepsilon^*$ and $\Delta\sigma^*$ can be calculated only by taking the principal strain and stress applied to the specimen, respectively. Applicability of eqs (3) and (4) to low-cycle fatigue failure life at high temperature was examined in the following.

4. HIGH-TEMPERATURE BIAXIAL LOW-CYCLE FATIGUE TESTS FOR CRUCIFORM TEST SPECIMENS

The material tested was 1Cr–1Mo–0.25V with a chemical composition by wt% ratio of 0.29C, 0.30Si, 0.74Mn, 0.006P, 0.003S, 0.03Cu, 1.12Cr, 0.31Ni, 1.16Mo, 0.25V, 0.0018Sb and remainder Fe. A Young's modulus

FIG. 8. High-temperature biaxial low-cycle fatigue test apparatus.

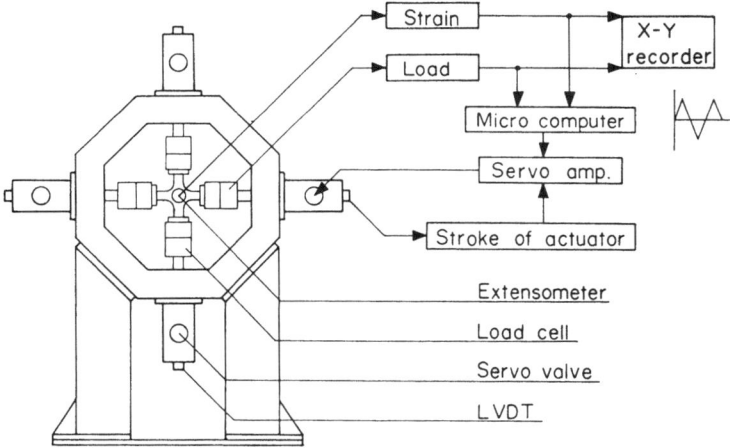

FIG. 9. High-temperature biaxial low-cycle fatigue test apparatus.

of 205 GPa and a yield stress of 410 MPa in the cyclic constitutive relation at 823 K in air were recorded.

Figure 8 [10] shows the overview of the biaxial low-cycle fatigue testing machine. The testing machine has four electrohydraulic actuators in x- and y-directions and a load capacity is 5 tons. The cruciform test specimen is accurately subjected to biaxial stresses in the range of the strain ratio ϕ or the stress ratio λ ranging from -1 to 1 at high temperatures by using the following control system. Figure 9 [10] is the block diagram of the testing machine. The most important work was the construction of the software and control circuit which constantly maintain the central point of the cruciform test specimen as shown in Fig. 10 [10] during the test. Basically, analogue control for the stroke in the actuators was adopted and an additional digital feedback control circuit was used for performing the strain-controlled test. The strain in the central portion of the specimen is thus maintained in each cycle by changing the function wave in the servo-amplifiers through a microcomputer. The measurement of strain in the central portion was by using a pair of X-Y differential transducer-type devices. We can see that the strain in the central portion of the specimen remains constant even if the material behavior is that of cyclic hardening or softening during the test. A specially designed electric furnace was used for heating the central portion of the cruciform specimen. Since we cannot measure the stress directly in the cruciform specimen, we also constructed an iso-maximum

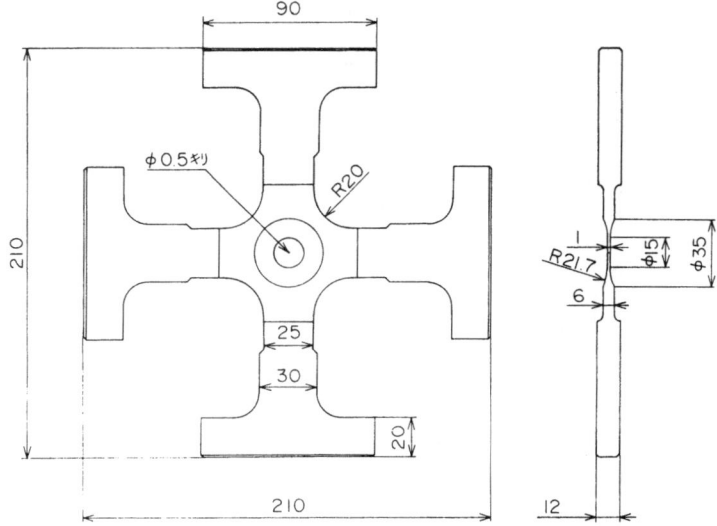

FIG. 10. Cruciform test specimen (mm).

principal stress and iso-von Mises' equivalent stress diagrams as shown in Figs. 11(a) and (b) [10] analytically by using the cyclic constitutive equation mentioned above.

Figure 12 [10] shows the low-cycle fatigue failure lives, N_f and N_c, of a centrally circular notched cruciform specimen under triangular strain waveform with a strain rate of 10^{-3} s^{-1} at 823 K in air. The test was performed under a strain ratio of $-1 \leqq \phi \leqq 1$ at a fixed maximum principal total strain range, $\Delta\varepsilon_1$, of 0·5% and 0·78%. Here N_f is defined as the number of cycles at which the tensile load amplitude in X- and Y-directions of the specimen decreases to three-quarters of the maximum value, and N_c is the number of cycles when the cracks reach about 100 μm. We can see that the low-cycle fatigue crack initiation life and the failure life of the material depend on the biaxiality of the stress. It is especially noticeable in the strain ratio of $-1 \leqq \phi \leqq -0.5$–0 as shown in Fig. 7.

Figure 13 [10] shows the comparative strain range versus the number of cycles to failure curves based on the strain parameters: (a) $\Delta\varepsilon_1$ (maximum principal strain), (b) $\Delta\bar{\varepsilon}$ (von Mises' equivalent strain), (c) $\Delta\gamma_{max}$ (maximum shear strain), (d) $\Delta\{(\gamma^*_{max}/2) + 0·2\varepsilon^*_n\}$ (Γ^*-plane theory) and (e) $\Delta\varepsilon^*$ (the authors). Lines of factor of 2 in failure life were drawn

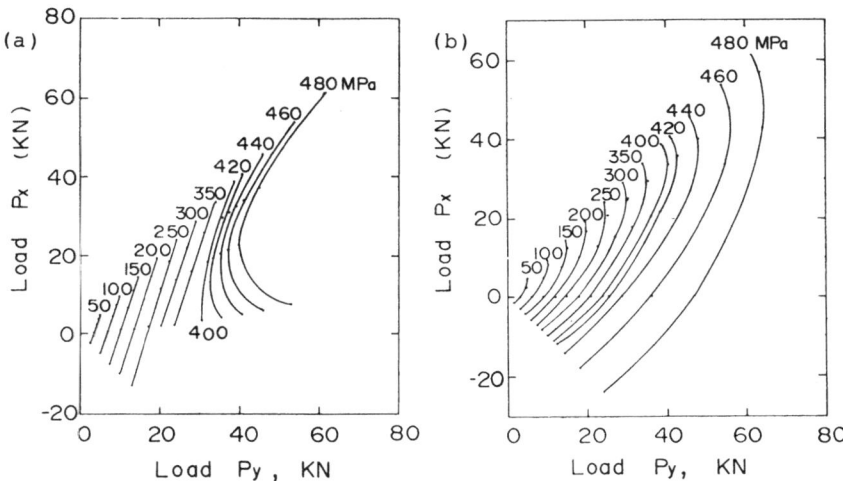

FIG. 11. (a) Iso-maximum principal stress curves and (b) iso-von Mises' equivalent stress curves.

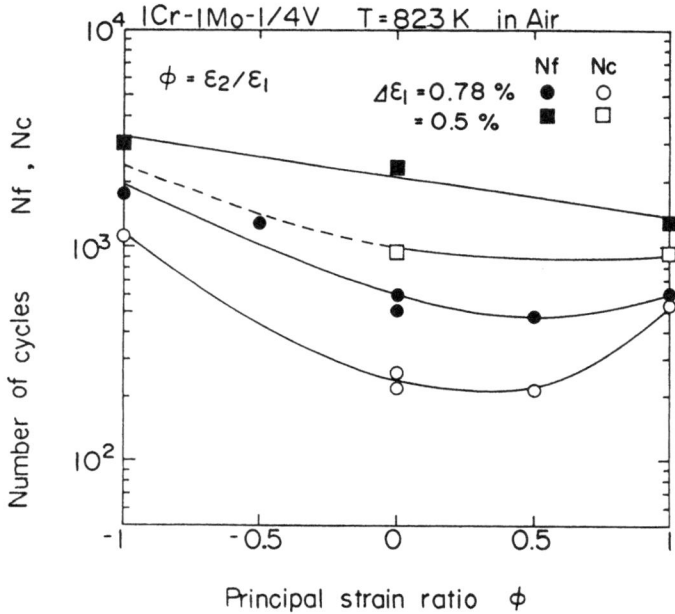

FIG. 12. Effect of principal strain ratio on low-cycle crack initiation life and failure life ($\Delta\bar{\varepsilon} = 0.78\%$, 0.5%).

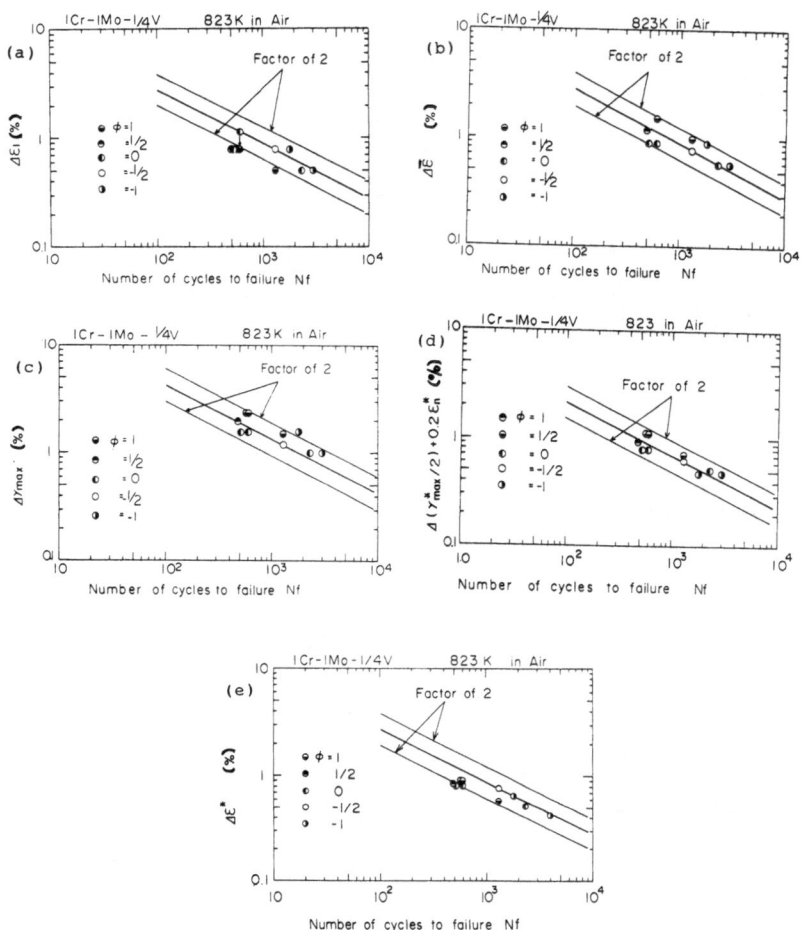

FIG. 13. Experimental relations between the comparative strain range and the number of cycles to failure. (a) Maximum principal strain range $\Delta\varepsilon_1$; (b) von Mises' equivalent strain range $\Delta\bar{\varepsilon}$; (c) maximum shear strain range $\Delta\gamma_{max}$; (d) Γ^*-plane strain; (e) equivalent strain range based on COD $\Delta\varepsilon^*$.

along the curve, of which the slope is -0.5 and the data point under $\phi = -0.5$ (push–pull) falls on the same lines. We found from those comparisons that there is no great difference between the parameters of (a) to (e) and the classical strain parameter of the von Mises' equivalent strain $\bar{\varepsilon}$ is adequate for the correlation of the actual fatigue failure life for a wide range of the strain ratio. However, it is seen that the strain

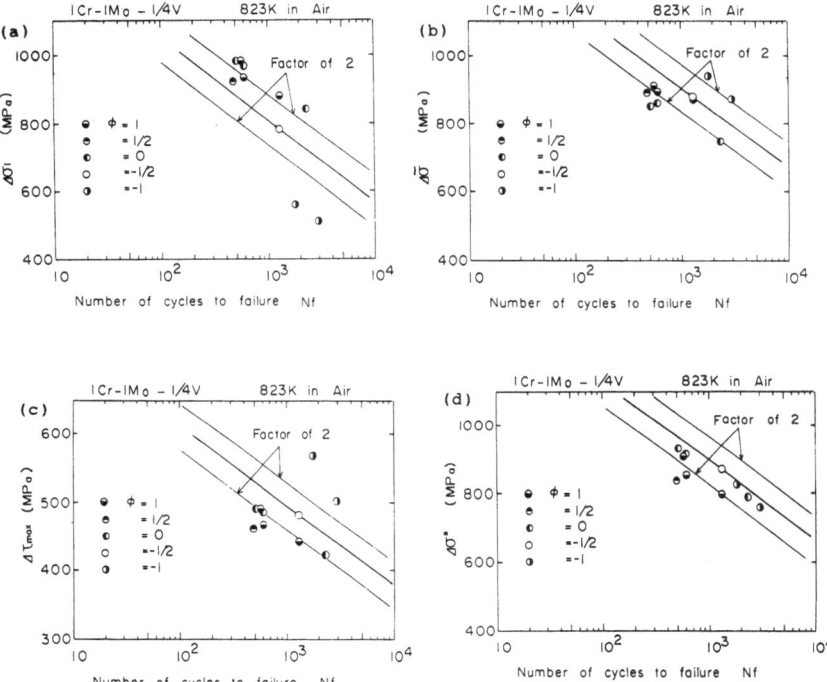

FIG. 14. Experimental relations between the comparative stress range and the number of cycles to failure. (a) Maximum principal stress range $\Delta\sigma_1$; (b) von Mises' equivalent stress range $\Delta\bar{\sigma}$; (c) maximum shear stress range $\Delta\tau_{max}$; (d) equivalent stress range $\Delta\sigma^*$.

parameter based on Γ^*-plane theory and the ε^* based on the COD concept are more adequate for this type of correlation. This was also confirmed in the test data for 304 type solution heat treated austenitic stainless steel under $-1 \leqq \phi \leqq 0.5$ [11].

In the stress basis, on the other hand, the situation changed. Figure 14 [10] shows a data comparison on the comparative stress parameters: (a) $\Delta\sigma_1$ (maximum principal stress), (b) $\Delta\sigma$ (von Mises' equivalent stress), (c) $\Delta\tau_{max}$ (maximum shear stress) and (d) $\Delta\sigma^*$ (the authors). $\Delta\sigma^*$ was calculated from eqn. (4) through Fig. 11a. We found from those comparisons that maximum principal stress, von Mises' equivalent stress and maximum shear stress cannot properly correlate the high-temperature biaxial low-cycle fatigue data while only equivalent stress σ^* based on the COD concept can correlate the data properly. This was also confirmed from the comparison of data under push–pull and reversed torsion for type 304 solution heat treated austenitic stainless steel [11, 12]. The cause of poor correlation is that the principal stress criterion does not include an effect of the stress parallel to the crack which has an effect of increasing the crack propagation rate while von Mises' equivalent stress does include the effect of the parallel stress. The $\Delta\sigma^*$ parameter was also applicable to the data of the low-cycle fatigue test with stress hold times in each cycle [4] and those of the tests with and without hold time under varying principal stress axes [13].

ACKNOWLEDGEMENT

This first series of studies was supported by the Research Institute of Science and Engineering, Ritsumeikan University and the second series of studies was performed through special Coordinate Funds of the Science and Technology Agency of the Japanese Government. The authors wish to thank Professor M. Tokizane, Ritsumeikan University, for carrying out the experiment in the first series of studies.

REFERENCES

1. GARUD, Y. S., *J. Test. Eval.*, **9** (1981), 165.
2. BROWN, M. W. and MILLER, K. J., ASTM, STP, **770** (1982), 482.
3. OHNAMI, M., SAKANE, M. and HAMADA, N., *J. Soc. Mater. Sci., Japan*, **35** (1986), 230 (in Japanese).
4. HAMADA, N., SAKANE, M. and OHNAMI, M., *Fatig. Engng Mater Struc.*, **7** (1984), 85.
5. NISHINO, S., HAMADA, N., SAKANE, M., OHNAMI, M., MATSUMURA, N. and TOKIZANE, M., *Fatig. Engng Mater. Struct.*, **9** (1986), 65.
6. MURA, T., SHIRAI, H. and WEERTMAN, J. R., *Proc. 2nd Int. Symp. 7th Canad. Fract. Conf. Defects, Fracture and Fatigue*, 1982, p. 67.
7. NISHINO, S., SAKANE, M. and OHNAMI, M., *Proc. 29th Japan Congr. Mater. Res.*, 1986 p. 53.
8. LOHR, R. D. and ELLISON, E. G., *Fatig. Engng Mater. Struct.*, **3** (1980), 1.
9. HAMADA, N., SAKANE, M. and OHNAMI, M., *J. Soc. Mater. Sci., Japan*, **34** (1985), 214 (in Japanese).
10. SAKANE, M., ITSUMURA, T. and OHNAMI, M., Paper of the *JSME*, No. 864–2 (1986), 13 (in Japanese).
11. SAWADA, M., SAKANE, M. and OHNAMI, M., Paper of the *JSME*, No. 864–2 (1986) (in Japanese).
12. HAMADA, N., SAKANE, M. and OHNAMI, M., *Bull. JSME*, **28** (1985), 13–41.
13. OHNAMI, M., SAKANE, M. and HAMADA, N., ASTM, STP 853 (1985), 622.

12

Creep–Fatigue–Hot Corrosion Interactions in High Temperature Structural Alloys

J. K. TIEN, W. L. KIMMERLE and E. A. SCHWARZKOPF

Center for Strategic Materials, Henry Krumb School of Mines, 918 Mudd Building, Columbia University, New York, NY 10027, USA

ABSTRACT

The prolonged exposure in hot gases under extremes of variable loading cycles presents a most difficult problem for systematic scientific study. Several phenomena observed in the heat resistant structural alloys that prove the existence of creep–fatigue (CF) and creep–fatigue–corrosion (CFC) interactions are discussed herein. It is found, for example, that fatigue fail-safe behavior is eliminated where either creep or hot corrosion or both interact with high cycle fatigue effects. The individuality of high temperature structural alloys is underscored and is shown to be intrinsic with the nature of the dislocation pinning particles. Strong pinning as with fine oxide strengthening particles, result in a frequency dependent anelastic phenomenon that in essence results in cyclic strengthening. These creep–fatigue complications found in actual engineering materials systems, coupled with severe service conditions, present interesting challenges to the development of constitutive life prediction equations for high temperature structural alloys.

1. INTRODUCTION

The adverse service conditions in high temperature gas turbine or jet engine applications generally expose components to loading that requires resistance to creep, fatigue, corrosion or superimposed creep–fatigue interactions and creep–fatigue–corrosion interactions. Much of the earlier work on high-temperature structural alloys was concerned with static

183

creep or conventional fatigue testing. In many cases, however, these testing methodologies represent a poor simulation of actual service conditions where time dependent deformation can occur under aggravated conditions of cyclic loading (i.e. creep–fatigue interactions).

Such cantonized testing precludes the development of a scientific understanding or even discovery of phenomena triggered by creep on fatigue or vice versa. These interactively induced effects would complicate life predictive models. Numerous predictive models are in use to account for creep and fatigue lives separately [1–6]. The creep (and stress rupture) models are based on thermally assisted deformation concepts and many recognize dislocation contributions and the effects of back or resisting stress caused by strong pinning points. Parametric equations have also been developed to forecast fatigue lives. For example, fatigue models range from the simpler ones which relate strain range and cycles at failure, to more advanced and complicated models which account for temperature, frequency, and wave shape dependencies. The best known of the early models is the simple, yet elegant Coffin–Manson (or Manson–Coffin) equation.

Where interactive complications are minimum, dependable fatigue–creep parametric equations which recognize thermal assistance are available. For example, strain range partitioning through wave shape dependencies, recognizes creep components [6]. There are also models based on linear damage rules. These also partition the damage into creep and fatigue components [7].

Damage accumulation controlled creep–fatigue or fatigue–creep models apply well for cases where the dislocation substructures are not significantly affected by the sequence details of application of, for example, athermal fatigue damage and creep damage. Athermal damage can be fatigue at very high frequencies at any temperature or fatigue at any frequency at low homologous temperatures. Creep damage need not be just thermally assisted deformation at constant applied stress. More generally, it can result from a loading scheme and thermal environment that provides plenty of time exposure at higher homologous temperatures. Low frequency fatigue at elevated temperatures and or fatigue loading with long dwell times at elevated or high homologous temperatures, fall into this category. Interestingly, even for the mechanically well-behaved and tough steels, experimental creep–fatigue life results are often inconsistent with the predictions of models based on the linear damage rules [7–9].

Other papers in this conference will be discussing some aspects of

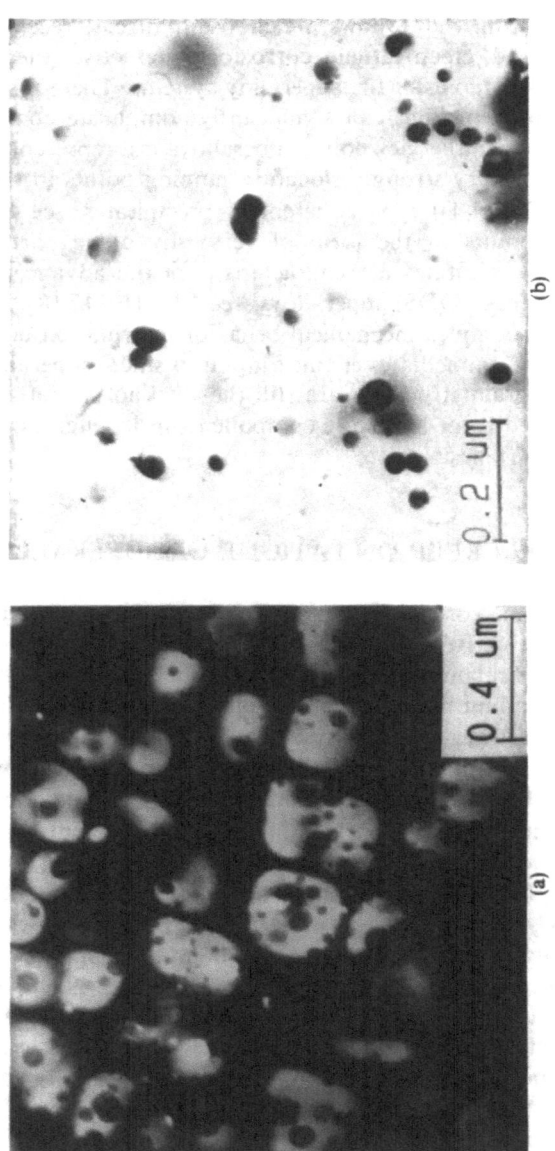

FIG. 1. (a) Dark field TEM of MA6000 revealing the cuboidal morphology of the (Ni_3Al) γ' precipitates. (b) Bright field TEM of MA6000 showing the fine and hard oxide dispersoids (Y_2O_3).

the constitutive aspects of creep–fatigue and creep–fatigue–corrosion interactions [10–16]. In what follows, we choose to discuss recent results on creep–fatigue and creep–fatigue–corrosion interactive effects that were found from the studies of the superalloy systems. These results are illustrative of phenomena that can significantly complicate constitutive protocols. From a scientific viewpoint, superalloys are representative of microstructures with very strong dislocation pinning points in the form of the very fine $Ni_3(Al, Ti)$ type coherent γ' precipitates, see Fig. 1a. Stronger pinning points in the form of very tiny oxide particles in addition to the γ' precipitates are characteristic of the advanced oxide dispersion strengthened (ODS) superalloys, see Fig. 1b [17,18]. On the more practical side, complex mechanical behavior information on super-alloys would be of immediate engineering use since superalloys in various forms and grain structures are still the workhorse materials for the critical stator and rotor hardware components in jet engines, rockets and landbase gas turbines.

2. IMPACT OF CREEP ON FATIGUE OF SUPERALLOYS

That thermal assistance or creep is deleterious to the fatigue behavior of superalloys is loudly advertised in Fig. 2 [19–20]. Fatigue tests were performed on a typical nickel-base superalloy under stress controlled, trapezoidal wave loading between a maximum, or upper, stress that was held constant and a lower applied stress that was varied to decrease the stress range and increase the mean stress. Testing of this type allows one, with a minimum of tests, to plot the gamut of creep–fatigue conditions on a single S–N type plot. The upper points of Fig. 2 represent essentially fatigue conditions as these tests employ high stress ranges and low mean or creep stresses. The lower points represent pure creep conditions as these tests employ small stress ranges and thus large mean stresses. When this sort of fatigue testing is done at low homologous temperatures (ambient for superalloy) with a frequency of 1 Hz, a S–N curve with conventional fail-safe behavior is generated as shown in curve A in Fig. 2. This can be viewed as a standard fatigue curve for initiation controlled superalloys with no creep interaction or mean stress effect. However, when the same tests are performed at elevated temperatures, a creep–fatigue interaction is evident as underscored by curve B of Fig. 2. That there is creep–fatigue interaction is seen in the data by the fact that the higher temperature curve has different (short) lives at all stress ranges.

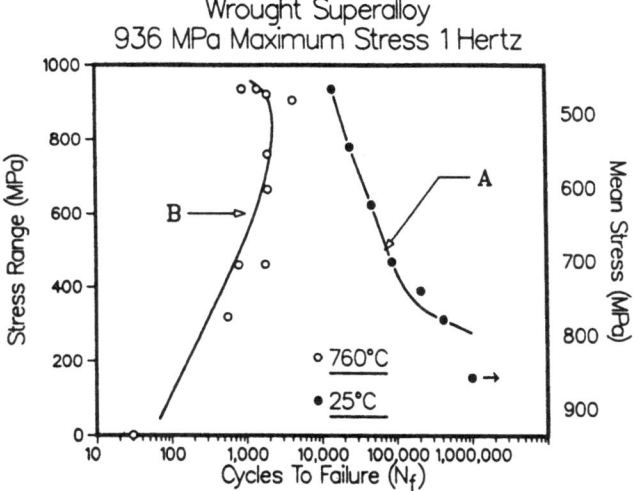

F IG. 2. S–N(A) and S–N inversion (B) curves of a typical superalloy [20].

We see an inversion at the lower stress range, or more importantly, at higher mean stress, indicating a strong impact of creep on fatigue.

The explanation for this inversion is intuitive once one accepts that creep is bad for fatigue. At high stress ranges failure is due to cyclic deformation while at lower stress ranges, failure is due to creep and stress rupture. Fractographic results, Fig. 3, are consistent with the inversion. Above the nose of the inverted curve, both initiation and propagation modes are intragranular, indicative of fatigue failures in superalloys, while below the nose, fracture surfaces reveal that initiation is intergranular with propagation becoming more and more integranular at lower stress ranges. Indeed, the bottom-most point is a stress rupture result from a creep test.

The deleterious effects of creep on fatigue in superalloys is discussed in detail elsewhere [21]. Figure 3 summarizes the situation fairly succinctly. Creep of superalloys results in grain boundary sliding effects with their attendant wedge crack formation, coalescence and eventual intergranular crack growth. As illustrated in Fig. 4, this situation can be aggravated by the role of the constricted or planar slip bands (a slip mode characteristic of superalloys) produced by fatigue loading [22]. Even without grain boundaries, creep, perhaps through coalescence of voids formed at inclusions and pores, can have similar inversion impact on directionally

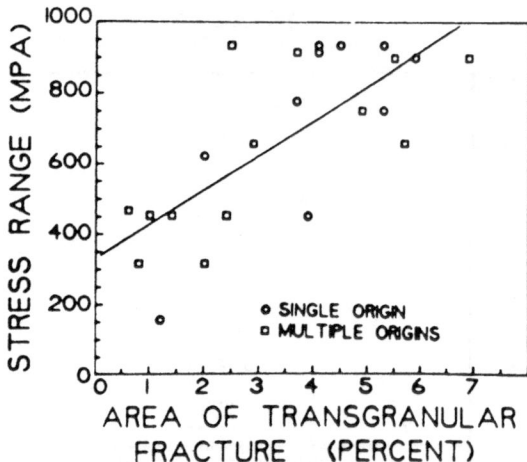

FIG. 3. Cyclic stress range versus area of transgranular fracture for the tests of Fig. 2 [20].

solidified or monocrystalline high temperature structural materials systems. Figure 5 shows the S–N inversions of an advanced Ni_3Al type intermetallic material in directionally solidified (columnar grained) form, the DS curve, and in the monocrystalline form, the SC curve [23].

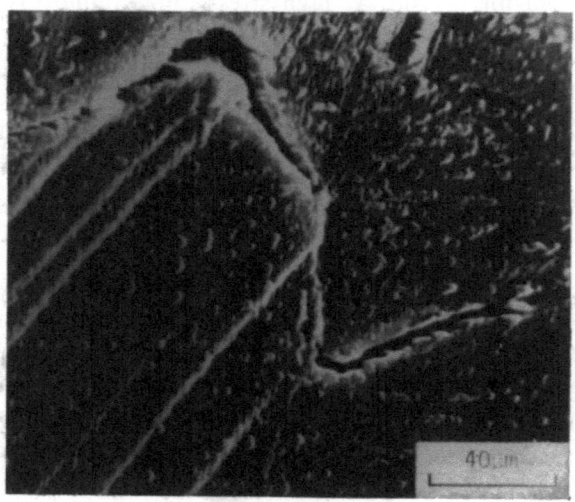

FIG. 4. SEM side-view of the gauge length of a typical polycrystalline superalloy following the higher frequency cyclic creep rupture. Test conditions; 707 MPa-41 MPa at 760°C [22].

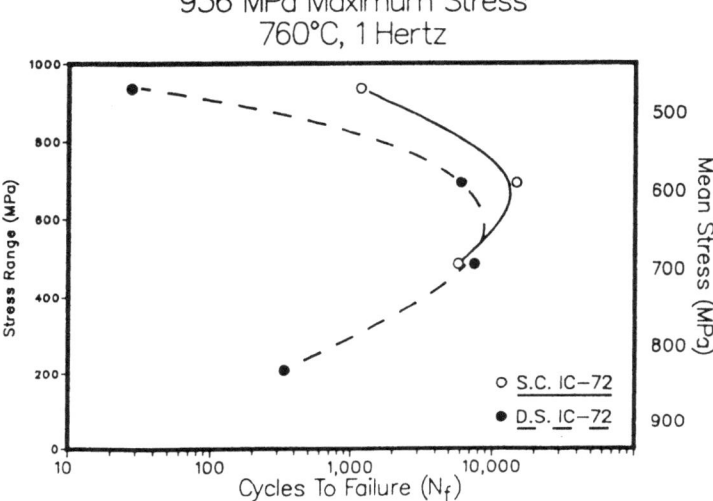

FIG. 5. Single crystal and directionally solidified intermetallic Ni$_3$Al type material revealing inversion of the S–N curve for fatigue test at 760°C [23].

3. IMPACT OF HOT CORROSION ON CREEP–FATIGUE OF SUPERALLOYS

A phenomenon that gives intergranular failures similar to those after elevated or high temperature creep–fatigue loading is the interaction of fatigue with corrosion. This not only gives intergranular failures with creep, but also without creep. Although the appropriate curves in Fig. 6 look like Fig. 2, the dive or inversion of these curves is due to the effects of corrosion and not due to increases in mean stress (or variable creep), since the curves in Fig. 6 were generated at constant mean stresses. The elimination of the fatigue fail-safe effect, Fig. 6, is apparently due to hot-salt corrosion-creep effects. At high enough cycles or long enough times, the grain boundaries even in these high nickel and chromium containing superalloys, are attacked reminiscent of time dependent stress-corrosion cracking, Fig. 7 [24,47].

The creep–fatigue inversion effect is even more pronounced if one intentionally introduces the hot corrosion component to the constant maximum stress, creep–fatigue distinction test. This is shown as comparisons of creep–fatigue and creep–fatigue–corrosion curves in Fig. 8 for

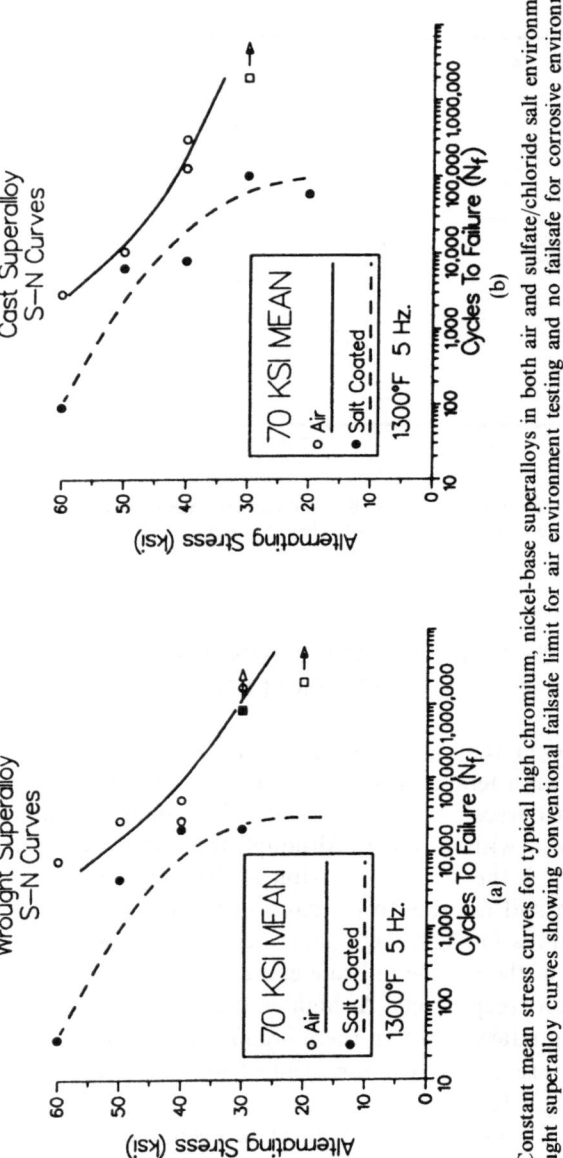

FIG. 6. Constant mean stress curves for typical high chromium, nickel-base superalloys in both air and sulfate/chloride salt environments. (a) Wrought superalloy curves showing conventional failsafe limit for air environment testing and no failsafe for corrosive environment testing. (b) Cast superalloy curves similar to the wrought superalloy case [24].

FIG. 7. Example of a thoroughly corroded nickel-base superalloy creep specimen tested at 870°C with Na_2SO_4 deposition in air: (a) side view of gauge length; (b) top view of fracture surface. Note extreme intergranular mode of attack. [47].

a typical cast superalloy and a typical wrought superalloy [24]. Interestingly, however, the bottom or creep points in the wrought alloy curve coincide. This is consistent with the fact that this alloy, Udimet 720, an originally landbase gas turbine blading alloy, was designed to be creep-corrosion resistant [25].

4. THE ANELASTIC PHENOMENON FROM CREEP–FATIGUE INTERACTION IN ODS SUPERALLOYS

For fatigue loading, cycling between zero and a maximum positive stress, cyclic creep acceleration (increase in the minimum strain rate and decrease in the rupture life) based on 'time on load' has been reported for a number of steels [26–29] and microstructurally simple systems [30–35]. The basic explanation for this type of behavior is that during the off-load periods, the hardened microstructure that formed during the on-load periods is allowed to recover [26,27]. Thus, when the load is reapplied, a period of primary creep occurs once again until work hardening results in the steady state creep rate observed during static creep. These repeated periods of primary creep during the initial portion of the on-load cycle causes exhaustion of the creep ductility and the

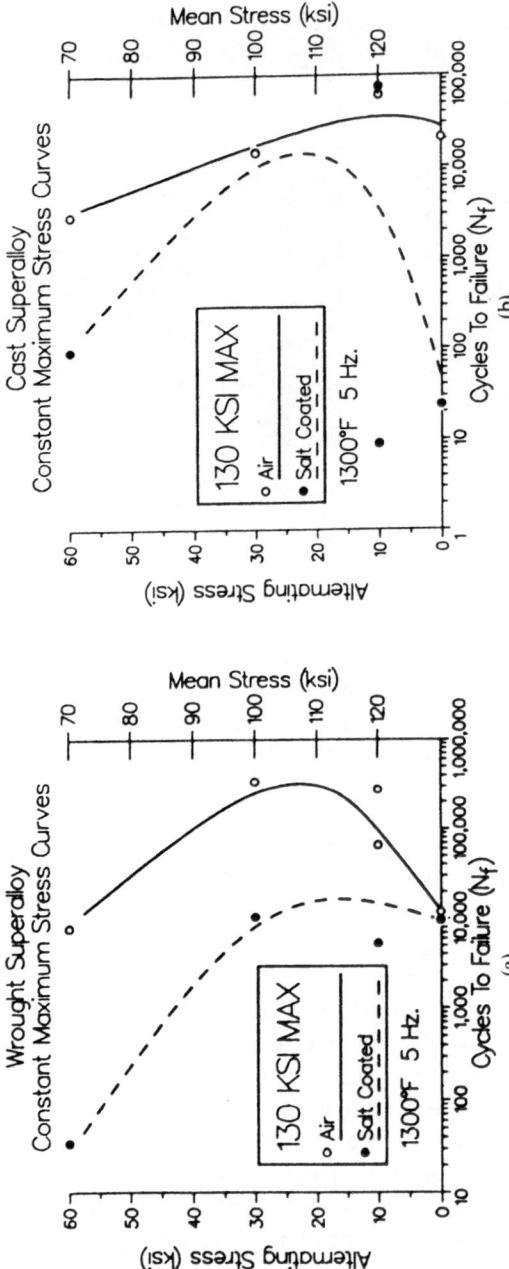

FIG 8. Constant maximum stress curves for typical high chromium, nickel-base superalloys in both air and sulfate/chloride salt environments. (a) Wrought superalloy curves showing large effect of salt at all conditions except at pure creep. (b) Cast superalloy curves decrease in lives to corrosive environment at all conditions including at pure creep [24].

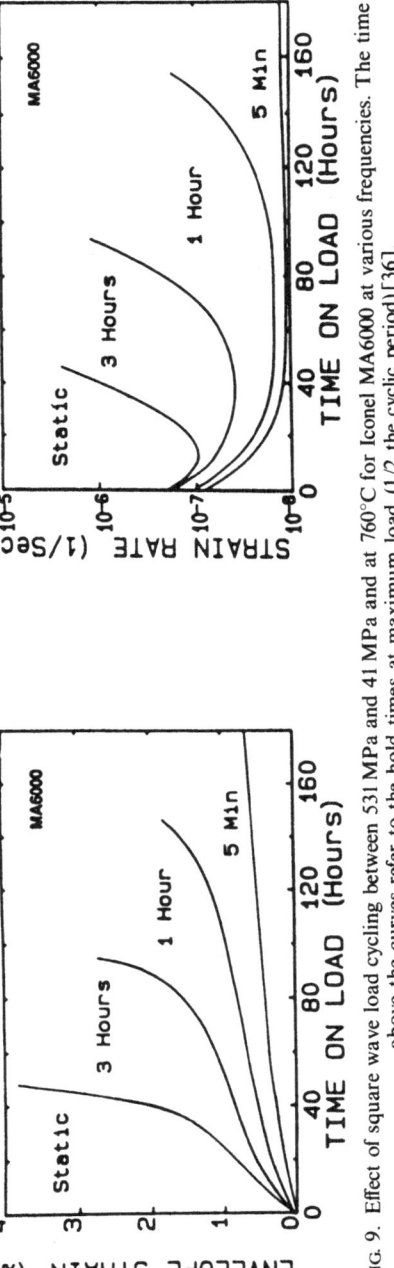

Fig. 9. Effect of square wave load cycling between 531 MPa and 41 MPa and at 760°C for Iconel MA6000 at various frequencies. The time above the curves refer to the hold times at maximum load (1/2 the cyclic period) [36].

observed creep acceleration. Similar creep acceleration, driven by the annealing phenomenon, has also been reported for the γ' strengthened superalloys under conditions where off-load annealing was encouraged [36–37].

Contrary to this behavior of γ' strengthened superalloys, a cyclic creep deceleration effect as a result of load cycling, uniquely underscoring creep–fatigue interaction, has been reported for two oxide dispersion strengthened (ODS) alloys, Inconel MA754 [38] and Inconel MA6000 [39], and PbSn eutectic [40]. Cyclic creep deceleration has also been reported as a contributing factor to the strength in SiC whisker reinforced aluminum composites [41]. It is noted that like the hybrid ODS alloys and SiC reinforced materials, PbSn eutectics are also strengthened by hard pinning points in the form of the fine, closely spaced and incoherent eutectic rods.

A significant apparent cyclic creep deceleration was observed in the ODS alloys in the frequency range from 1 Hz to $6 \, h^{-1}$, with the effect becoming more pronounced as the frequency increased. An example of this cyclic creep behavior for the Inconel MA6000 is given in Fig. 9 [39]. Similarly, Fig. 10 reveals this apparent creep deceleration for the PbSn eutectic. The time above each of the curves is the hold time at maximum load, i.e. one-half the cyclic period. The effect of load cycling on the minimum strain rate and stress rupture life are clearly evident in this figure. This unique cyclic deceleration effect for periods of load application ranging from hours to seconds was attributed to the storage and recovery of anelastic strain, Fig. 11 [38, 39, 46]. This type of strain recovery has also been documented many years ago in hot-pressed silicon nitride [48].

A model has been proposed to describe load controlled low cycle fatigue (cyclic creep) behavior of these ODS alloys which are strengthened by fine yet very hard pinning particles [42]. The model takes into account that the recoverable strain stored during the on-load half-cycle (the anelastic strain) is responsible for what appears to be a net frequency dependent cyclic creep deceleration. The net effect is a frequency dependent decrease in the minimum strain rate and increase in the cyclic rupture life, or cycles to failure. To obtain the envelope cyclic creep rate, $\dot{\varepsilon}_{cyc}$ the instantaneous nonrecoverable creep rate as shown in eqn. 1 is averaged over the on-load time, $1/2\nu$, where ν is the cyclic creep frequency

$$\dot{\varepsilon}(t) = \dot{\varepsilon}_{sta}[1 - B\exp(-t/\tau_1) - (1-B)\exp(-t/\tau_2)]^n \qquad (1)$$

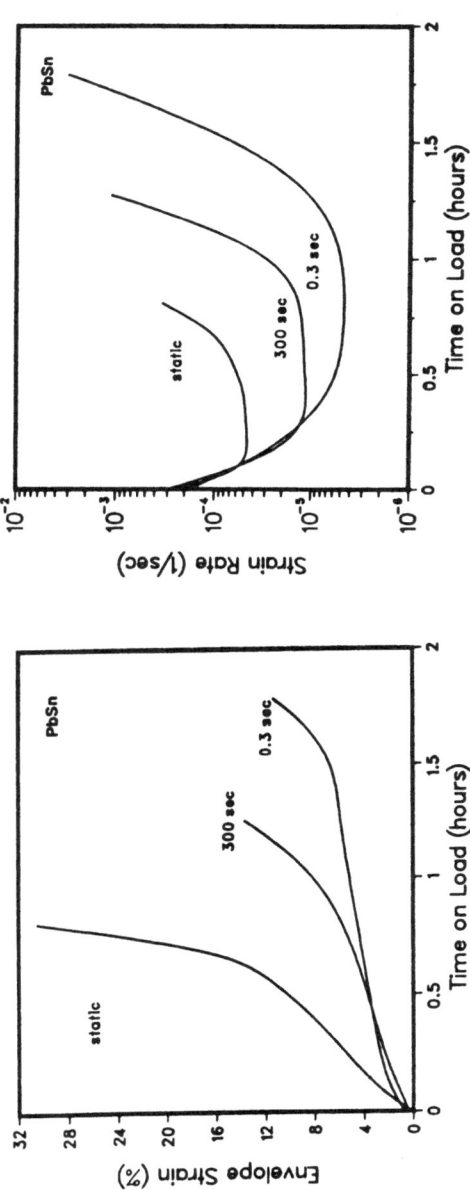

FIG. 10. Effect of square wave load cycling between 0·138 MPa and 13·8 MPa at approximately 0·7T_m for PbSn eutectic at various frequencies. The time above the curves refer to the hold times at maximum load (1/2 the cyclic period) [40].

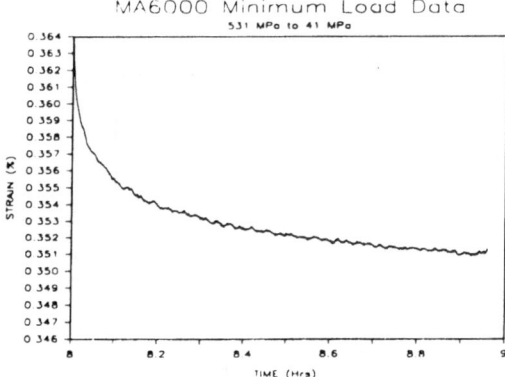

FIG. 11. A typical anelastic relaxation curve during the minimum stress half cycle in a cyclic creep test with square wave loading. The load was cycled between 531 MPa and 41 MPa with a frequency of $0.05\,h^{-1}$ at $760°C$ [46].

$$\dot{\varepsilon}_{cyc}(v) = 2v\dot{\varepsilon}_{sta} \int_0^{1/2\,v} [1 - B\exp(-t/\tau_1) - (1-B)\exp(-t/\tau_2)]^{20}\,dt \qquad (2)$$

where τ_1 and τ_2 are time constants, and B is a measure of the magnitude for each of the exponentials, v is the cyclic frequency, n is the apparent stress exponent, and $\dot{\varepsilon}_{sta}$ is the static strain rate.

Accordingly, the storage of anelastic strain during the initial portion of the on-load cycle delayed 'nonrecoverable creep and resulted in the apparent cyclic creep deceleration. Detailed TEM analysis of the crept ODS alloys has revealed that the mechanism driving the anelastic recovery process is departure side pinning of the dislocations to the hard pinning particles (Fig. 12). This increases the stress required for dislocation escape to a value approaching the ambient temperature Orowan stress levels [43–45].

It has been proposed that this attractive interaction between the oxide particles and the mobile dislocation is a consequence of void space around the particles. The dislocations instead of climbing over the particle barriers, as would be expected with thermal assistance are sucked into the void space and can escape only by progressively bowing between the particles with the last contact point at the dislocation departure side of the particle. This type of pinning guarantees not only a higher resisting stress at the higher temperatures of a magnitude like that expected at ambient temperature, but also provides a creep strain storage

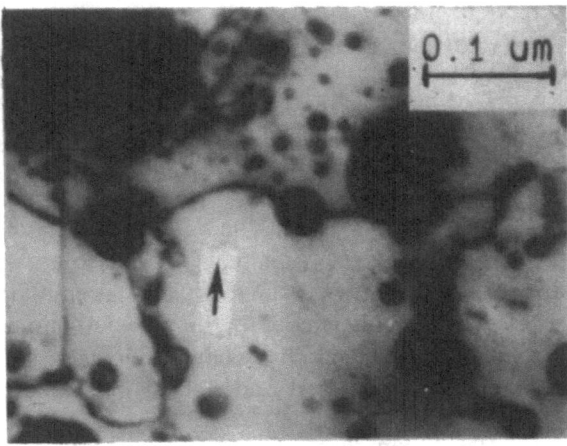

F IG. 12. TEM of crept MA6000 shows a dislocation bowed between two oxide dispersoids. The dislocation appears to be stuck on the departure side of the oxide particle.

mechanism. Upon unloading, the bowed segments will creep back (anel-astically relax), resulting in a recovering of some of the creep strain. On loading, a certain amount of bowing is required before the dislocation can depart from the particles and result in nonrecoverable creep. If the frequency of load–unload is high, then most of the creep strain observed at on-load will be recovered resulting in the observed apparent creep deceleration effect.

5. CONCLUDING REMARKS

In this paper, we reported and discussed several phenomena observed in the heat resistant structural alloys that are the consequence of creep–fatigue and creep–fatigue–corrosion interactions. The hot corrosion in-teraction in essence results in high cycle fatigue effects where fail-safe behavior is eliminated. Constitutive equations eventually must recog-nize such complications that are cycle dependent.

At the elevated and high temperatures, thermal assistances are not helpful to fatigue resistance. Indeed, S–N curves, obtained while holding the stress constant, tend to invert instead of approaching a fatigue limit. It should be noted that this type of test condition is consistent with certain application missions of high temperature structural alloys, be they polycrystalline, directionally solidified or monocrystalline super-alloys, ODS superalloys or even structural intermetallics.

Generalization of creep–fatigue behavior of high temperature struc-tural alloys can be imprudent. Individuality is evident at least as a function of the nature of the dislocation pinning points. For example, for the coherent precipitate γ' strengthened superalloys, loading and unload-ing at high temperatures results in a shortening of the fatigue life. In contrast, when the dislocation motion is retarded by not only slow passage through the coherent γ' precipitates, but also attractively pinned to such strong pinning points as fine oxide strengthening particles intentionally added into the superalloys, then a net enhancement of life results. This phenomenon, which pivots about recoverable strain storage and anelastic relaxation, is, as expected, very cyclic frequency depen-dent.

These creep–fatigue complications found in actual engineering materials systems present interesting challenges to the development of constitutive life prediction equations for high temperature structural alloys based on linear damage or life fraction or strain partitioning concepts. As discussed, the difficulties are not only those expected from complex loading conditions in adverse and hot environments, but are also those further complicated by the individuality of behavior exhibited by the materials themselves.

ACKNOWLEDGEMENTS

We would like to thank the Materials Research Division of NSF and USAF Office of Scientific Research for their funding of some of this work. It is recognized that IN738, Inconel MA754 and Inconel MA6000 are trademarks or designations of the INCO family of companies. Udimet 720 is a designation of Special Metals Corporation.

REFERENCES

1. GAROFALO, F., *Fundamentals of Creep and Creep Rupture in Metals*, New York, The MacMillan Co., 1965.
2. LI, J. C. M., *J. Appl. Phys.*, **32** (1961), 525.
3. COFFIN, L. F., *Trans. Am. Soc. Mech. Engrs*, **76** (1954), 923.
4. MANSON, S. S., *Exp. Mech.*, **5, 6, 193** (1965), 199.
5. HOWSON, T. E., D. Eng. Sci. Thesis, Columbia University, 1981.

6. MANSON, S. S., HALFORD, G. R. and HIRSCHBERG, M. H., NASA TMX-67838 1971.
7. PLUMTREE, A. and DOUGLAS, M. J., *Advances in Fracture Research* (D. Francois, ed.), New York, Pergamon Press, 1982, p. 2423.
8. PLUMTREE, A. and LEMAITRE, J., *Advances in Fracture Research* (D. Francois, ed.), New York, Pergamon Press, 1982, p. 2379.
9. PLUMTREE, A. and PERSSON, N. C., *Fracture 1977* (D. M. R. Talbin ed.), University of Waterloo Press, Waterloo, Canada, 1977, p. 821.
10. IIDA, K., this volume, pp. 1–13.
11. PETTIT, F. S., in Iida, K., this volume, pp. 135–147.
12. OHNAMI, M., SAKANE, M., NISHINO, S. and ITSUMURA, T. in Iida, K., this volume, pp. 165–182.
13. NARUMOTO, A., in Iida, K., this volume, pp. 219–231.
14. ENDO, T., in Iida, K., this volume, pp. 233–244.
15. MCEVILY, A. J., MINAKAWA, K. and NAKAMURA, H., in Iida, K., this volume, pp. 291–301.
16. KOMAI, K. and MINOSHIMA, K. in Iida, K., this volume, pp. 373–388.
17. HOWSON, T. E., STULGA, J. E. and TIEN, J. K., *Met. Trans. A*, **11A** (1980), 1599.
18. HOWSON, T. E., MERVYN, D. A. and TIEN, J. K., *Met. Trans. A*, **11A** (1980), 1609.
19. CHEN, C. L., FRITZEMEIER, L. G., XIE, X. and TIEN, J. K., *Met Trans. A*, **13A** (1982), 1951.
20. PAULSON, R. R., FRITZEMEIER, L. G. and TIEN, J. K., *Met. Trans.*, **14A** (1982), 727.
21. TIEN, J. K., NAIR, S. V. and NARDONE, V. C., *Flow and Fracture in Metals*, ASM Press, 1985.
22. MATEJCZYK, D. E., D. Eng. Sci. Thesis, Columbia University,1983.
23. BELLOWS, R., HO, C. and TIEN, J. K., *Met. Trans. A*, submitted for publication.
24. SCHWARZKOPF, E., STEFFANI, J. and TIEN, J. K., unpublished research, Columbia University.
25. WHITLOW, G. A., BECK, C. G., VISWANATHAN, R. and CROMBIE, E. A., *Met. Trans. A*, **15A** (1984), 23.
26. EVANS, J. T. and PARKINS, R. N., *Acta Met.*, **24** (1976), 511.
27. MORRIS, D. G. and HARRIES, D. R., *J. Mater. Sci.*, **13** (1978), 985.
28. DAY, M. F. and CUMMINGS, W. M., *J. Mech. Eng. Soc.*, **10** (1968), 36.
29. KLOOS, K. H., GRANACHER, J., BARTH, H. and RIETH, P., *Advances Fracture Research* (D. Francois, ed.), New York, Pergamon Press, 1982, p. 2355.
30. SHETTY, D. K. and MESHII, M., *Met. Trans.*, **6A** (1975), 349.
31. BRADLEY, W. L., NAM, S. W. and MATLOCK, D. K., *Met. Trans. A*, **7A** (1976), 425.
32. COUTINHO, C. B., MATLOCK, D. K. and BRADLEY, W. L., *Mater. Sci. Engng.* **21** (1975), 239.
33. SHETTY, D. K. and MESHII, M., *Mater. Sci. Engng*, **32** (1977), 283.
34. MESHII, M., UEKI, M. and CHIOW, H. D., *Strength of Metals and Alloys* (Proc. of the 5th Int. Conf.), New York, Pergamon Press, 1979, vol. 1, p. 245.
35. SHELDON, G. P. and YESKE, R. A., *Met. Trans. A*, **9A** (1978), 5.

36. SULLIVAN, C. P., WEBSTER, C. A. and PIEARCEY, B. J., *J. Inst. Met.,* **96** (1967), 274.
37. WEBSTER, C. A. and PIEARCEY, B. J., *Trans. ASM,* **59** (1966), 847.
38. MATEJCZYK, D. E., ZHUANG, Y. and TIEN, J. K., *Met. Trans. A,* **14A** (1983), 241.
39. NARDONE, V. C., MATEJCZYK, D. E. and TIEN, J. K., *Met. Trans. A,* **14A** (1983), 1435.
40. WEINBEL, C. R., POLLICK, S., KANG, S. and TIEN, J. K., unpublished research, Columbia University.
41. NARDONE, V. C. and STRIFE, J. R., unpublished research, United Technologies Research Center.
42. NARDONE, V. C., MATEJCZYK, D. E. and TIEN, J. K., *Met. Trans. A.,* **16A** (1985), 1117.
43. NARDONE, V. C. and TIEN, J. K., *Scripta Metall.,* **17** (1983), 467.
44. NARDONE, V. C., MATEJCZYK, D. E. and TIEN, J. K., *Acta Metall.,* **32**, No. 9 (1984), 1509.
45. COOPER, A. H., NARDONE, V. C. and TIEN, J. K., *Superalloys 1984* (M. Gell, C. S. Kortovich, R. H. Bricknell, W. B. Kent, and J. F. Radavich, eds), Warrendale, Pennsylvania, Metallurgical Society of AIME, 1984, p. 359.
46. KIMMERLE, W. L., NARDONE, V. C. and TIEN, J. K., *Met. Trans. A,* in print.
47. ANING, D., D. Eng. Sci. Thesis, Columbia University, 1975.
48. ARONS, R. M. and TIEN, J. K., *J. Mater. Sci.,* **15** (1980), 2046.

13

Material Characterization and Material Requirement of the High-temperature Components of the High-temperature Gas-cooled Reactor

MASAKI KITAGAWA, HIROSHI HATTORI, AKIRA OHTOMO, YOSIHITO NARITA and HISAO IWAMATSU

Research Institute, Ishikawajima-Harima Heavy Industries Co., Ltd., Toyosu 3-1-15, Koto-ku, Tokyo, Japan.

ABSTRACT

The design temperature of the high-temperature gas-cooled nuclear reactor (abbr. as HTR) system is as high as 950°C. Because 950°C is the highest of the design temperatures for the various nuclear reactors, it was expected that various new material behaviors may be encountered. These phenomena should be considered in the design of the high-temperature components of the HTR.

During the development of the helium heat exchanger for the HTR direct steel making system, the following material properties were considered to be important R&D items at the high operating temperature:

(1) Creep–environment interaction under cyclic loading;
(2) tribology problem at the metal-to-metal interface;
(3) high hydrogen permeability.

In order to overcome these problems in the HTR design, material characterization was performed at the level of laboratory experiments, and inputs from the experience of designing intermediate heat exchangers were also considered.

1. INTRODUCTION

Material for high-temperature applications must possess high-temperature strength as well as hot corrosion resistance. Creep resistance was considered to be the most important property for the high-temperature applications, and hence materials with high creep resistance were developed. However, the thermal stress is usually high in high-temperature equipment, and here high low cycle fatigue strength is also required. This requirement opposes the requirement for high creep resistance. Environmental effects on the creep and fatigue strength is another important property for high-temperature applications.

At very high temperatures, thermally activated behaviors are usually accelerated, and therefore, it is possible that some new requirements may become important for the component design. The authors are experienced in developing heat exchangers for high-temperature gas-cooled reactors (abbr. as HTR) whose operating temperature is the highest among nuclear reactors. In this paper, the design considerations involving the following problems of high-temperature gas cooled reactors are discussed; namely, the creep–fatigue–environment interaction, wear, self welding and hydrogen permeation. The methods of the overcoming of those problems are also discussed.

2. STRUCTURAL STRENGTH PROBLEMS AT THE HTR TEMPERATURES AND THE STRENGTH DESIGN

Figure 1 summarizes the general structure of the structural design codes for the nuclear components, including the applicable temperatures and the considered failure modes. At the HTR temperatures, the enhancement of thermally activated phenomena such as creep and structural changes and the thermal stress activated phenomena such as the thermal ratchet and thermal fatigue are the most important design consideration. The failure due to self welding and wear increase in importance at these temperatures. [1]

2.1. Creep Properties in the HTR Environments [2,3]

The coolant for the HTR is pure helium gas. Figure 2 compares the creep curves of the Inconel 617 at 1000°C in air and helium gas of 99·995% purity. It is seen that the creep in helium gas of 99·995% purity is much higher than in air and in very pure helium (the purity of which cannot be measured quantitatively because of experimental reasons, but it was estimated much higher than 99·995%). Figure 3 shows the carbon distribution in the specimens tested in 99·995% helium gas at 1000°C which clearly indicated that decarburization occurred in the helium gas.

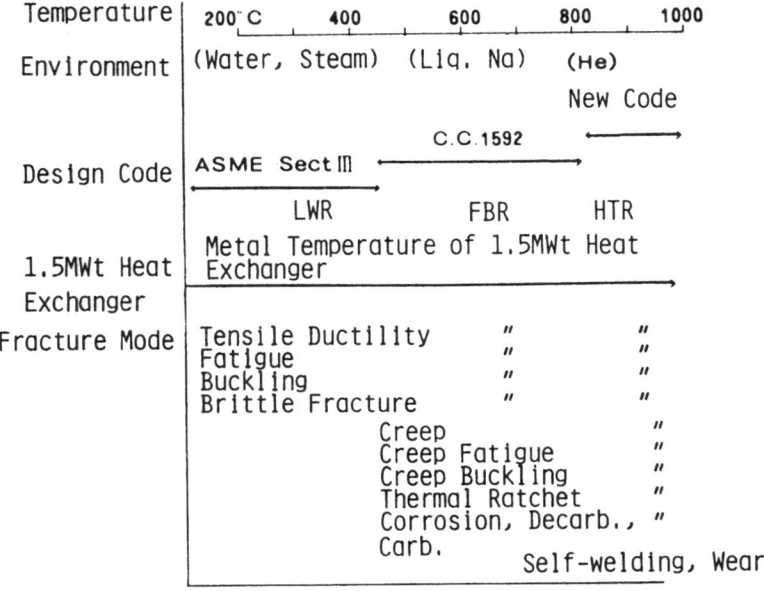

FIG. 1. Design codes for nuclear components and the considered failure mode.

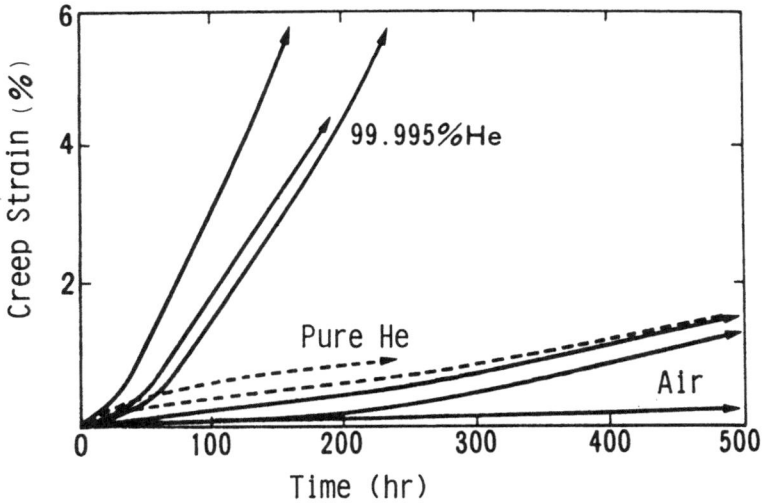

FIG. 2. Comparison of creep curves in air and He.

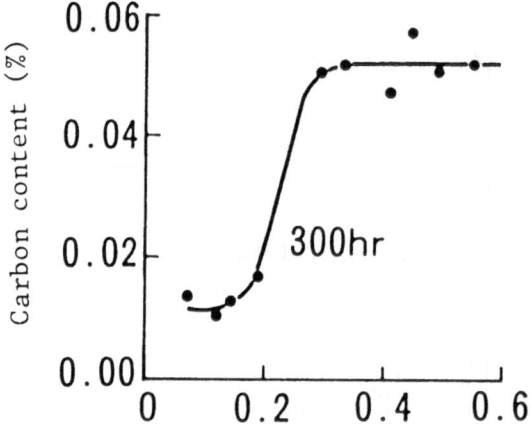

FIG. 3. Carbon distribution after the creep test in helium gas.

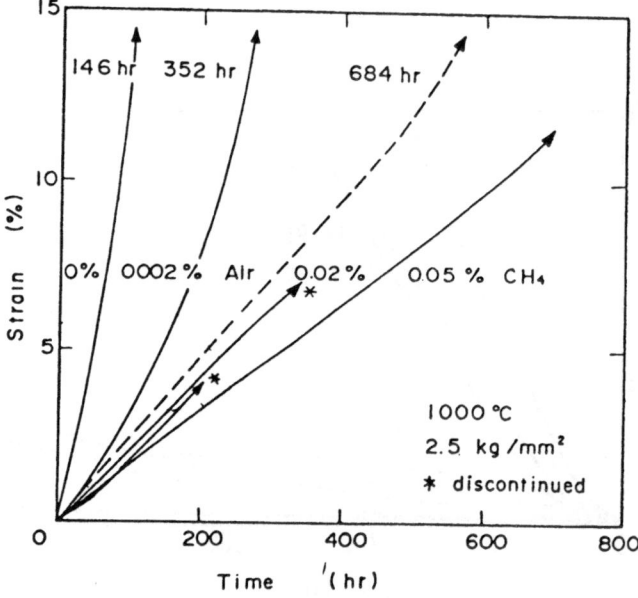

FIG. 4. Effect of methane gas addition on creep strength of Inconel 617 in helium gas.

Figure 4 shows the creep curves in the helium gas with various amounts of the methane (CH_4) for the purpose of preventing decarburization. It is seen that the creep strength is improved by additional of a small amount of CH_4. It was concluded that the creep resistance was decreased due to the decarburization caused by the oxygen in the helium gas.

Materials with high stability in low oxygen environments should be developed. Note that most of the high temperature materials were developed assuming that the materials were to be used in the normally oxidizing environments. Whether carburization or decarburization will occur depends on a combination of the material and environment (impurity level in the helium gas) factors. JAERI B gas, which is considered to most closely represent the oxidation and carburization capability of the actual HTR helium gas possesses a high carburizing capability.

As is well known, the creep strength is strongly dependent on the grain size and the plastic deformation enhances recrystallization. The de-

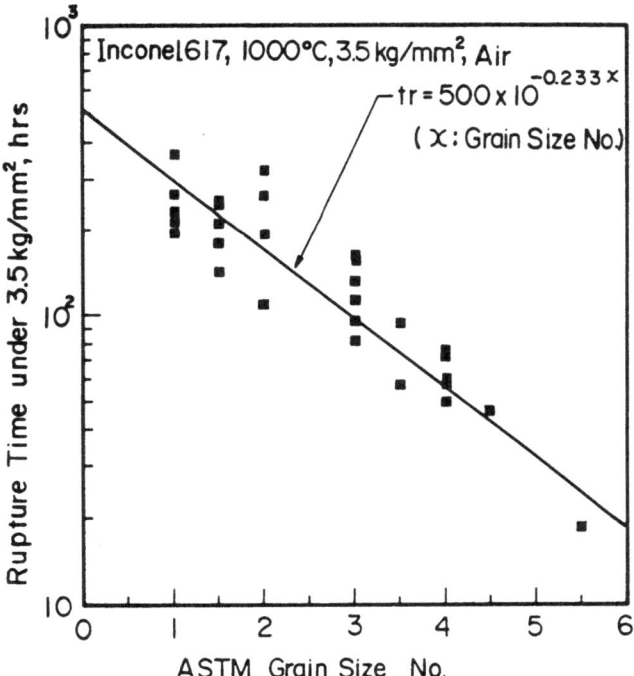

FIG. 5. A relationship between creep rupture time and grain size of Inconel 617.

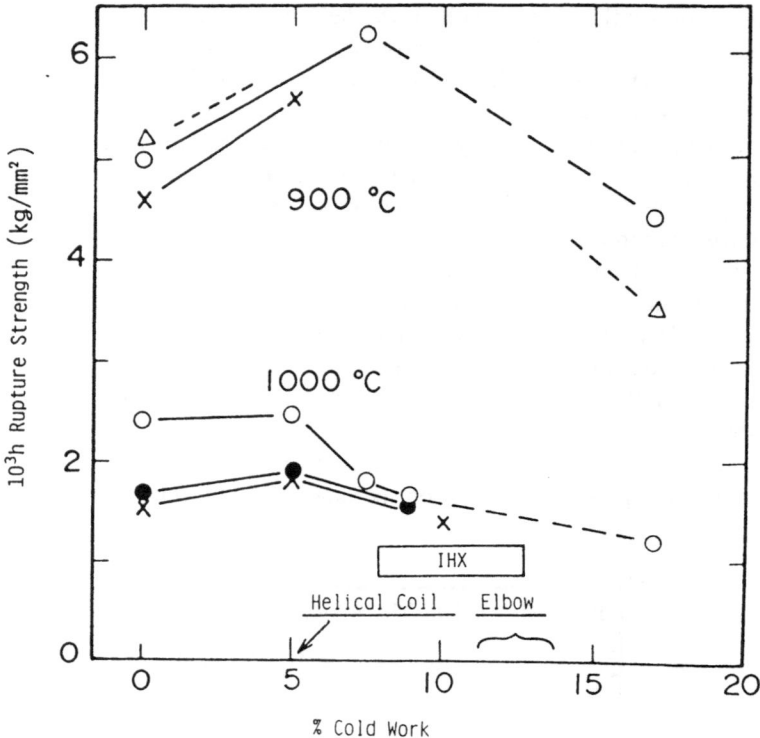

Fig. 6. Variation of creep rupture strength of Inconel 617 with cold work.

pendency of the creep strength of Inconel 617 on the grain size and on the prior plastic deformation are shown in Figs. 5 and 6. In the development of new alloys, a limit on the grain size is desirable and helpful in achieving the desired combination of creep strength and ductility. The solution heat treatment for the stabilization of the structure is also desirable when any prior plastic deformation exceeds the limit. In our development, the materials of grain size of ASTM No. 0·5–3·0 and with plastic deformation under 5% were allowed to be used mainly from the viewpoint of creep resistance, and secondly, to optimize the fatigue ductility.

2.2. Creep–Fatigue Problems in the HTR Environment [2, 3]

Although all the possible countermeasures are taken to reduce the thermal stress, the heat exchanger tubes are subjected to the most severe

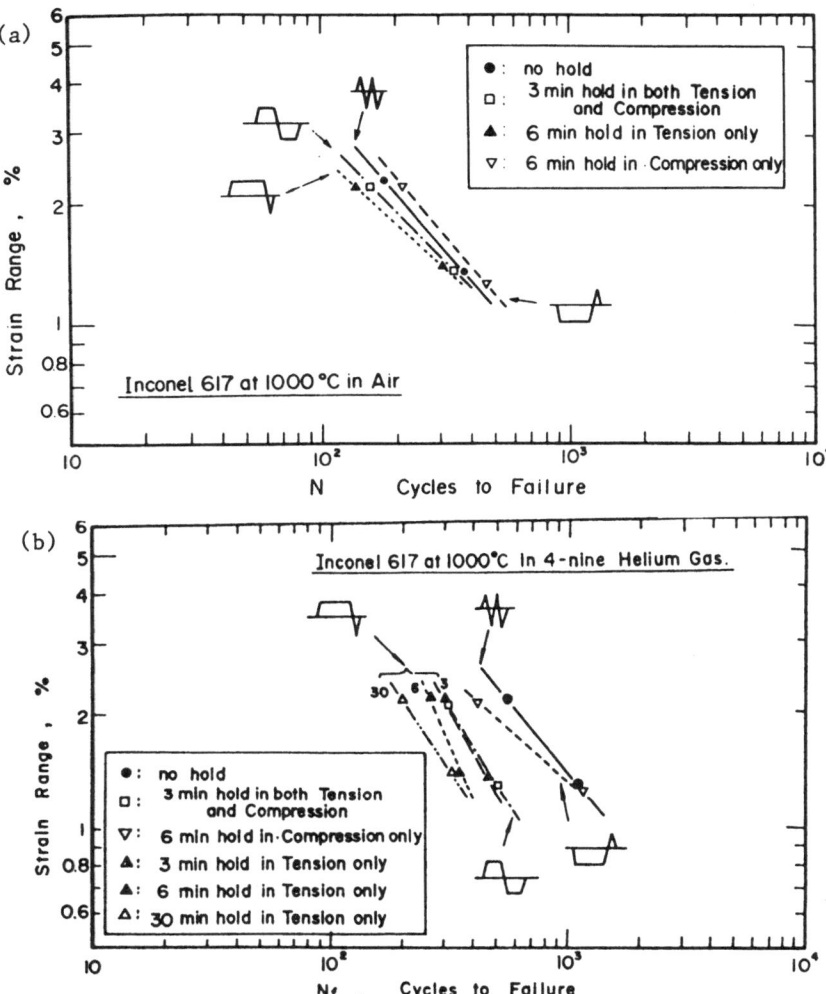

FIG. 7. The effect of hold time on the low cycle fatigue life of Inconel 617 at 1000°C (a) in air and (b) in helium gas of 99·995% purity.

cyclic thermal stress condition. For example, the tubes of the helical coil type heat exchangers are subjected to cyclic thermal stress due to the temperature difference of the centerpipe and the bundle of the heat exchanging tubes. The tubes are also subjected to constant stress due to

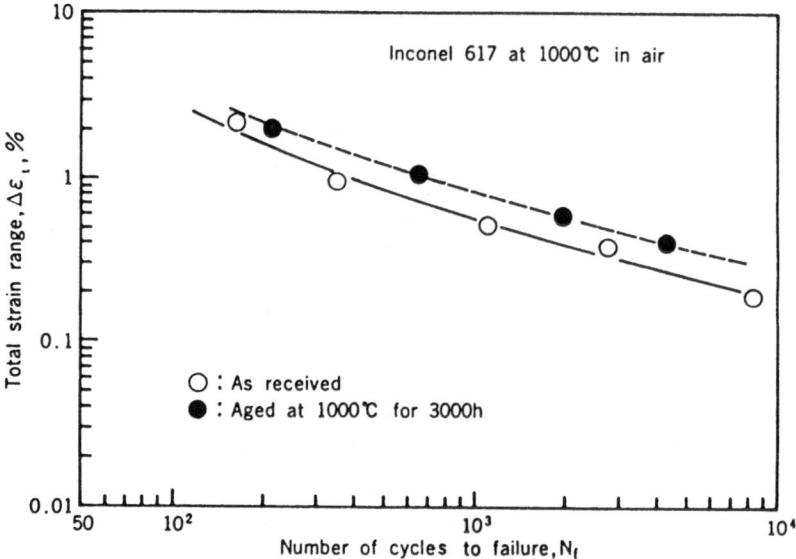

FIG. 8. Comparison of low cycle fatigue strengths of solution treated and aged Inconel 617.

the internal pressure. Therefore, failure due to creep–fatigue interaction is an important problem.

Figure 7 shows the creep–fatigue strength of the Inconel 617 in air and of 99·995% purity helium gas. The fatigue strength in the helium is higher than that in the air. This is considered to be due to the fact that there are fewer active impurities in the helium gas than in air, preventing the oxidation of the slip plane or the newly cracked surface. However, the creep–fatigue strength in the helium gas is nearly equal to the creep–fatigue strength in air. This may be due to two reasons; (1) the failure mode of the creep–fatigue at this temperature is grain boundary separation which is less affected by the environment compared to the surface fatigue failure; (2) since the environment effect is cumulative with time, the long test time in the creep–fatigue interaction test supplies enough active impurity to the failure site.

In order to improve the strength at these temperatures, it is important to increase the strength of the grain boundary because all the failure mode is that involving grain boundary separation. Figure 8 compares the fatigue strength of the solution treated Inconel 617 and aged Inconel 617. It is seen that the fatigue strength of the aged material is higher than that of the solution treated material. This may be related to the carbide precipitation on the grain boundary which increases grain boundary sliding resistance. This technique of increasing the fatigue strength by carbide precipitation on grain boundaries is employed successfully in a newly developed nickel-base alloy for HTR application.

2.3. Failure under Cyclic Thermal Strain with Ratchet Strain [4]

In the components used at very high temperatures, failure associated with large deformation is often encountered. In the case of the helical coil type heat exchanger as shown in Fig. 9, the stub portions (a piece between centerpipe and tube) are subjected to internal pressure and cyclic bending due to the thermal expansion and contraction. Consequently, the diameter of the stub is increased due to the internal pressure and cyclic bending (creep ratcheting). Although conventional design procedure recommends a plastic deformation limit of 1%, the limitation is sometimes difficult to meet because of the severe loading conditions at

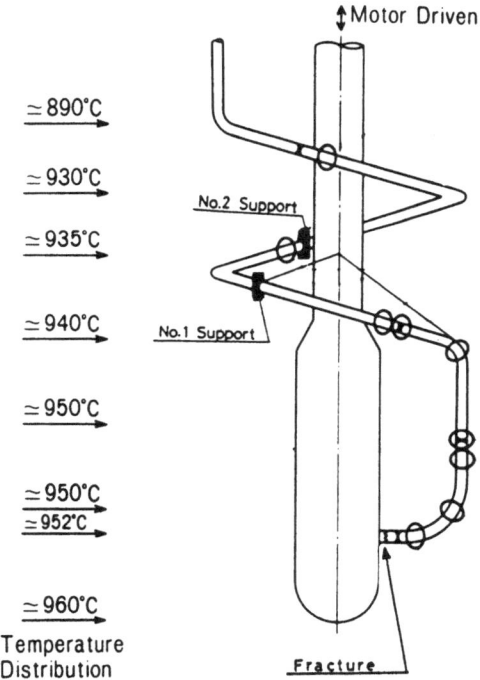

FIG. 9. Fracture model of helical type heat exchanger.

very high temperatures. In such cases, life criteria which include the effect of the ratcheting strain are required to insure structural integrity. Figure 10 shows the fracture criterion developed for the life estimation of the accelerated lifetime test of the partial model of the HTR intermediate heat exchanger and the results of the life estimation compared with the experimental data. The surprisingly good agreement between experiment and prediction may be partially due to the appropriate fracture criterion which includes the effect of the above mentioned ratchet strain.

Criteria

$$\phi_f + \phi_c + \phi_D = 1$$

Where ϕ_f is fatigue damage, ϕ_c is creep damage, and ϕ_D is ductility exhaustion damage.

Based on the experimental results, above equation was simplified as follows.

$$\int \frac{dt}{t_{rc}} + 0.9 \int \frac{d\varepsilon}{\varepsilon_r} = 1$$

Life estimation results

Tested	Time to fracture (hr)		
Tube No.	Experiments	Estimated (inner S.)	Estimated (outer S.)
1	695	1033	793
2	665	1053	793
3	> 700	1117	787
4	440	557	421
5	372	445	341

$\phi_c / \phi_v = \frac{1}{6} \sim 2$

FIG. 10. Failure criteria for creep–fatigue condition with ratchet strain.

3. TRIBOLOGY PROBLEMS IN HTR TEMPERATURES AND COUNTERMEASURES

Vibrational motion and the cyclic thermal expansion of the tubes causes small cyclic relative motion between the support and the tubes. This motion can result in wear problems. Furthermore, the metal-to-metal contact often causes self welding at the support, which is undesirable because it increases the thermal stress in the tubes. R&D for preventing these problems is important.

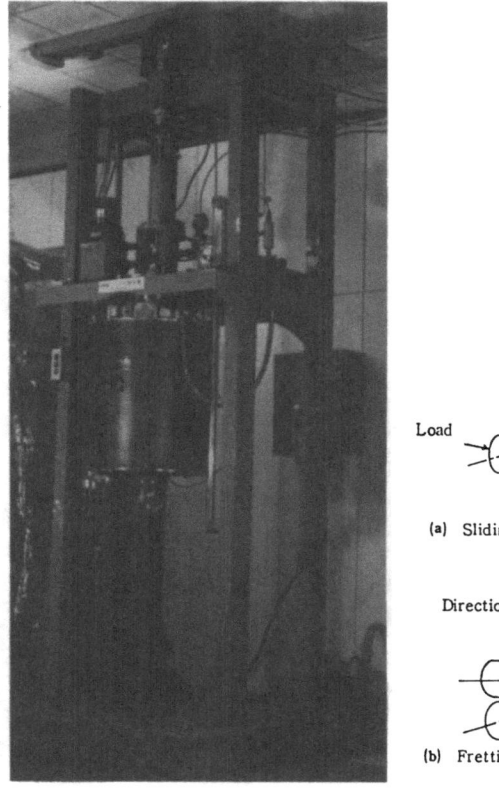

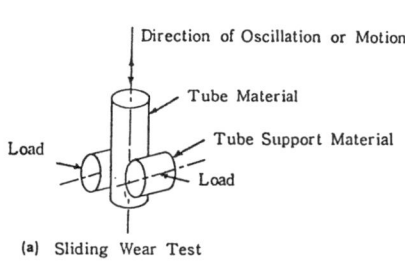

Direction of Oscillation or Motion

Tube Material

Load

Tube Support Material

Load

(a) Sliding Wear Test

Direction of Oscillation or Motion

Tube Material

Tube Support Material

(b) Fretting Wear Test

FIG. 11. Sliding and fretting wear apparatus.

3.1. Characterization of Wear and Friction

Figure 11 shows the experimental apparatus for the characterization of the wear (sliding wear and fretting wear) under relatively small cyclic deformation. Figure 12 shows an example of the surface of the specimen after the sliding wear test. The surface of the Inconel 617 to Inconel 617 contact is roughened significantly due to the cyclic sliding while the surface is very smooth if the surfaces of both metals are spray-coated with a cermet layer.

Figure 13 shows the results of the friction measurement after the sliding wear test. It is seen that the ceramic coated surface has much lower friction coefficients than those from the metal-to-metal experiments.

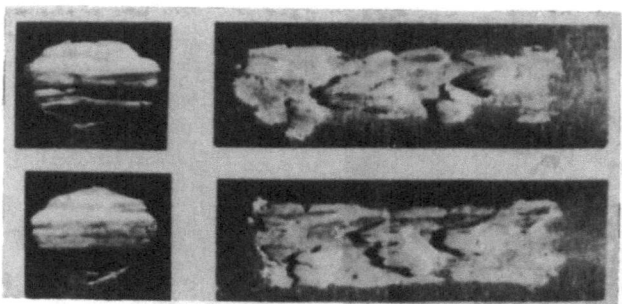

a) Inconel 617 to Inconel 617

b) Spray-coated by a cermet

FIG. 12. Specimen surface after the sliding wear test in helium gas at 1000°C.

In practice, the peeling-off of the coated ceramics is a problem. It is important to develop the technique to make a more adherent layer.

3.2. Characterization of Self Welding in the HTR Environment

Several combinations of metals and cermets (mainly zirconia compounds with some additives) were tested. As is similar to the results of the wear, the ceramic coated surfaces had the highest resistance to the self welding.

3.3. Material Design Against Wear and Self Welding

As is shown above, the ceramic coated surface has better resistance against wear and self welding. In order to get good resistance for the extended periods of service time, the stability and strength of the coated layer must be studied.

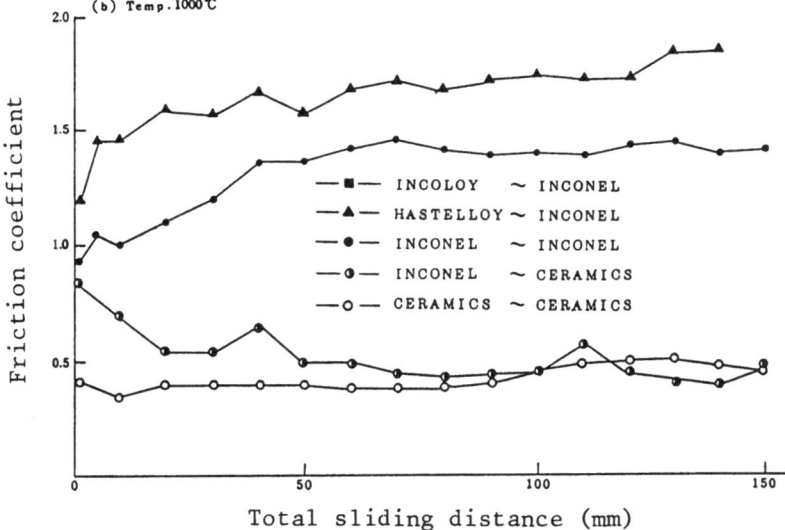

Fɪɢ. 13. Friction coefficient after the sliding wear test in helium.

4. HYDROGEN PERMEABILITY AND THE COUNTERMEASURES

One of the possible applications of the HTR is as the heat source in a direct steel making plant. The heat of the reactor is transferred to the intermediate heat exchanger and to the reducing gas heater. The reducing gas includes large amounts of hydrogen. The hydrogen at HTR temperatures can rather easily penetrate the tube wall behind the reactor core. This transferred hydrogen can in turn damage the internal structure of the reactor which is made of graphite. In order to assure the structural integrity, the hydrogen penetration through the wall should be reduced as much as possible. Furthermore, to determine the capacity of the purification system, it was necessary to characterize the hydrogen permeability of the alloys at the HTR temperatures.

4.1. Test Apparatus for Hydrogen Permeability

The apparatus consists of (1) supply system of the mixed gas of hydrogen, reducing gas, etc., (2) test section which simulates the heat exchanging tubes in the real plant, (3) hydrogen measurement system which is capable of measuring the 0·1 ppm density and (4) purification and circulation system.

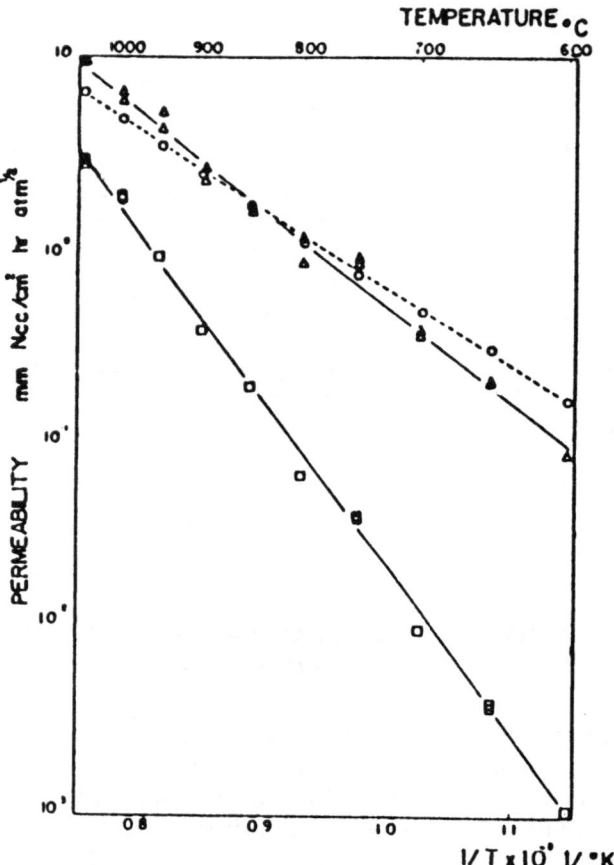

FIG. 14. Permeability to hydrogen of calorized and chromized Inconel 617. ○, As received; △, chromized (inner surface); □, calorized (inner surface).

4.2. Hydrogen Permeability and the Countermeasures

The temperature dependence of the hydrogen permeability of Inconel 617 and Incoloy 800 are shown in Figs. 14 and 15, respectively. As a countermeasure against the hydrogen permeability, the tube surface was chromized or aluminized by diffusion. The treated tubes as well as the as-received tubes were tested at temperatures between 600°C and 1050°C. The chromized tube showed almost no reduction in the hydrogen permeability. However, the aluminizing (calorizing) was found to reduce the hydrogen permeability significantly.

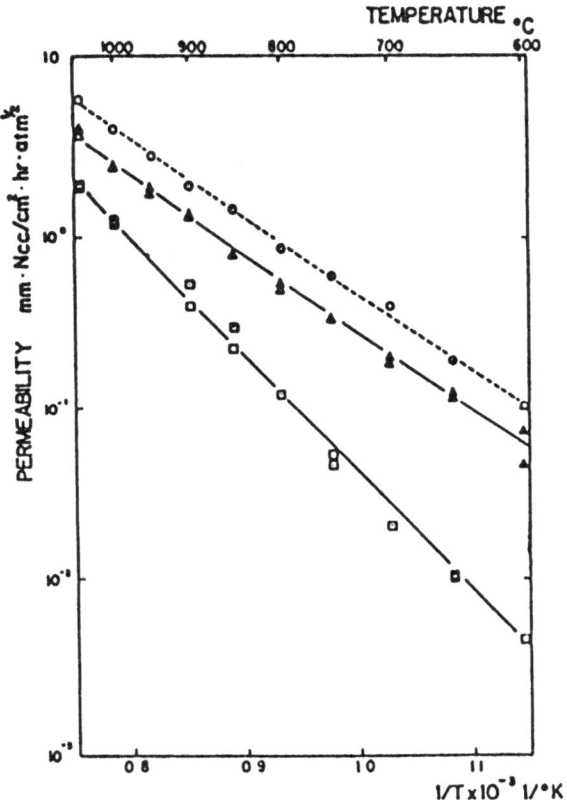

FIG. 15. Permeability to hydrogen of calorized and chromized Incoloy 800. ○, As received; △, chromized (outer surface); □, calorized (outer surface).

5. OTHER MATERIAL REQUIREMENTS FOR HTR APPLICATIONS

Figure 16 shows the relative importance of each high-temperature property for application in the HTR's intermediate heat exchanger tube. The relative importance of each property was decided by answers to inquiries from specialists of material strength, high-temperature plant designers and specialists in corrosion at HTR environment and alloy developers. Since the countermeasures against hydrogen permeability and wear and self-welding were given by the surface coating, those properties were not included in the inquiry.

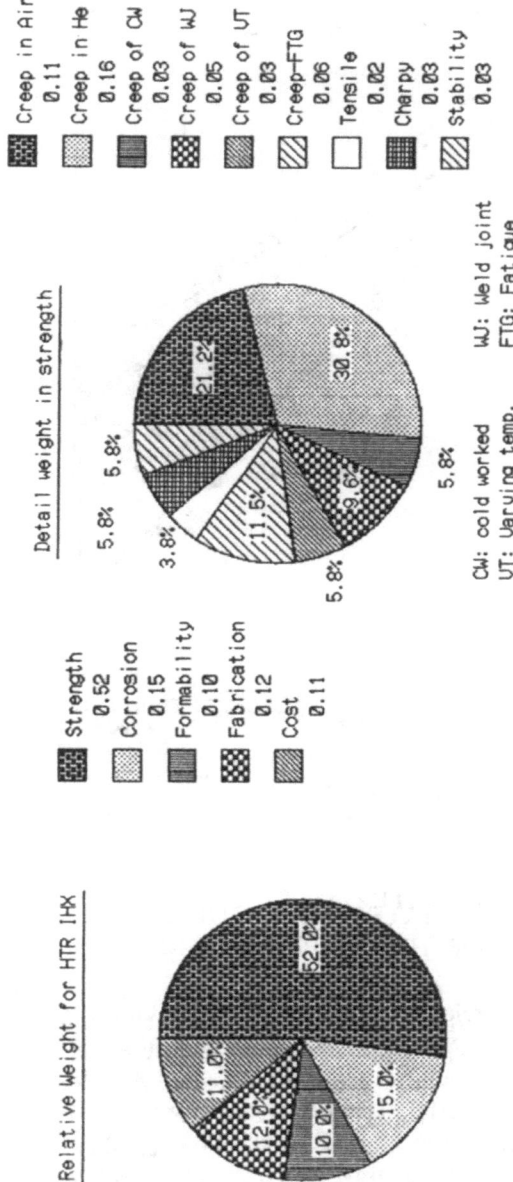

FIG. 16. Relative weight of importance for the selection (or development) of alloys for high temperature component of HTR.

It is seen that the importance of the strength is considered to be very high. The corrosion property comes next in importance. The formability and fabrication capability is also considered to be fairly important. The cost of the material is not of major concern, which is not surprising at the early stage of the development.

Figure 16 also shows the detailed weight of the strength properties for HTR intermediate heat exchanger tubes. Because of the high operating temperatures, it is seen that material with creep resistance is sought.

REFERENCES

1. KITAGAWA, M. *et al.*, *Elevated Temperature Design Symposium*, Mexico City, September 19–24, 1976, pp. 33–40.
2. KITAGAWA, M. *et al.*, *Proc. 8th SMIRT*, Berlin, 1979, F9/1.
3. HATTORI, H. *et al.*, *Proc. Int. Conf. Creep*, Tokyo, 1986, pp. 117–22.
4. KITAGAWA, M. *et al.*, *J. Nucl. Technol.*, **66** (1984), 675–84.

14

Creep, Fatigue and Environment Interactions in Cr–Mo Steels

A. NARUMOTO

Plate Laboratory, Technical Research Division, Kawasaki Steel Corp., Kawasaki-cho 1, Chiba, Japan.

ABSTRACT

The effects of temperature, strain rate, strain wave form and environment on the low cycle fatigue life of the Cr–Mo steels were investigated and the damage modes were assumed to be classified into creep mode and environment assisted fatigue mode. A life prediction method is proposed for each damage mode.

1. INTRODUCTION

Low cycle fatigue at elevated temperatures has been studied with respect to creep–fatigue interaction. This is used for fatigue life predictions in ASME Boiler and Pressure Vessel Code Case N-47. It is known, however, that environment also plays an important role in low cycle fatigue at elevated temperatures. The present paper describes the effects of strain rate, temperature, wave form as well as environment on the low cycle fatigue life of Cr–Mo steels. It also proposes a new prediction method that takes into account the creep, fatigue and environment interactions.

2. MATERIALS AND EXPERIMENTAL PROCEDURES

2.1. Materials

The steels used are four kinds of Cr–Mo steels with various Cr and Mo contents. Their chemical compositions are shown in Table 1. The materials were received in the form of rolled plate.

TABLE 1
Chemical compositions of steels used

	C	Si	Mn	P	S	Cu	Ni	Cr	Mo
A204B	0·25	0·24	0·73	0·010	0·008	0·01	0·03	0·04	0·49
A387-12	0·16	0·26	0·58	0·010	0·003	0·01	0·20	1·03	0·54
A387-22	0·13	0·08	0·51	0·009	0·008	0·12	0·12	2·41	1·01
A387- 5	0·09	0·24	0·44	0·007	0·007	0·03	0·07	4·65	0·47

2.2. Experimental Procedure

The fatigue test specimens employed were smooth round bars with a gauge section of 10 mm in diameter and 30 mm in length as shown in Fig. 1. Prior to the test, each specimen was polished in the longitudinal direction with No. 400 emery paper.

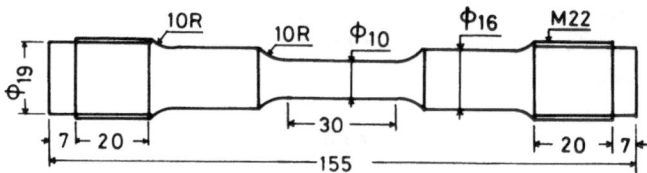

FIG. 1. Specimen dimensions (mm).

The low cycle fatigue test was carried out by controlling the axial strain in air and vacuum at elevated temperatures using a servo-hydraulic testing machine with induction heating equipment and a vacuum chamber which could reduce the pressure to less than 1×10^{-5} mm Hg. The axial strain was measured with an LVDT type extensometer whose gauge length was 25 mm. The strain wave forms were even–even, fast–slow, slow–fast and strain hold triangular as illustrated in Fig. 6. The strain rate was between 1%/s and $1 \times 10^{-3}\%$/s. The fatigue life, N_f, was defined as the number of cycles required to reduce the load from a stabilized level by 25%.

3. EXPERIMENTAL RESULTS

3.1. Effects of Strain Rate and Temperature

Figure 2 shows the effects of temperature and strain rate on the fatigue

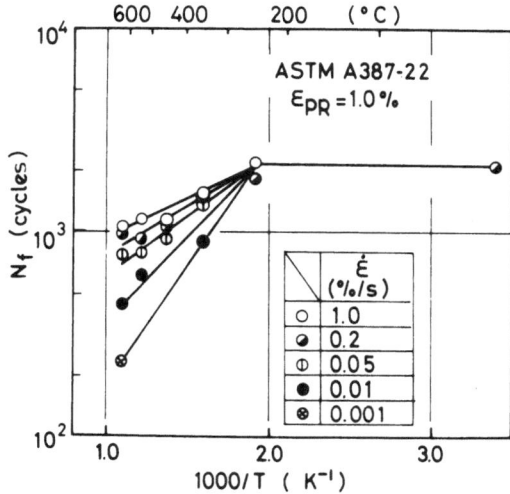

FIG. 2. Effects of temperature and strain rate on the fatigue life of A387-22 steel.

life under even–even wave forms for an ASTM A387-22 steel. The tests were performed at the total strain ranges, ε_{tR}, of 0·8% and 1·4%. The lives for the plastic strain ranges, ε_{PR}, of 0·5% and 1·0% were obtained by short range extrapolation or interpolation from the two experimental points. For both plastic strain ranges, no reduction of life was observed when $10^3/T$ was higher than 1·9 K⁻¹, i.e. the temperature was below 250°C. When the temperature was higher than 250°C, however, the fatigue life decreased as the temperature increased and/or the strain rate decreased. The same behaviors of decrease in life could be observed in the other steels which are listed in Table 1. Figure 3 shows micrographs of subcracks at the longitudinal cross-sections of the tested specimens. When the strain rate was 2×10^{-1}%/s and the temperature was 350°C (a), the crack initiated along a slip band and propagated at an angle of 45° with respect to the specimen surface as observed in the specimen tested at room temperature. At a low strain rate (b), however, a crack propagated in the direction normal to the surface and the crack width increased due to the oxidation of the crack surface. The thickness of the oxide layer increased as the temperature increased and/or the strain rate decreased (c,d). It was almost uniform through the crack depth up to the crack tip. This observation indicates that crack propagation is accompanied by oxidation at the crack tip. The growth of the oxide layer is coincident with the decrease in life at slow strain rate and/or high temperatures.

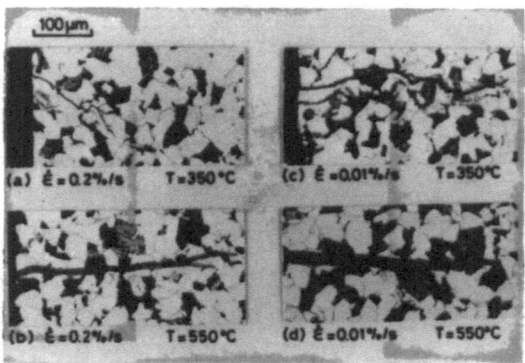

FIG. 3. Subcracks observed in the longitudinal cross-sections of tested specimens of A387-12 steel.

3.2. Effect of Strain Wave Form

The effect of strain wave form on the life is shown in Fig. 4. In this test, one cycle period was about 280 s, and the total strain range was 1·4%. The life under a fast–slow wave form coincided with that under an even–

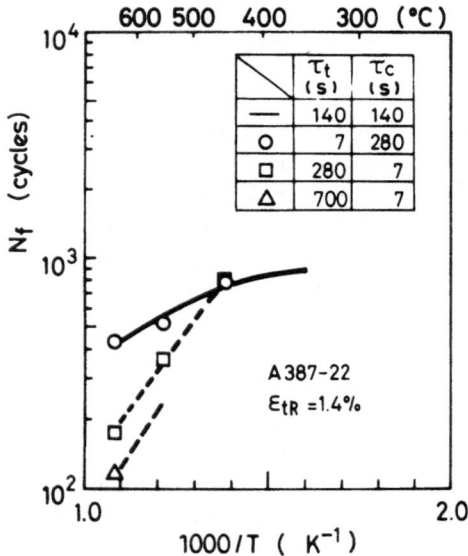

FIG. 4. Effect of strain wave form on the fatigue life.

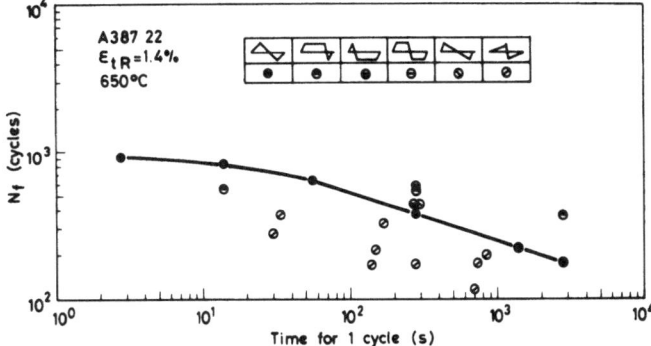

FIG. 5. Plots of fatigue lives under various wave forms against cycle period.

even wave form. On the other hand, the life obtained under a slow–fast wave form was also coincident with those under other wave forms at 450°C ($10^3/T = 1.4$). Above this temperature or in the so-called creep temperature region, however, the life under a slow–fast wave form decreased as the temperature increased much more than in the other wave forms.

Figure 5 shows the relation between the life and the time for one cycle under various wave forms for an A387-22 steel at 650°C. The lives obtained under slow–fast wave forms are significantly shorter than those under the other wave forms. Though the strain hold period has been considered to be damaging, the damage should be discussed in comparison to the life under even–even wave form with equivalent cycle period. Figure 5 indicates that the strain hold cycle is less damaging than the even–even wave form with the equivalent cycle period.

3.3. Effect of Environment

Figure 6 shows the relation between the compression time, τ_c, and N_f in air and vacuum at 650°C. The tension time, τ_t, was 140 s, and ε_{tR} was 1·4%. In vacuum, N_f increased with the increase in τ_c and reached the value obtained at room temperature. The fact that no difference existed between ε_{PR}–N_f curves obtained at room temperature and in vacuum at elevated temperatures for the even–even wave form has been reported by Coffin [1]. In air on the other hand, though no environmental effect was observed when τ_c was shorter than the critical value, when τ_c exceeded the critical value a significant environmental effect on N_f was observed and the fatigue life began to decrease with the increase in τ_c.

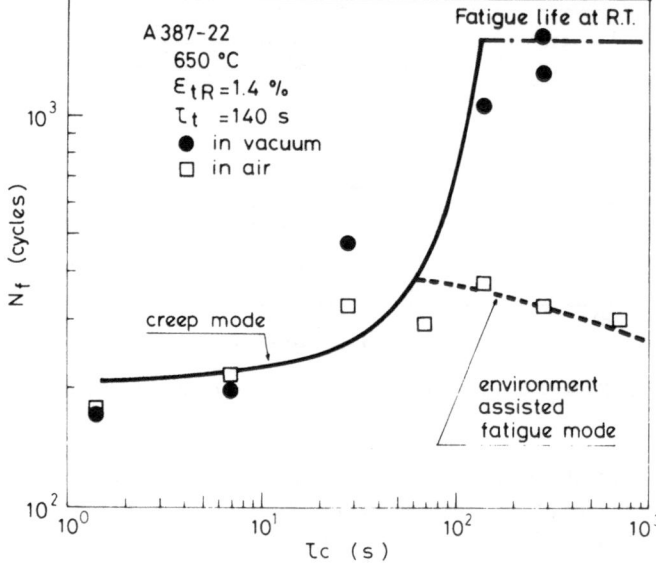

FIG. 6. Effect of environment on the fatigue life under sawtooth wave form.

4. DISCUSSION

4.1. Failure Mode

It is indispensable to the prediction of failure life to know that the damage mechanism and creep–fatigue interaction mechanism are assumed in ASME B&PV Code Case N-47. Experimental results, obtained in this work, however, maintain the following hypotheses for damage mechanism:

1. Failure modes are classified into creep mode and environment assisted fatigue mode.
2. In the creep mode, damage is accumulated only in the tensile period and in compressive creep the tensile creep damage recovers. Then the creep damage per one cycle depends on $(\tau_t - \tau_c)$.
3. In the environment assisted fatigue mode, environmental damage is accumulated during whole cycle period.
4. Observed life is determined by one of the two modes which gives the shorter life.

On the bases of these hypotheses, the test results obtained in this work can be explained as follows.

In Figs. 4 and 5, failure lives under slow–fast wave forms were shorter than those under the other wave forms. This is attributed to the difference in failure mode, that is, the life under slow–fast wave form is determined by the creep mode though the environment assisted fatigue mode is observed in the other wave forms. If the creep crack initiates at the interior of the specimen, it is reasonable that there is no environmental effect on the creep mode failure as shown in Fig. 6. Figure 6 also shows the recovery of the tensile creep damage by compressive creep. In vacuum, failure life increased with increase in τ_c. When τ_c exceeds τ_t, there is no longer any creep damage and the failure life is almost equivalent to that obtained at room temperature which is regarded as being a result of only fatigue induced damage. In air, when τ_c exceeds a critical value, failure mode changes from creep mode to environment assisted fatigue mode, so the failure life begins to decrease with increase in cycle period.

Strain hold wave forms gave longer lives than that under even–even wave form with the same cycle period as shown in Fig. 5. It means that environmental damage for strain hold wave forms is less than that for even–even wave forms. This is attributed to oxidation. Since the oxide layer is not cracked during the holding period, the specimen surface is protected by the oxide layer. In the even–even wave form, on the other hand, alternating cracking of oxide layer and oxidation of newly revealed surface will accelerate the environmental damage. Therefore, evaluation of the environmental damage due to strain hold cycle pivoting about the cycle period will result in a conservative estimation.

It is of interest whether there exists interaction between two failure modes. Figure 7 shows the subcracks on the longitudinal cross-section of the specimen failed under different τ_c values. Failure life increases with increase in τ_c as already mentioned. The depth of the surface subcrack also increases with increase in τ_c, that is, increase in number of cycles. Therefore, it is suggested that at least the surface crack depth depends on only environment assisted fatigue damage and there is no interaction between the two modes.

4.2. Life Prediction for Environment Assisted Fatigue Mode

The environmental damage induced in one cycle, ϕ_e, is defined by the following equation:

$$\phi_e = 1/N_f - 1/N_0 \tag{1}$$

A. Narumoto

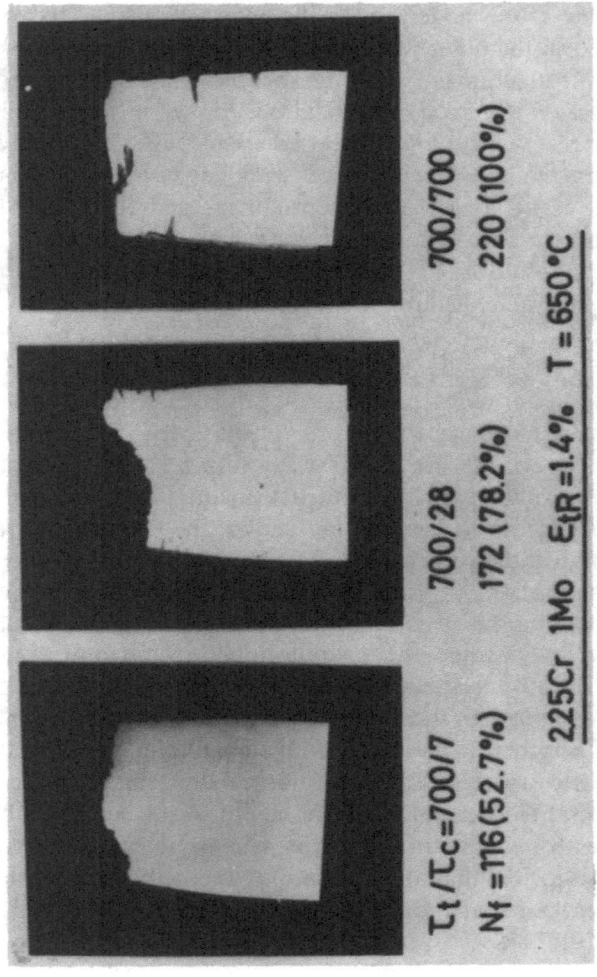

Fig. 7. Comparison of the surface subcrack depth observed in the longitudinal cross-sections of the specimen tested under slow–fast wave forms with varying compression times.

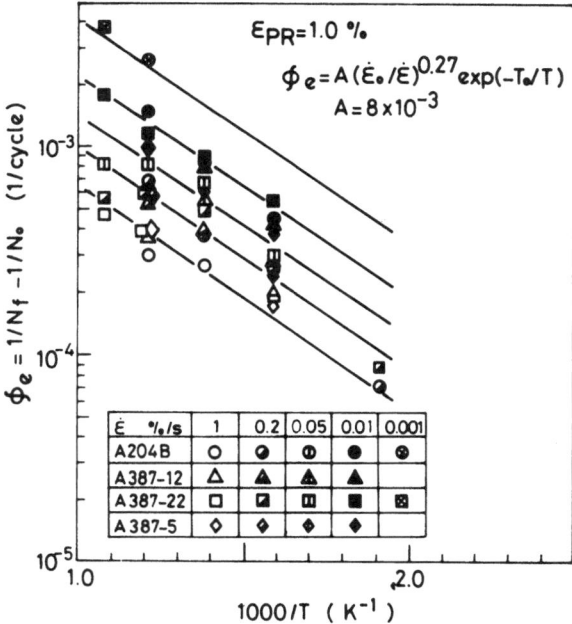

FIG. 8. Effects of temperature and strain rate on environmental damage in one cycle at ε_{PR} of 1.0% for several Cr–Mo steels.

where N_f is the life at a given strain rate (or time for one cycle) and temperature, and N_0 is the life at room temperature without environmental damage. The relations between ϕ_e and temperature for low alloy steels are shown in Figs. 8 and 9 for the plastic strain ranges of 1.0% and 0.5%, respectively. The value of ϕ_e was not significantly affected by Cr and Mo contents and is expressed by the following equation:

$$\phi_e = A \; (\dot{\varepsilon}_0/\dot{\varepsilon})^{0.27} \exp\left(-\frac{T_0}{T}\right) \tag{2}$$

where $\dot{\varepsilon}_0$ is the reference strain rate equal to $1\%/s$, T_0 is 2500 K, $\dot{\varepsilon}$ is the strain rate in $\%/s$, T is the test temperature in K and A is the plastic strain range dependent constant. The value of A was 8×10^{-3} for the plastic strain range of 1.0% and 4×10^{-3} for 0.5%. The environment assisted fatigue life for any combination of plastic strain range, strain rate and temperature can be predicted from the life obtained at room temperature using eqns. (1), (2) and the Manson–Coffin relationship. At

A. Narumoto

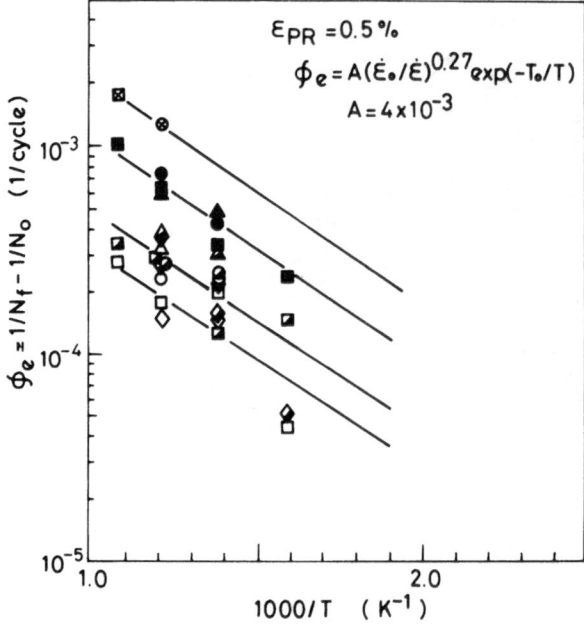

FIG. 9. Effects of temperature and strain rate on environmental damage in one cycle at ε_{PR} of 0·5% for several Cr–Mo steels. Key as for Fig. 8.

very low strain rate and/or at very high temperature, this prediction method can be described as follows:

$$\phi_e \gg 1/N_0 \tag{3}$$

$$N_f \propto \phi_e^{-1} \tag{4}$$

$$N_f \propto \nu^{0.27} \tag{5}$$

where ν is the frequency. Equation (5) is equivalent to the frequency modified life concept which was proposed by Coffin [2]. The method proposed above is able to cover the transition region between room temperature and high temperature, and does not require the experimental constants which are used in the frequency modified life method.

The predicted lives for an environment assisted fatigue mode are shown in Fig. 10 in comparison with the experimental ones. The figure includes the data obtained under constant strain range at various strain rates and temperatures and under various strain ranges at a constant strain rate and temperature for several Cr–Mo steels listed in Table 1. The predicted life agreed with the experimental one within a factor of 1·5.

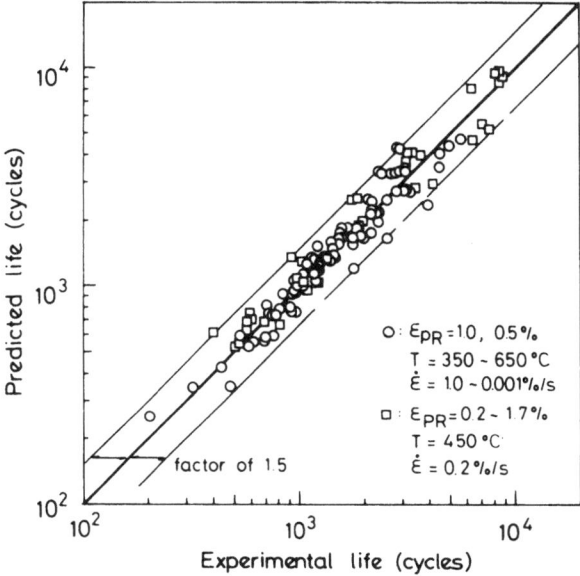

FIG. 10. Comparison of predicted life with experimental life for environment assisted fatigue mode.

4.3. Life Prediction for Creep Mode

The creep damage mode was observed in the slow–fast wave forms. As τ_c increased, the tensile creep damage decreased. Figure 11 shows the comparison between the static creep and fatigue test results obtained under slow–fast wave forms at 550°C and 650°C using the Larson–Miller parameter which is given by the following equation:

$$P_{LM} = T\{20 + \log t_r\} \tag{6}$$

where τ_r is the creep rupture time in hours. In the present study, the creep stress and the rupture time for the fatigue test were defined by the maximum stress in a stabilized hysteresis loop and eqn. (7), respectively.

$$t_r = N_f(\tau_t - \tau_c)/3600 \tag{7}$$

A considerably good agreement was observed between the static creep and fatigue test results. Thus the fatigue life for the creep damage mode can be predicted using eqn. (7) and the P_{LM} value which is obtained from the static creep test under a stress equal to the maximum cyclic stress.

Figure 12 shows the experimental and predicted lives versus τ_c for the

A. Narumoto

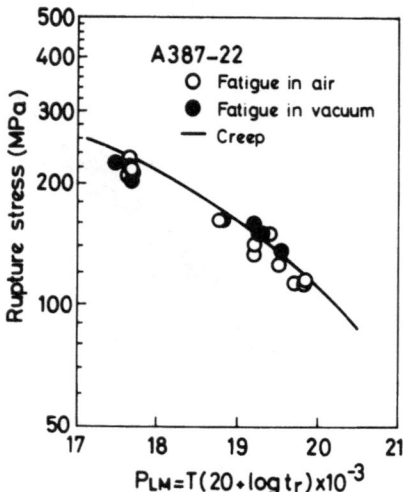

Fig. 11. Comparison of fatigue test results with static creep test results.

A387-22 steel tested at 650°C. The values of τ_t were 28, 140 and 700 s. Solid lines and dotted lines indicate predicted lives for creep mode and environment assisted fatigue mode, respectively. The experimental points fell on the predicted lines for creep mode or that for environment assisted fatigue mode whichever gave the shortest life.

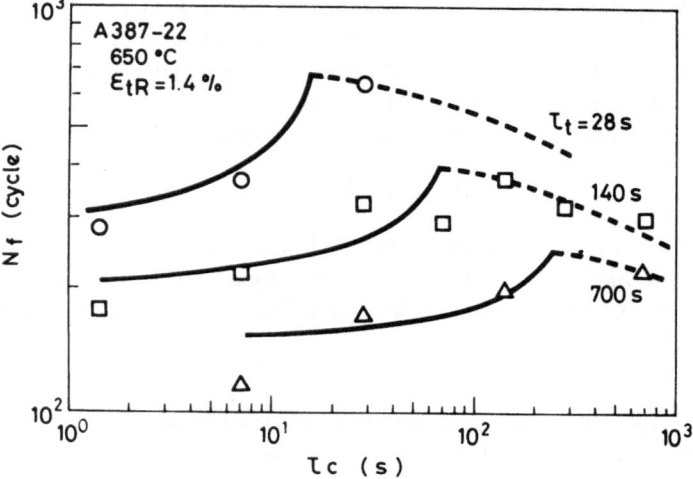

Fig. 12. Comparison of predicted life and experimental life obtained in various sawtooth wave forms.

5. CONCLUSIONS

The conclusions obtained in the present study are as follows:

(1) The fracture mode in low cycle fatigue at elevated temperatures for low alloy steels can be classified into two modes. They are the creep mode and the environment assisted fatigue mode. The fracture mode which gives the shortest life determines the life.

(2) The compressive creep recovers the tensile creep damage.

(3) The strain hold period is less damaging than the even–even wave form with an equivalent cycle period.

(4) A life prediction method for the two fracture modes is proposed.

REFERENCES

1. COFFIN, L. F. Jr *Proc. Int. Conf. Fatigue*, NACE-2, Houston, Tex., 1972, pp. 590–600.
2. COFFIN, L. F. Jr *Proc. Air Force Conf. on Fatigue and Fracture of Aircraft Structure and Materials*, AFFDL TR7 0-144, 1970, pp. 301–312.

15

Material Degradation and Life-time Prediction of Fossil Power Plant Components

T. ENDO

Materials and Strength Research Laboratory,
Takasago R and D Center Mitsubishi Heavy Industries, Co., Ltd,
Shinhama 2-1-1, Arai-cho, Takasago, Hyogo, Japan.

1. INTRODUCTION

The recent deterioration problem of materials for fossil power plant components due to being put into long term service at high temperatures constitute a subject of the utmost concern to power companies and plant manufacturers. The requirement imposed upon these components is that they can serve longer than originally expected, in spite of their frequent starts and stops. Accordingly materials from which these components are made must be capable of serving satisfactorily throughout their design lives.

Here we are considering HP–IP turbine rotors as representative of the items being exposed to such environments and consider the current conditions and the problems of their deterioration assessment and life prediction technologies.

2. PROBLEMS INVOLVED IN MATERIAL DEGRADATION ASSESSMENT AND LIFE PREDICTION TECHNOLOGIES OF HP–IP TURBINE ROTOR

2.1. Changes of Material Quality Due to the Age of Manufacture of Cr–Mo–V Turbine Rotor Steel

The majority of the fossil power plants currently operating under the conditions described above were installed during the periods from 1950s to 1960s. As regards turbine rotors, they have experienced considerable

T. Endo

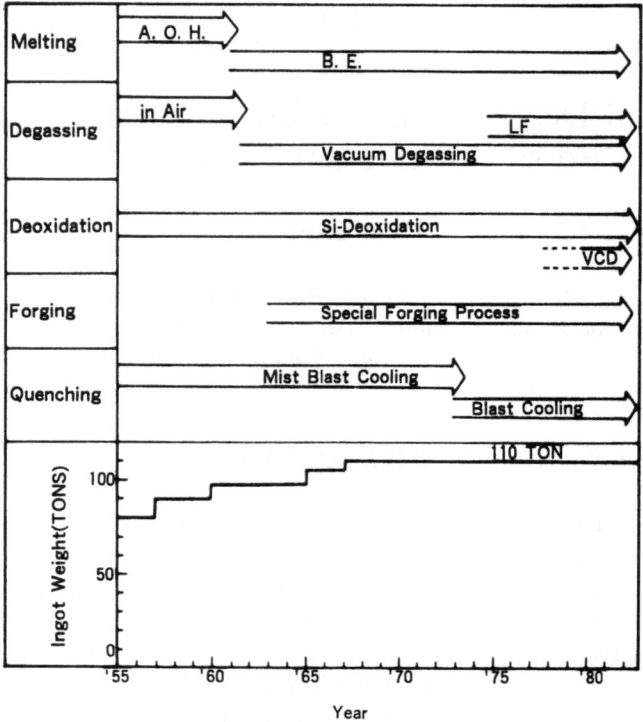

F IG. 1. Chronological changes of manufacturing processes of Cr–Mo–V rotor forgings.

changes in their production history. This will include some problems in applying the currently available Cr–Mo–V steels to all the experiments for life-time assessments. Therefore, it becomes necessary that we should conduct the survey of manufacturing history of old rotors and accumulate data of material tests obtained from old turbine rotors in actual operation.

Figure 1 shows changes in the manufacturing process of Cr–Mo–V rotor materials, while Fig. 2 gives a figure of changes in the main impurity elements. The rotor steels produced in the periods of 1950s and 1960s contained much of such trace elements as P, S, Sn *et al.* Much difference is recognized in the mechanical properties according to the chronological difference in manufacturing processes. Figure 3 shows the shift of FATT after long-term heating obtained from specimens taken from high-temperature parts of rotors in actual service.

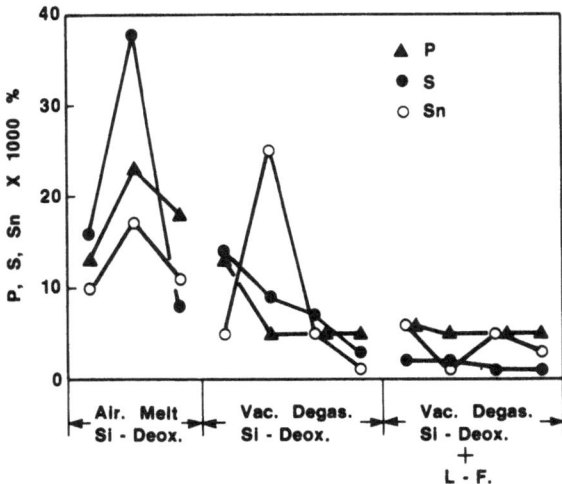

FIG. 2. Changes of impurity content according to manufacturing processes.

As is evident in the figure, wide differences exist in the shifts according to operation hours of individual rotors, and it is recognized that differences are also caused by different levels of metal temperature. The shifts of FATT were affected the most by the P and Sn impurity contents of old rotor materials.

Next comes the problem about creep rupture strength and ductility, and here again individual properties of rotor materials are found to have an important role.

Creep rupture strength will be increased by raising the quenching temperature, but this will adversely affect creep rupture ductility by reducing it greatly. In view of this, the general approach since the middle of the 1960s has placed importance on keeping high ductility by reducing the quenching temperature. However, much is still left unsolved about the mechanism of reduction of creep rupture ductility, especially the effects of trace elements. Figure 4 illustrates the respective creep rupture ductilities, creep–fatigue lives and crack propagation characteristics of two different Cr–Mo–V rotor steels containing different impurity elements. Forging A with much impurities (P, S and As) exhibits lower rupture ductility and a remarkable reduction of creep–fatigue life. It is also apparent that while these two rotors do not show any difference in low cycle fatigue crack propagation, there is a great difference in creep

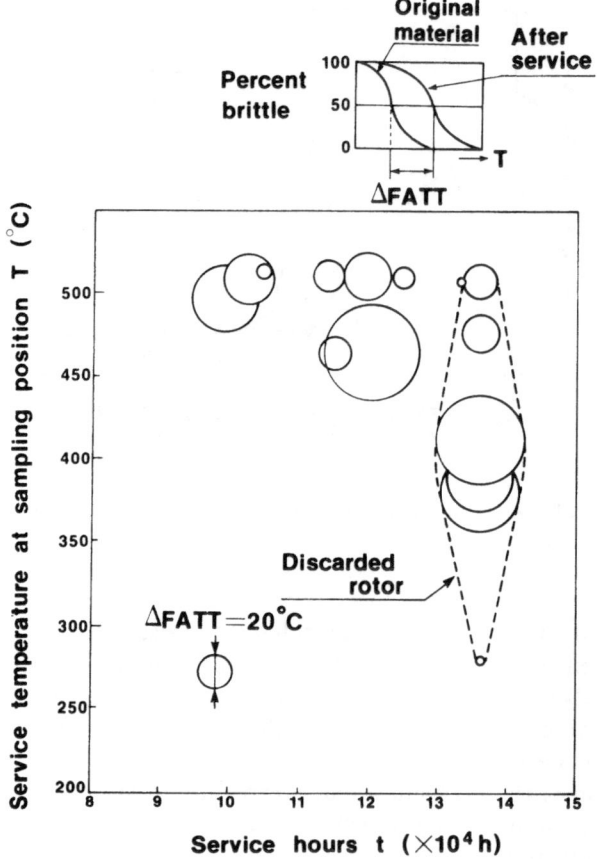

F IG. 3. Shift of FATT due to long term exposure in service [4].

crack propagation rate. As can be seen from above, it is important to take into account, in making life-time assessments of rotors, the fact that mechanical properties vary according to their chronological or compositional difference. So, we are making continuous efforts to acquire material data by post exposure examination of rotors at every opportunity, because we are not always familiar with the respective properties of all rotors now in operation.

2.2. Degradation of HP–IP Rotor and its Material Strength after Long-term Service

Following are various examples of deterioration noted in HP–IP

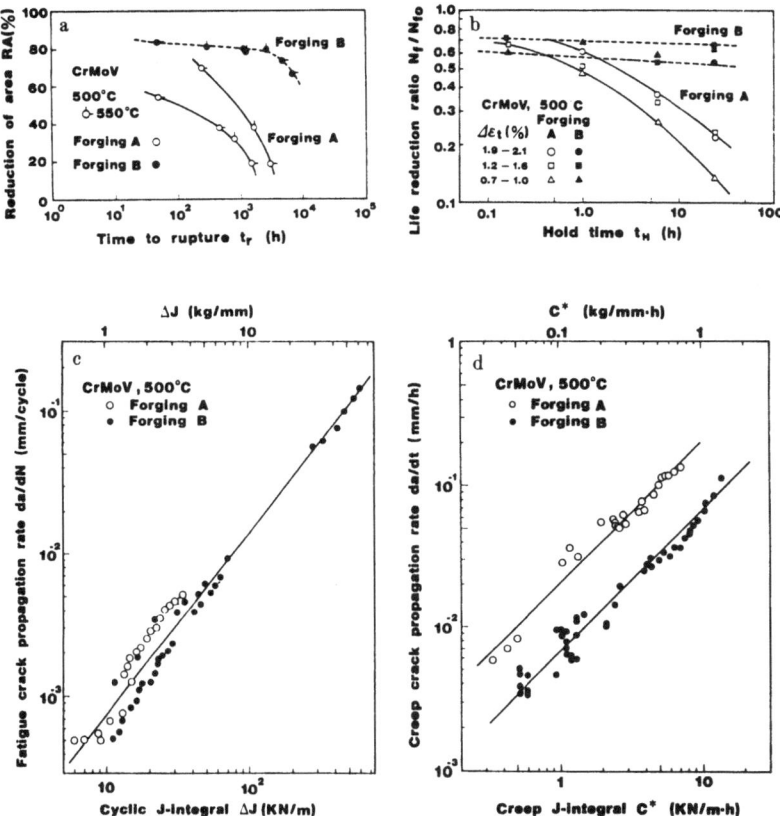

FIG. 4. The comparison of mechanical properties of two Cr–Mo–V rotor steels at elevated temperatures [5]. (a) Difference in rupture ductility; (b) Effect of hold time in creep–fatigue tests; (c) L.C.F. crack propagation rate; (d) creep crack propagation rate.

rotors which had been in operation for many hours. It can be said that very few material tests have been performed on the HP–IP rotors which have used more than 100 000 h. Naturally, what is being introduced here cannot be said to cover all the aspects of deterioration coming from long service, but it shows the results obtained through our studies on HP–IP rotors which had been in operation with metal temperatures less than 500°C.

Figure 5 shows the creep rupture strength of various parts of the HP–IP rotor operated for 140 000 h. This scarcely shows any change due to long term exposure to high temperatures. In the case of a different type of rotor it is reported that a remarkable deterioration has been detected

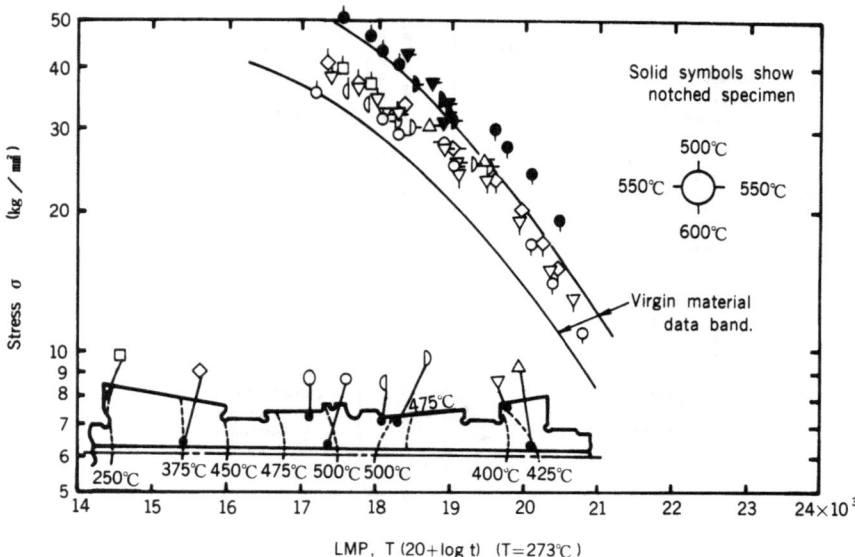

FIG. 5. Creep rupture test results for replaced HP–IP rotor used for 140 000 h.

after exposure at a metal temperature of 538°C [1]. For these Cr–Mo–V rotor materials, the change in the creep rupture strengths due to long exposures to high temperature has been related to the softening phenomenon, and various investigations are under way in this direction.

The cause of softening can be explained as a result of growth of fine carbides which change their shape during long-term heating. This makes softening an effective means for detecting processes of deterioration by creep. Softening accelerates once the applied stresses exceed a certain amount.

Goto [2] has worked out a method to show the unified relations between softening, applied stresses, temperature and exposure time, and this method is being used as a parameter for deterioration evaluations. Figure 6 indicates the result of hardness measurements on the surface of the above mentioned HP–IP rotor and the trend curve has been obtained by Goto's unified method. We may add that deterioration is not observed here in the form of softening.

Next item is the thermal fatigue that is generated at the notch root on the surface of the rotor exposed to high temperature steam. The HP–IP rotor produces transient thermal stresses at starts and stops. Currently, even in old fossil power plants, load following operations are being done daily or weekly, and attentive monitoring of the outside surface of rotors

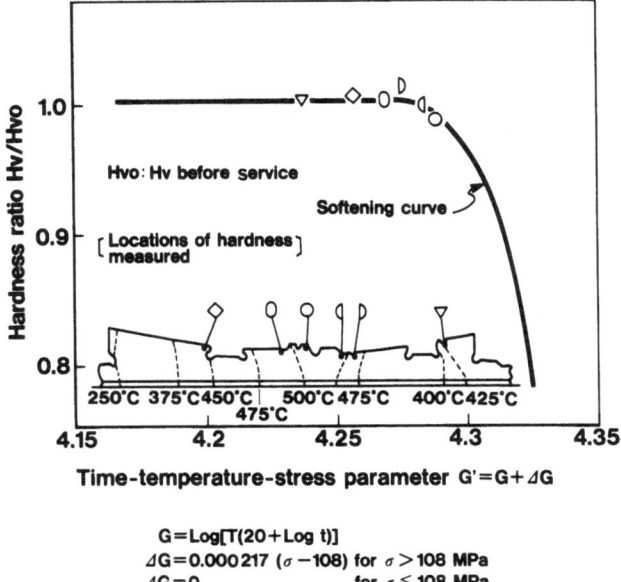

$$G = Log[T(20 + Log\ t)]$$
$$\Delta G = 0.000\,217\ (\sigma - 108)\ \text{for}\ \sigma > 108\ \text{MPa}$$
$$\Delta G = 0 \qquad\qquad\qquad \text{for}\ \sigma \leqq 108\ \text{MPa}$$

FIG. 6. An example of hardness measurement of rotor surface after 140 000 h.

becomes all the more necessary to detect any initiation of thermal fatigue cracks.

One example of deterioration noticeable in the above mentioned HP–IP rotor used for 140 000 h is a slight softening that can barely be noticed at the bottom of the surface. This sort of softening is different from the phenomenon due to creep, and is a result of cyclic softening of rotor steels due to cyclic plastic strains.

Figure 7 illustrates this. Softening is barely noticeable at a depth less than 1 mm at the notch root where cyclic strains are obvious. Figure 8 gives the stress contour which has been obtained by FEM elastic–plastic stress analysis conducted on dummy grooves and T-root grooves on the HP rotor, and it can be seen that the location of plastic deformation zones and the softening zone given in Fig. 7 are in good correspondence.

When, in this way, detection of a slight softening phenomenon on the surface becomes feasible, preventive measures will be available and life-time to surface cracking can be extended by adopting 'skin-peeling'. Figure 9 shows an example of low cycle fatigue test, which testifies to the efficacy of 'skin-peeling'. The specimen illustrated as a solid line shows a test piece where the notch root of the groove was left untouched, while

T. Endo

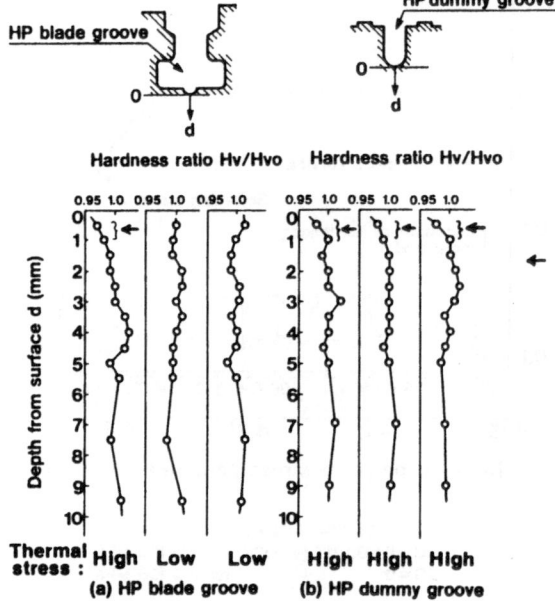

FIG. 7. An example of surface softening observed in a service rotor.

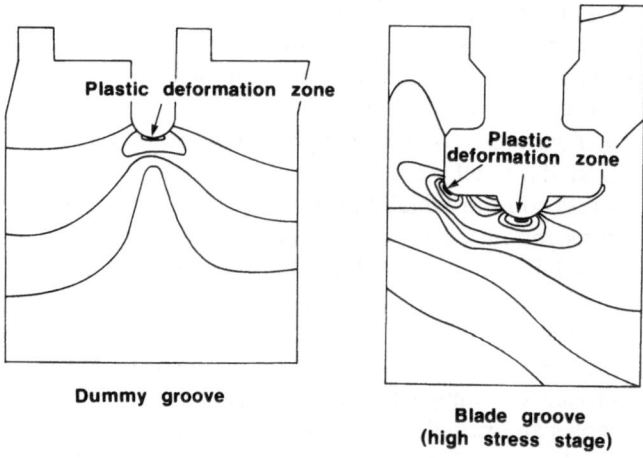

FIG. 8. An example of elastoplastic stress analysis around notch root [4].

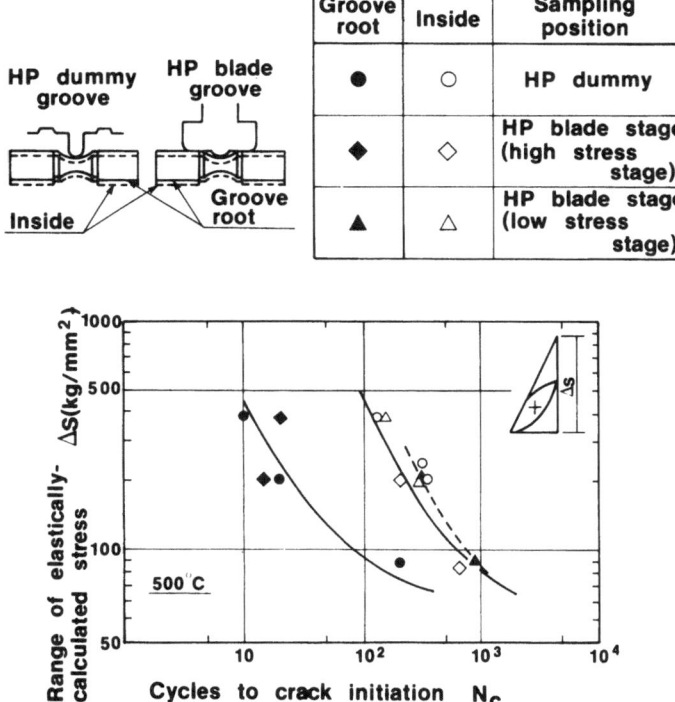

Groove root	Inside	Sampling position
●	○	HP dummy
◆	◇	HP blade stage (high stress stage)
▲	△	HP blade stage (low stress stage)

FIG. 9. An example of the effect of 'skin-peeling' on the low cycle fatigue life.

the dotted line shows the specimen whose surface has been machined off and has been taken at a depth 3 mm deeper than the notch root.

The test has been done on two kinds of specimens, one from a HP dummy groove and a T-root groove which were subjected to high stresses and the other from a T-root groove subjected to low stresses. The figure shows that by performing this 'skin-peeling' of the plastic deformation zone, low cycle fatigue life can be greatly extended. From this it can be said that measures should be easy to take for coping with a deterioration phenomenon in peripheral regions.

2.3. Development of Nondestructive Evaluation System for Detection and Monitoring of Material Degradation

Now that representative cases of deterioration detected in HP-IP rotors have been shown in Section 2.2, we will proceed to discuss the

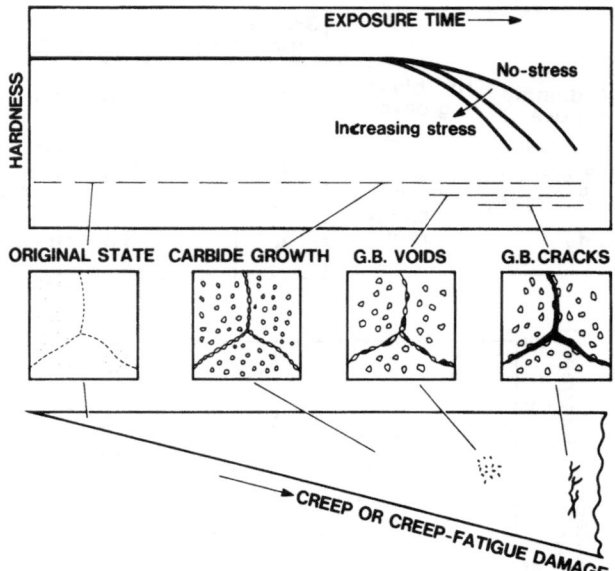

FIG. 10. Schematic expression of progress of creep or creep–fatigue damage.

assessment system for detecting these cases. Generally speaking, if we try to illustrate schematically the progress of damage of creep or creep–fatigue, it will take the shape of Fig. 10. This means that a change will take place in the microstructure of the metal regardless of whether stress is applied or not. In the case of Cr–Mo–V rotor steel, which is the subject of our study here, the carbide is chiefly $M_{23}C_6$ which will grow into coarser $M_{23}C_6$ through long exposure at elevated temperature.

If further stress were to be applied to it, the damage would increase its speed and according to the level of stress it would proceed to generate creep voids and grain boundary cracking. Deterioration assessment requires development of hardware for its detection and software for passing judgement upon it. It is indispensable that hardware be performed by nondestructive means, and for this purpose its development is based upon the above mentioned post-exposure material tests to find out at what stage of descriptions given in Fig. 10 the present deterioration stands.

Among the currently available methods of nondestructively detecting deterioration, the most effective will be:

(1) hardness measurement;
(2) detection of creep voids by means of replicas.

FIG. 11. Inspection devices for turbine rotor bore.

Other methods, so far tried, seem not to have proved very successful. As regards the steam turbine rotor, the peculiarity of the assessment positions must be considered for the devices. Figure 11 shows typical devices for turbine rotor bore inspection [3].

2.4. Technological Problems for Predicting Rotor Life

It has to be admitted that a number of technological problems still remain in performing life prediction of high-temperature components. Even when attention is confined to the life-time of rotors in the high-temperature regions, various problems can be thought of in the life prediction technology:

(1) creep–fatigue life assessment method;
(2) notch creep rupture evaluation method;
(3) prediction of crack growth in creep regime;

being among the number.

In such an actual situation, it cannot be helped that some percentage of error should exist in the life prediction values worked out for respective parts of the rotor. Consequently, the calculated lives will indicate the time when the above mentioned detection of deterioration should be performed.

It would be not too much to say that the materials of old fossil power plants are being taxed with work to their limits. We have introduced our recent studies and problems concerning Cr–Mo–V rotor steel, which are representative of the high-temperature parts of fossible power turbines.

3. SUMMARY

Current needs required for the limited use of HP–IP rotor steels are summarized as follows:

(1) Understanding is important concerning the changes of material properties due to chronological changes of manufacturing processes, chemistry, etc.
(2) Periodic monitoring is also important on the material degradation which is represented as softening due to creep or cyclic strain, temper embrittlement and creep voids.
(3) Nondestructive evaluation system development is a highly important item to apply to the damage detection techniques.

REFERENCES

1. MURAMATSU, M. *et al.*, *Karyoku-genshiryoku Hatsuden*, **35**, No. 8 (1984).
2. GOTO, T. *Proc. 2nd Int. Conf. Creep Fracture of Engng Materials and Structures at Swansea, UK*, 1984.
3. KADOYA, Y., GOTO, T. *et al.*, ASME Paper, 85–JPGC–Pwr10, 1985–10.
4. KADOYA, Y., ENDO, T. *et al.*, EPRI Seminar, Life assessment of turbogenerator rotors for fossil power plants, 1984–9.
5. ENDO, T. and SAKON, T., *International Conference on Creep*, Tokyo, 1986–4.

16

Structural Materials for Cryogenic Use

K. ISHIKAWA, K. HIRAGA and T. OGATA

National Research Institute for Metals, Tsukuba Lab.,
Sengen 1-2-1, Sakura-mura, Niihara-gun, Ibaraki, Japan.

ABSTRACT

This paper presents an overview on cryogenic structural materials. Material properties at liquid helium temperature are described for application in superconducting technology which requires highly reliable structural materials. The candidate materials are nonmagnetic high strength austenitic steels and titanium alloys of which the mechanical properties, i.e. strength, toughness, fatigue and creep are strongly affected by composition and microstructure.

1. INTRODUCTION

Cryogenic technology is basic and essential in the development of advanced technology in energy, information and medicine. It is more difficult to achieve a liquid helium temperature (4 K) environment, although liquefaction of helium gas was achieved 70 years ago. Great progress, however, in research and development of fusion energy and space technology has been made and rapid advances in R&D have also been made.

With the evolution of the R&D from experimental to a practical scale, severe conditions are imposed on the reliability of cryogenic systems and the safety of the apparatuses. Similar requirements for structural materials have stimulated R&D effects in new materials technology which must have improved properties compared to existing materials.

Structural materials must be studied from a different standpoint at cryogenic temperature environment.

The primary importance of the structural material for cryogenic use are its strength, namely tensile, fracture, fatigue and creep where limit analysis is recommended to design the most efficient refrigerators for achieving cryogenic temperatures. The more important cryogenic systems of recent years are superconducting machinery which works at liquid helium temperature. The structural materials for such apparatus are selected for their superior mechanical and physical properties. New structural materials suitable for the system are indispensable since there are distinct limitations in the properties of existing materials.

The necessity for developing new materials, which satisfy the requirements, are stressed during the stage of laboratory research. Much progress has been made in R&D of new structural materials, mainly in Japan and the results are appreciated [1]. Cryogenic materials are also being evaluated for their respective needs and this paper will review such evaluation and related problems.

2. STRUCTURAL MATERIALS FOR CRYOGENIC USE

At present, the cryogenic materials are classified into metallic and nonmetallic materials (organic composites). From the practical viewpoint based on energetic studies, the former are more advanced than the latter. In the future, the advantages of both materials will be exploited in the R&D of designing efficient cryogenic systems.

Austenitic stainless steels, typically AISI 304, are presently important structural materials for cryogenic use and are superior to any other materials. The mechanical and physical properties, together with data scattering of the respective properties, of austenitic stainless steels at low temperatures are shown in Fig. 1. The physical properties of various materials do not scatter as much but are more temperature dependent. On the other hand, there is a large dispersion in the mechanical properties at low temperatures due possibly to the difference in the impurity contents and processing.

3. EVALUATION AND PROBLEMS

3.1. Tensile Properties

Tensile properties are of primary importance for the evaluation of structural materials both at low and high temperatures. Tests are carried

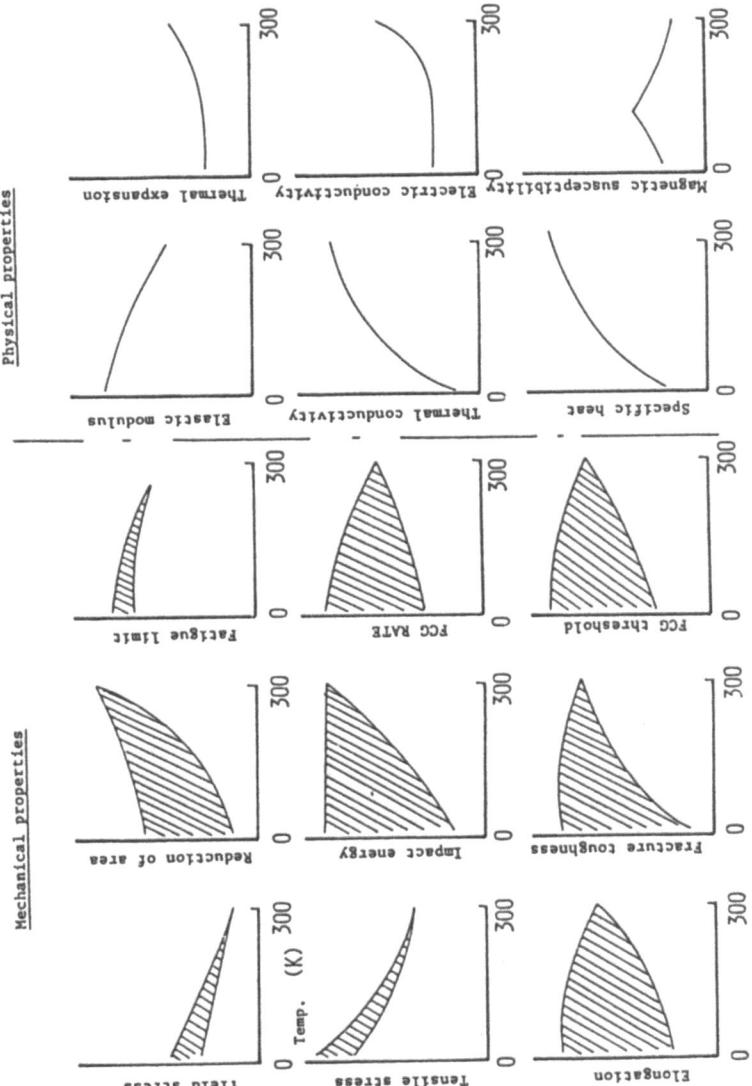

Fɪɢ. 1. Temperature dependence of mechanical and physical properties of austenitic stainless steel.

K. Ishikawa, K. Hiraga and T. Ogata

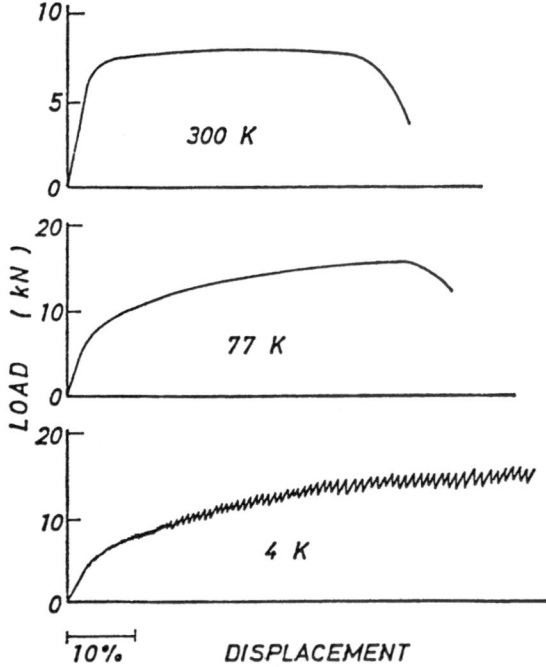

FIG. 2. Load–displacement curves of high manganese austenitic iron alloy at low
temperatures.

out using simple loading of round specimens. In Fig. 2, the load–
displacement curves of high manganese austenitic iron alloy are shown.
The test temperature affects the shapes of the flow curves. At the liquid
helium temperature (4 K), there is observed discontinuous behavior in
the flow curves [2]. This behavior is not dependent on the crystal
structure of the materials [3] but is affected by the strain rate as shown in
Fig. 3. At higher strain rate, the curve becomes smoother. The lower the
test temperature, the less the specific heat and thermal conductivity of
metallic materials. Therefore, a large temperature rise of the specimen is
produced adiabatically through small plastic deformation. Thus, the flow
curve produces a repeated load drop due to temperature increase. This
effect is conspicuously noted in tensile stress and elongation but not in
yield stress. Actually, the temperature increase reaches 150 K for stainless
steels [4].

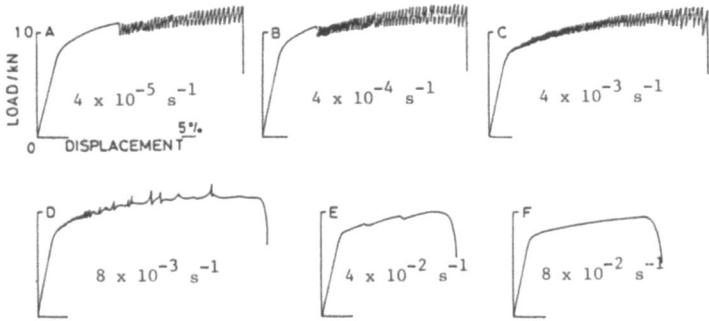

F IG. 3. Effect of strain rate on flow curves of AISI 310S at 4 K.

3.2. Toughness Properties

The K_{IC} test is important for low toughness materials and voluminous data have been compiled for various materials. Cryogenic materials, however, are basically very tough and thus the K_{IC} test is not applicable.

Historically, Charpy impact energy is used to evaluate highly tough materials and the J_{IC} test is expected to replace the K_{IC} test. the former method is standardized from high to liquid nitrogen temperatures and voluminous data have been obtained for various materials. Moreover, the J_{IC} test can be used for materials selection for lower temperature use. At liquid helium temperature, however, the method cannot be standardized because of difficulty in sustaining liquid helium temperature.

Recently, a simple method has been developed for maintaining the test temperature at 4 K by minimizing the heat flux from the surroundings. The J_{IC} test of ASTM E-812 is applied for evaluating the toughness of the materials at 4 K and the problems in determining the critical J_{IC} values are discussed.

3.3. Fatigue Properties

Fatigue properties are essential for the safety of the systems and the prediction of the life but evaluation is incomplete and the study is continuing. The S–N curves to failure are the basic data required with low cycle fatigue, fatigue crack propagation (FCP) and threshold properties of FCP receiving much attention. Voluminous data of FCP for various ductile materials at liquid helium temperature have been compiled and with little differences in materials. Material dependence in the high strain gauge of low cycle fatigue, however, is noted with little dependence in the low strain range of high cycle fatigue. Many problems remain to be resolved in fatigue properties at cryogenic temperatures since there is a shortage of data in this field. Moreover, though stress-number to failure (S–N Curve) and ΔK_{th} values are important parameters, further studies are necessary for quantitative evaluations.

3.4. Creep Properties

Creep properties have been ignored, since the creep was thought to take place at the temperatures close to absolute zero and would thus be negligible, if any. Recently, the importance of creep properties at cryogenic temperature has been indicated, since the superconducting machinery is subject to a large electromagnetic force and the structural stability of a superconducting magnet is affected by strain at micro level. For instance, AISI 304LN stainless steel creeps at a strain rate of $5.6 \times 10^{-11} \, s^{-1}$ after 200 h at liquid helium temperature at yield stress, and the total strain will reach about 4% after 24 years. Thus creep strain is a significant factor in the design of superconducting magnets.

Figure 4 shows the creep behaviors of austenitic stainless steels at 77 K where the creep rate after 200 h cannot be ignored. Also, strengthened copper alloys do not necessarily exhibit lower creep strain rate at lower temperatures [5]. Furthermore, the relaxation problem of superconductors subject to stronger electromagnetic force and a cyclic creep problem under alternating current are yet to be resolved.

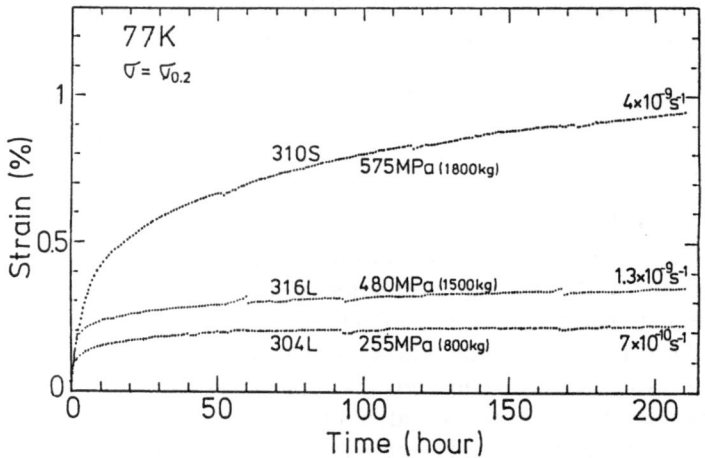

FIG. 4. Creep curves of austenitic stainless steels at 77 K.

4. R & D OF NEW STRUCTURAL MATERIALS

The research and development of new structural materials for cryogenic use are being actively pursued and the material evaluation is made from the viewpoint of the application. Japanese steel makers have developed high strength and toughness materials for cryogenic use, i.e. high manganese steels for the superconducting magnets of fusion reactors. These materials are characterized by the addition of manganese and nitrogen where the former makes them nonmagnetic and the latter stronger and tougher at cryogenic temperatures. On the other hand, materials with high strength at both the ambient and the cryogenic temperature, are also important for supporting structural members in cryogenic systems. Such materials are precipitated austenitic superalloys, which are originally refractory materials, with high strength and toughness at liquid helium temperature.

REFERENCES

1. REED, R. P. and HOROUCHI, T., *Austenitic Steels at Low Temperatures*, New York, Plenum Press, 1983.
2. OGATA, T. and ISHIKAWA, K., *Tetsu-to-Hagane*, **71** (1985), 1647.
3. ISHIKAWA, K. and OGATA, T., *J. Japan Inst. Metals*, **50** (1986), 28.
4. OGATA, T. *et al.*, *Tetsu-to-Hagane*, **71** (1985), 1390.
5. TIEN, J. K. and YEN, C. T., *Adv. Cryo. Engng. Mater.*, **20** (1984), 319.

17

Cryogenic Fatigue Design of Austenitic Stainless Steels for Superconducting Magnet Applications

JUICHI FUKAKURA, KENICHI SUZUKI and HIDEO KASHIWAYA

Suehiro-cho 2–4, Tsurumi-ku, Yokohama, Kanagawa, Japan

ABSTRACT

The 4 K fatigue life curves on the mechanical failure of domestic structural materials for superconducting magnet applications, such as type 304L and type 316L stainless steels, are obtained under axial strain control. These curves are compared with ones at 300 K and 77 K, and with the literature data on foreign materials. Then, the paper focuses on the permeability increase with cyclic straining, due to martensitic transformation. The 4 K fatigue life curves on the permeability limit of type 304L and type 316L, which are the metastable austenitic stainless steels, are obtained. The fatigue tests on the permeability at 300 K and 77 K are also conducted for comparison. Finally, a cryogenic fatigue design criterion to prevent the mechanical failure and the permeability increase of type 304L and type 316L is proposed referring to the approach in the ASME Boiler and Pressure Vessel Code Section III.

1. INTRODUCTION

Recently, superconducting magnets have been used for experimental nuclear fusion reactors, NMR–CT, various kinds of accelerators and other applications. Support structures of these magnets include liquid He inside them and are subject to various kinds of mechanical and electromagnetic forces. Strength data, however, especially those of fatigue strength of the domestic structural materials at cryogenic temperatures, have scarcely been accumulated in our country. This restricts the development of a structural design standard for cryogenic apparatuses.

On the other hand, in the metastable austenitic stainless steels, which have been generally applied to cryogenic structures, part of the paramagnetic structure changes to ferromagnetic structure by mechanical strain and this tendency is intensified at low temperatures. Since the increase of magnetism is undesirable from the point of view of achievement of high magnetic spatial homogeneity or minimization of mutual magnetic force reactions, inspection of magnetic permeability has been performed for the virgin materials before the application to structures.

In operation of the apparatuses, however, repetition of straining by magnetic force reactions or heating cycles induces martensitic transformation in the material. With respect to the situation described above, the authors have conducted a study to clarify the fatigue properties and magnetic characteristics at cryogenic temperatures of domestic type 304L and type 316L stainless steels[1,2]. In this report, fundamental fatigue and magnetic characteristics are examined from the viewpoint of martensitic transformation by strain cycling.

2. EXPERIMENTAL PROCEDURES

2.1. Materials and Specimens

Testing materials of type 304L and type 316L were supplied as plates of 65 mm thickness. Tensile specimens and fatigue specimens were prepared from the midthickness of the plate with the specimen axis aligned with the rolling direction. Specimen geometries are shown in Fig. 1. Chemical compositions and mechanical properties are summarized in Tables 1 and 2 respectively.

2.2. Tension Test

Tension tests at 4 K were conducted at a crosshead speed of $10^{-3}/s$ in a liquid He cryostat, which was specially designed to make tension tests of four specimens possible at one time. Strain was measured by a strain gauge attached to the specimen for strain $\leq 1\%$, and by an extensometer attached to the specimen for strain $> 1\%$.

2.3. Fatigue Test

Figure 2 shows the cross-sectional view of the cryostat for fatigue testing at 4 K [1]. In the design of the cryostat bath, special attention was paid to maintaining high rigidity of loading columns and preventing heat leak by adopting double cylinder loading columns, by which liquid He

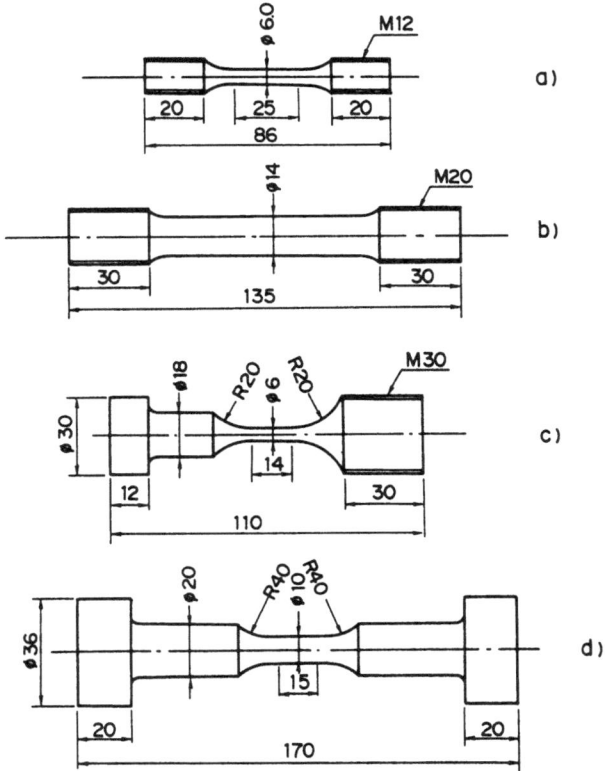

FIG. 1. Specimen configurations; (a) for 4 K tension; (b) for 77 K and 4 K tension; (c) for 4 K fatigue; (d) for 77 K and 4 K fatigue.

TABLE 1
Chemical compositions (wt %)

Material	C	Si	Mn	P	S	Ni	Cr	Mo	N
SUS 304L	0·013	0·61	0·9	0·028	0·008	9·3	18·5		0·07
SUS 316L	0·025	0·47	0·8	0·027	0·01	12·25	16·35	2·12	

consumption was controlled to relatively a low amount of less than 1·5 litre/h. Fatigue testing up to a strain range of 2·5% was successfully possible without specimen buckling. Fully reversed axial strain controlled fatigue testing was conducted in liquid He in a cryostat bath

TABLE 2
Mechanical properties

Material	Temp. (K)	σ_Y (kgf/mm^2)	σ_B (kgf/mm^2)	E_l (%)	RA (%)
SUS 304L	300	23	60	65	75
	77	36	150	45	61
	4	43	174	34	50
SUS 316L	300	22	54	65	73
	77	32	126	49	61
	4	44	147	48	62

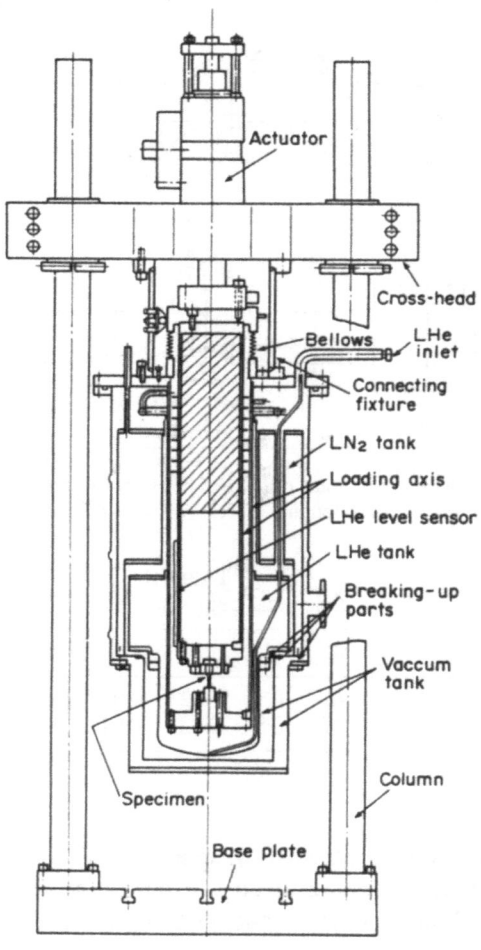

FIG. 2. Cross-sectional view of the cryostat for 4 K fatigue testing.

which was installed in a closed loop electrohydraulic machine of 50 kN loading capacity. Axial strain controlled repeated tension cycling was applied to several specimens to investigate the effect of mean strain on magnetic permeability. Strain measurement and control were based on the output of an extensometer attached to the specimen. The strain versus time wave was a symmetrical triangle with a strain rate of 0·4%/s.

2.4. Magnetic Permeability Measurement

The magnetic permeability of the specimen was measured at room temperature by an instrument made by the Severn Engineering Co., which determines the permeability of the object by comparing its magnetic force to that of a standard specimen.

3. TEST RESULTS AND DISCUSSION OF FATIGUE STRENGTH

3.1. Stress Response in Strain Cycling

Figure 3 shows an example of hysteresis curves of type 304L at 4K. Saw-tooth-like fluctuation of stress was observed in the early stage of cycling. This phenomenon is thought to be due to the intermittent repetition of specimen temperature rise and fall because of the small specific heat of the material at 4 K [3]. This phenomenon, however, disappeared with an increase of the number of cycles, with stress–strain characteristics ap-

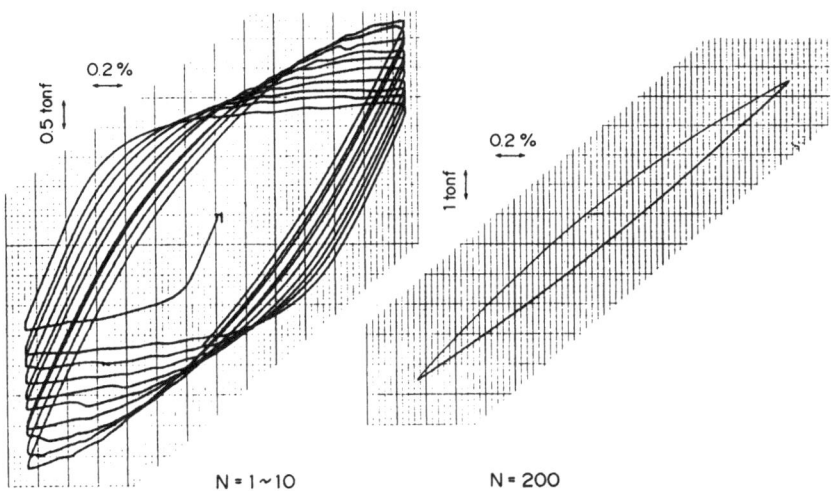

FIG. 3. Hysteresis curves of type 304L at 4 K ($\Delta \varepsilon_t = 2·52\%$).

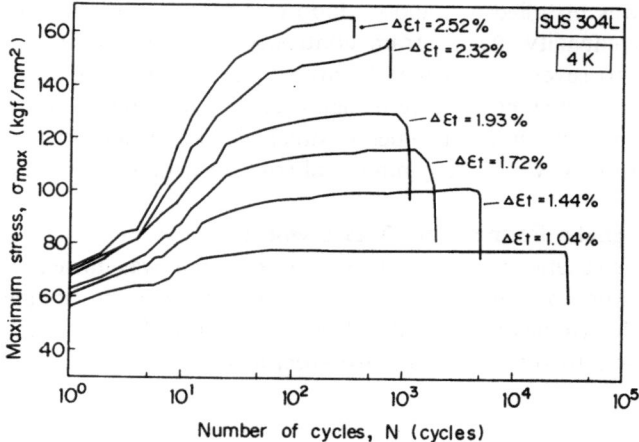

FIG 4. Change of maximum stress of type 304L with number of cycles at 4 K.

proaching elastic behavior by strong strain hardening. This stress fluctuation was not observed in the hysteresis curves at 77 K and room temperature. This tendency was the same in type 316L. Figures 4 and 5 show the appearance of the changes of maximum tensile stress (σ_{max}) of type 304L with strain cycling at 4 K, 77 K and room temperature

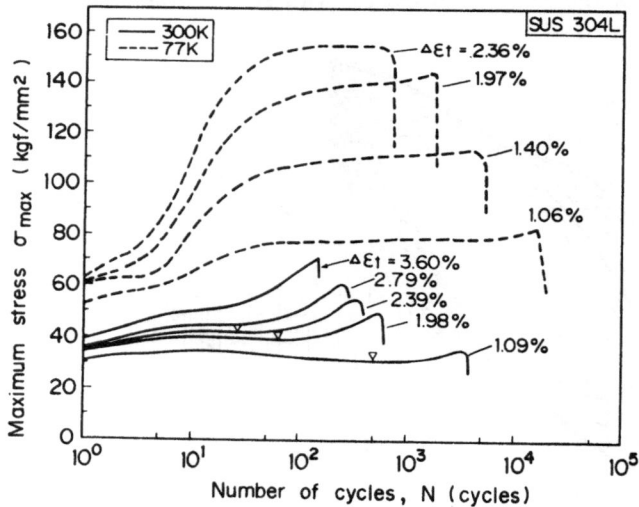

FIG 5. Change of maximum stress of type 304L with number of cycles at 300 and 77 K.

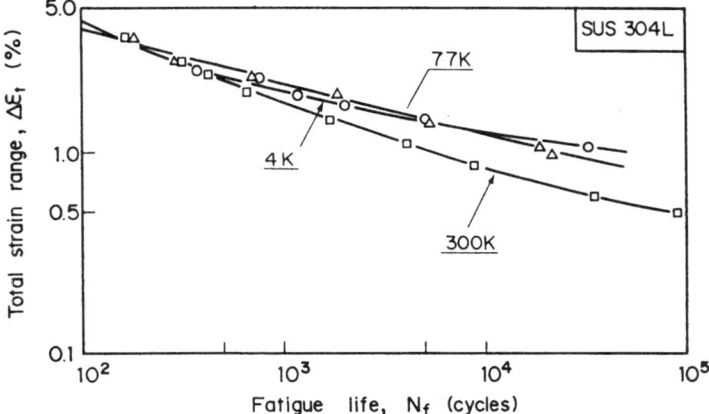

F IG. 6. Fatigue life curves of type 304L on mechanical failure.

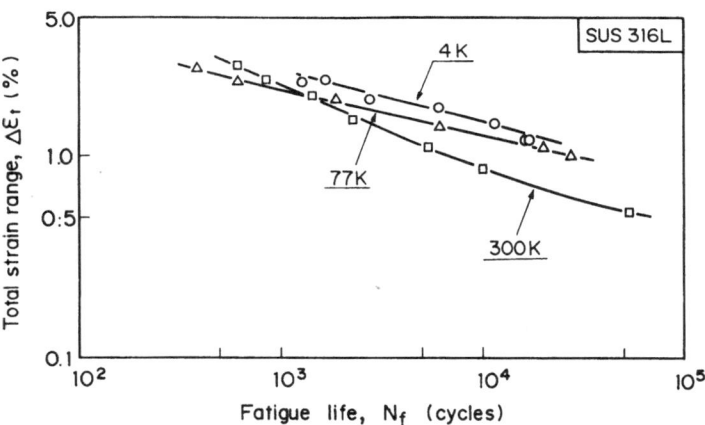

F IG. 7. Fatigue life curves of type 316L on mechanical failure.

respectively, for example. At 4 K and 77 K, σ_{max} increases sharply in the early stage of cycling and then increases gradually, approaching constant values. At room temperature, on the other hand, the increase of σ_{max} is gradual, then σ_{max} decreases, and finally increases again in the last stage of cycling. Behavior of stress decrease and increase is more prominent in type 304L than in type 316L, and the transition point in $\sigma_{max} - N$ curve which is denoted by a symbol in the figure is situated at longer life with a decrease of applied strain range ($\Delta\varepsilon_t$). The final stage hardening is thought to be due to martensitic transformation induced by strain cycling.

3.2. Fatigue Strength

Figures 6 and 7 are the fatigue test results of domestic type 304L and type 316L, respectively, from room temperature to 4 K. Here, fatigue life (N_f) is defined as the number of cycles where σ_{max} decreases to 3/4 of its maximum value. In type 304L, fatigue life curves at 4 K and 77 K are only slightly displaced, and their fatigue strengths are higher than those at room temperature in the long-life range by about 500 cycles. In the case of type 316L, the 4 K strength is higher than the 77 K strength and difference in 4 K and room temperature strength seems to be magnified in long-life range.

4. TEST RESULTS AND DISCUSSION OF MAGNETIC PERMEABILITY

Since magnetic permeability (μ) of type 304L and type 316L was 1·02–1·05 in as received condition, the maximum value of 1·05 was adopted as the representative standard value for the initial state of the specimen prior to testing in the investigation of increase of μ by mechanical straining. Figures 8 and 9 show the μ values of the two kinds of stainless steels after the application of strain cycling at fixed $\Delta\varepsilon_t$ at room temperature. In the case of type 304L, μ exceeds the value of 1·3 after the

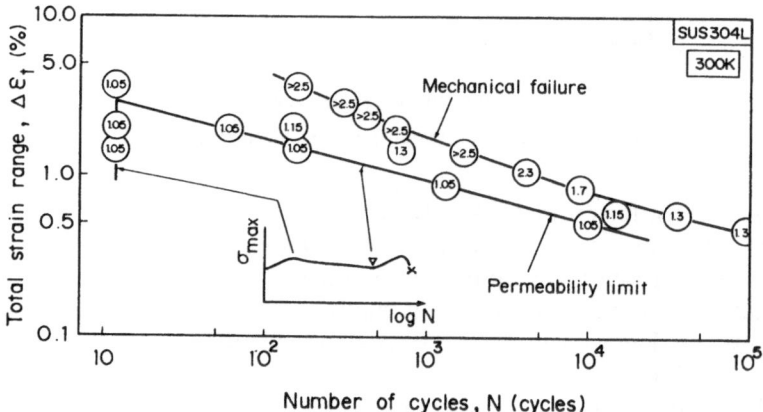

FIG. 8. Permeability change of type 304L with cyclic straining at 300 K.

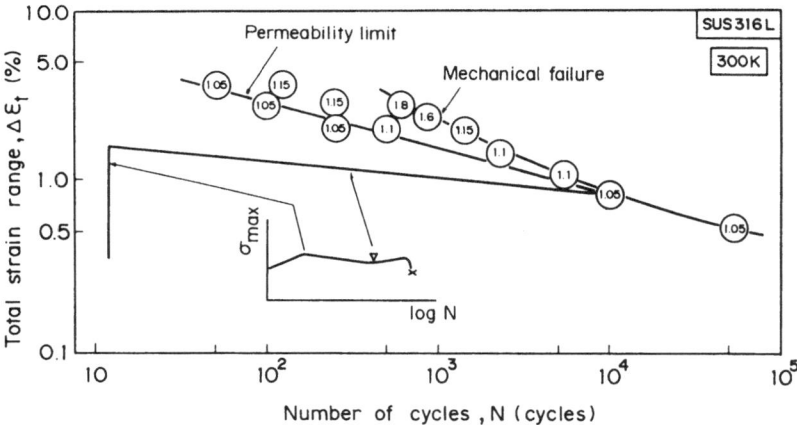

F IG 9. Permeability change of type 316L with cyclic straining at 300 K.

mechanical fatigue failure in type 304L, showing the tendency for μ to decrease with decrease of $\Delta\varepsilon_t$. The relationship between $\Delta\varepsilon_t$ and the transition number of cycles (N_σ) which is illustrated in the insert in the figure is presented as straight line in full logarithmic scale. N_σ increases with decrease of $\Delta\varepsilon_t$ and permeability measurement along this line indicated the μ value of 1·05, showing the permeability increase in the straining process between N_σ and N_f. So the fatigue life for permeability, N_μ, can be defined as the maximum number of cycles up to which no change in μ is observed. At room temperature, $N_\mu = N_\sigma$ in type 304L. The results for type 316L, however, are different, indicating $N_\mu > N_\sigma$.

Test results at 4K on type 304L and type 316L are shown in Figs. 10 and 11 respectively. Although the μ value after mechanical failure of the two steels exceeds the measurement limit of the instrument, 2·5, μ does not change from the virgin value 1·05 up to 10^4 or 2×10^4 cycles in the case of $\Delta\varepsilon_t = 0·3\%$. But a small increment of $\Delta\varepsilon_t$ to 0·4% gives a comparatively large μ of 1·2–1·5 even after 200 cycles. This tendency holds in the case of repeated tension cycling as shown in Fig. 11. As illustrated schematically in the inserts in Table 3, the hysteresis curve of $\Delta\varepsilon_t = 0·3\%$ at 4K is essentially a straight line, without any noticeable plastic strain range ($\Delta\varepsilon_p$) by shakedown to elastic behavior due to the high strain hardening characteristics of the materials, while $\Delta\varepsilon_p = 0$ in the case of $\Delta\varepsilon_t = 0·4\%$. Thus it should be concluded that the magnetic permeability did not increase as long as the specimen remained in elastic deformation behavior and the fatigue life curve for the permeability limit

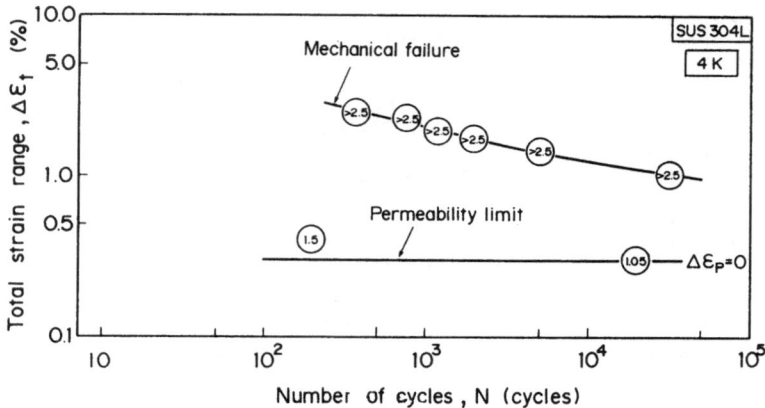

FIG 10. Permeability change of type 304L with cyclic straining at 4 K.

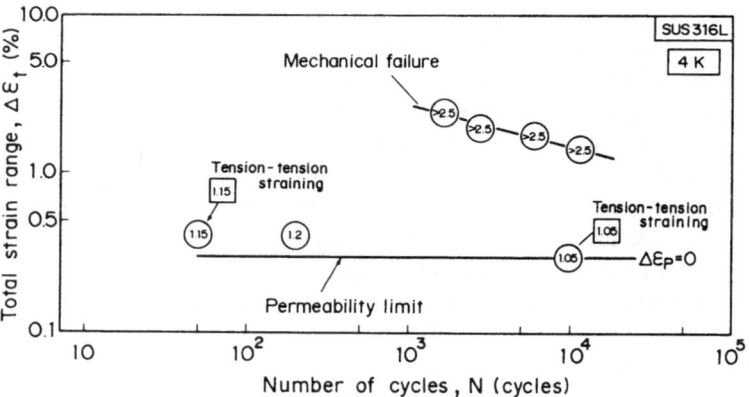

FIG 11. Permeability change of type 316L with cyclic straining at 4 K.

of these materials at 4 K should be regarded as a straight line at $\Delta\varepsilon_t = 0.3\%$, parallel to the abscissa. From Figs. 12 and 13, which show the experimental results of the two steels at 77 K, a horizontal straight line at $\Delta\varepsilon_t = 0.2\%$ is regarded as the fatigue limit for permeability of the steels on which $\Delta\varepsilon_p = 0$ for the hysteresis curve.

Fatigue curves for the magnetic permeability limit of the steels from room temperature to 4 K are summarized in Fig. 14. At room temperature, the fatigue life curves for the two steels are mutually parallel with the curve of type 316L located in the longer life range. Lowering the

TABLE 3
Summary of stress–strain relations

| Material | SUS 304L | | | SUS 316L | | |
Items Temp(K)	300	77	4	300	77	4
S_y (kgf/mm^2)	22	36	43	23	32	44
S_u (kgf/mm^2)	60	150	174	54	126	147
$\frac{2}{3}S_y$ (kgf/mm^2)	15	24	29	15	21	29
$0{\cdot}9S_y$ (kgf/mm^2)	20	32	39	21	29	40
$\frac{1}{3}S_u$ (kgf/mm^2)	20	50	58	18	42	49
ε_A(%)	0·07	0·10	0·11	0·08	0·11	0·12
ε_B(%)	0·08	0·11	0·14	0·08	0·10	0·14
ε_C(%)	0·11	0·17	0·20	0·12	0·15	0·21
ε_0 (%)	—	0·10	0·15	—	0·10	0·15
$\Delta\varepsilon_P$(%)	—	~0·01	~0·04	—	~0·02	~0·04
$\Delta\varepsilon'_P$(%)	—	—	—	—	~0·02	~0·04
E (kgf/mm^2)	19 500	21 000	21 000	19 500	21 000	21 000

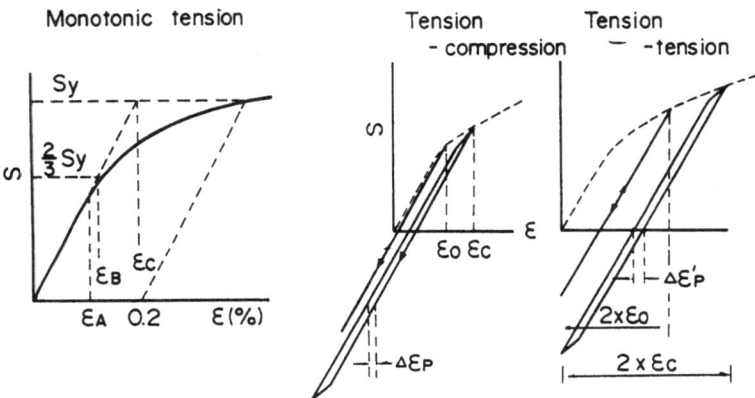

temperature to 77 K and 4 K, fatigue limits are expressed by horizontal lines of $\Delta\varepsilon_t = 0\cdot2\%$ and $\Delta\varepsilon_t = 0\cdot3\%$, respectively. The large strain limit for permeability at 4 K compared to that at 77 K will be attributed to the increase of $\Delta\varepsilon_t$ for shake down limit ($\Delta\varepsilon_p = 0$) by the increase of mechanical strength at low temperature.

Previous research [4–6] has investigated the ferromagnetic martensitic transformation in metastable austenitic stainless steels by mechanical working in the temperature range from LNG temperature (111 K) to 4 K, in relation to their tensile deformation characteristics at low tempera-

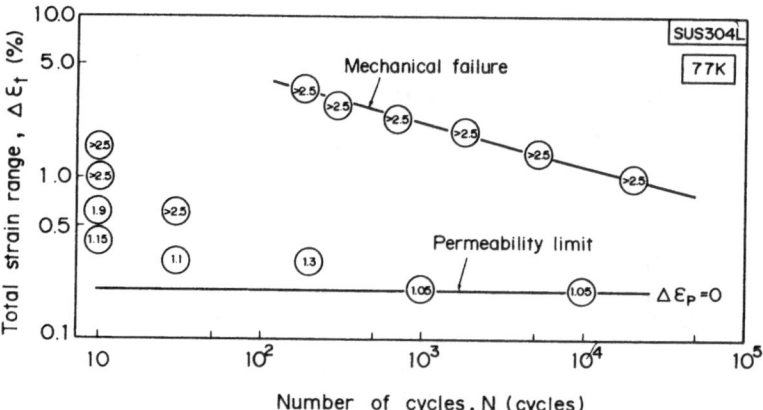

FIG. 12. Permeability change of type 304L with cyclic straining at 77 K.

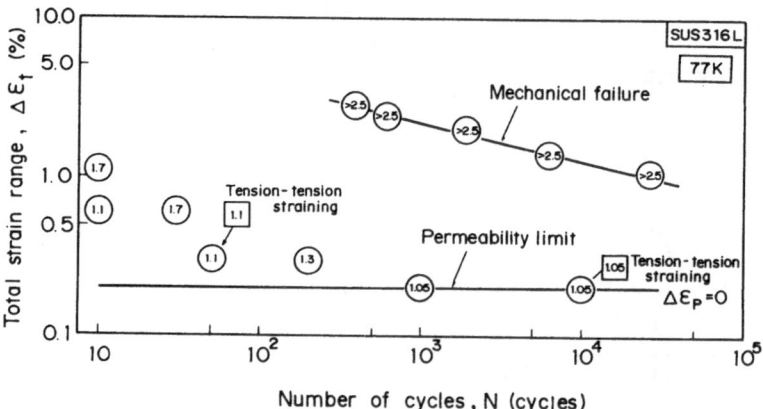

FIG. 13. Permeability change of type 316L with cyclic straining at 77 K.

tures. Although some papers discussed the effect of martensitic transformation, which is induced by the repetition of mechanical straining, on stress amplitude or mechanical fatigue life [7–9], relatively little data at cryogenic temperatures have been reported [7].

In this report, the fatigue life curves for the permeability limit were established for type 304L and 316L steels regarding the change of μ to be related to the amount of transformed martensite. As a result, it was confirmed that the upper bound of $\Delta\varepsilon_t$ in which $\Delta\varepsilon_p = 0$ corresponded to the permeability limit in strain cycling at 4 K and 77 K. This means that martensitic transformation occurs simultaneously with dislocation mul-

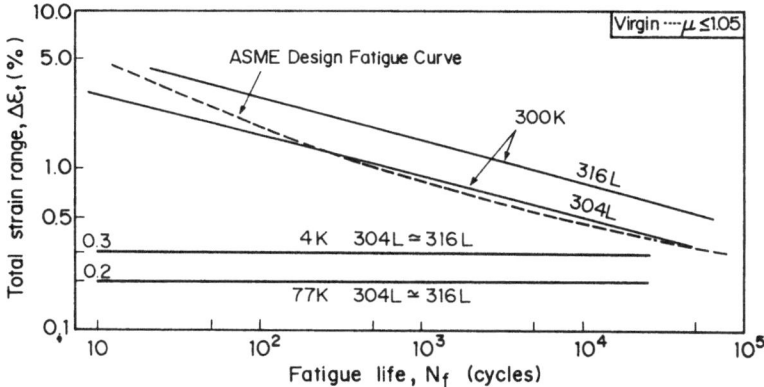

F ɪ ɢ. 14. Fatigue life curves on permeability limit ($\mu \leq 1.05$).

tiplication in plastic strain cycling. Since dislocation slip is inhibited by the formation of martensitic structure in austenitic structure, the materials begin to harden from the early stage of strain cycling as shown in Figs. 4 and 5. At room temperature, however, μ was unchanged till N_μ even in the case of plastic strain cycling. There were the cases where σ_{max} increases and decreases gradually with strain cycling at room temperature, and this behavior would be attributed to the dislocation multiplication and rearrangement in the austenitic phase. In the case of type 304L, σ_{max} increases again by inhibition of dislocation rearrangement by the martensite formation. A change of magnetic permeability could not be found, however, in type 316L at room temperature for a while after the beginning of last stage hardening. From the fact that the last stage hardening of this material was slight compared to that of type 304L, two possibilities may explain why the increase in μ was not detected until some amount of hardening. One is the limitation of μ measurement precision, and another is the possibility that the early stage martensitic transformation is to a paramagnetic form.

5. FATIGUE DESIGN OF STRUCTURAL MATERIALS FOR SUPERCONDUCTING MAGNETS

5.1. Design for Protection against Mechanical Failure

The ASME Boiler and Pressure Vessel Code Section III, which is one of the systematized structural design codes, has been tried for the design of the structures of superconducting magnets. This code gives the

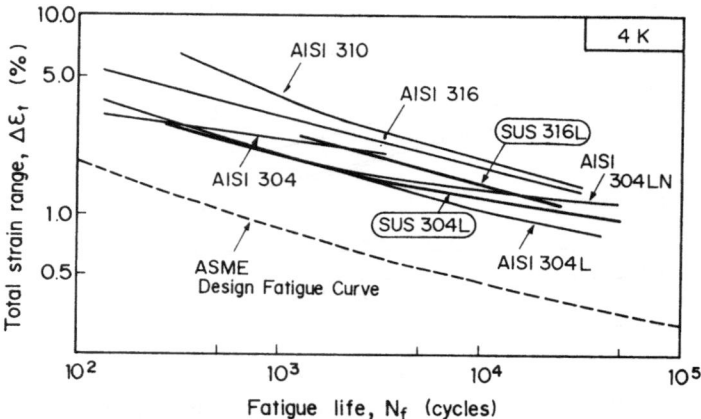

FIG. 15. Fatigue life curves on mechanical failure of various austenitic stainless steels at 4 K.

specified yield strength (S_y) and tensile strength (S_u) for austenitic stainless steels in the temperature range from 100 to 800°F. In the code the stress limitation of the range of the primary stress intensity + secondary stress intensity (S_n) should not exceed $3 S_m (= 2 S_y)$, where S_m is the allowable stress intensity, to protect from the low cycle fatigue failure of the components by ensuring that they shake down to elastic behavior. Concerning local structural discontinuity, fatigue analysis must be performed on the basis of the design fatigue curve in the code which includes the design margin of 20 and 2 on life and stress amplitude respectively, against the best fit curve of a large amount of fatigue test data in the temperature range from 100 to 800°F. Consequently, fundamental mechanical strengths and the design fatigue curve at 4 K and 77 K of the materials would be required in the first place to design the structures for superconducting magnets according to the procedures of ASME Code Sec. III. Although considerable data are available for static mechanical properties of austenitic stainless steels such as type 304L and type 316L at cryogenic temperatures, strain controlled fatigue data are rare at cryogenic temperatures and especially at 4 K. It is difficult to form a design fatigue curve from the data. Therefore, to examine the possibility of expanding the existing design fatigue curve to design at cryogenic temperatures, the design fatigue curve in the present code was compared with the $\Delta\varepsilon_t - N_f$ curves at 4 K and 77 K of various austenitic stainless steels in Figs. 15 and 16. The application of the design fatigue curve to

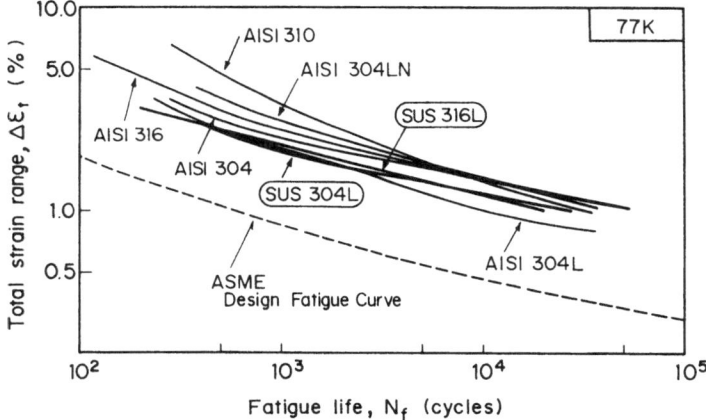

F IG. 16. Fatigue life curves on mechanical failure of various austenitic stainless steels at 77 K.

the data at cryogenic temperatures seems to be adequate, showing the design margin above 15 and 2 on life and stress amplitude respectively even for the minimum strength trend curve in the $\Delta\varepsilon_t - N_f$ curves in the figures. The accumulation of fatigue data at cryogenic temperatures is required to extract the design fatigue curve to make the most of the advantage of fatigue strength in the high cycle range.

In Section III of the ASME Code, the allowable design stress intensity S_m is determined from the yield strength and tensile strength at room temperature or design temperature. So S_m depends on the tensile properties at design temperature for the high temperature components, because mechanical strengths generally decrease with increase of temperature. If this design philosophy is applied in the design of cryogenic apparatuses, the advantage of the increase of mechanical strength at cryogenic temperature is not put to use since the S_m value is essentially determined by room temperature mechanical properties. Consequently, it is rational for S_m to be determined by the strength at design temperature.

5.2. Design for Protection of Magnetic Permeability

5.2.1. *The cause of magnetic permeability increase*

Generally, 4 cases are considered as possible causes of the permeability increase in austenitic stainless steels as shown in Fig. 17. (I) is permeability increase by cooling down with the peak value attained at the

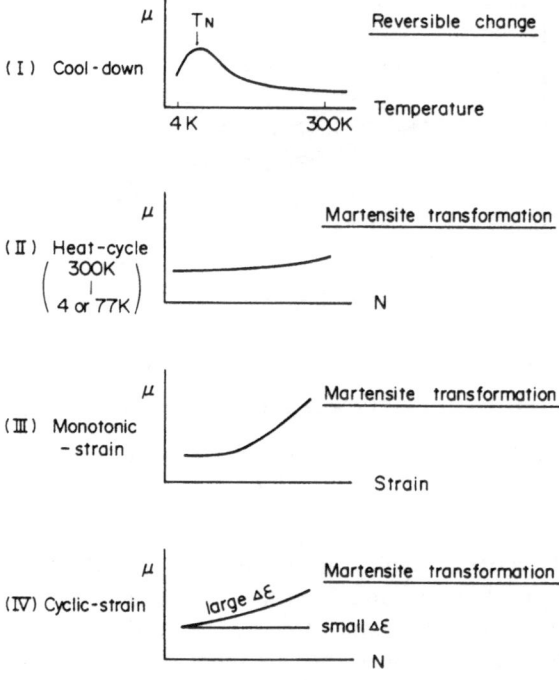

FIG. 17. Factors for permeability change.

temperature called the Neel point between 4 K and 60 K. This is a reversible process and μ recovers to the original value by heating the material. The permeability increase represented in (II)–(IV) is observed only in metastable austenitic stainless steels which in each case is due to the irreversible martensitic transformation. (II) shows the μ increase by martensite formation by heat cycling. There is a report [5] describing how the amount of martensite saturates with number of cycles. In (III) and (IV), the permeability increase is attributed to martensitic transformation by deformation and there seems to be some limiting value of strain below which martensitic transformation is inhibited as described above.

5.2.2. Stress limitation for protection against permeability increase

When type 304L and type 316L are adopted as structural materials for superconducting magnets, designers should take the effects of the μ increase into consideration. Although permeability increase by mechanisms (I) and (II) is generally controlled by chemical compositions of the materials, some stress limitation should be required on the stress caused by electromagnetic force, thermal deformation, etc., to protect against permeability increase. The permeability increase by the mechanisms (III) and (IV) was due to the martensitic transformation by monotonic or cyclic strain.

As shown in Fig. 14, the critical strain range for permeability at 4 K and 77 K is far smaller than the strain of the fatigue design curve for mechanical failure. Thus it is a severe limitation for cryogenic structures to restrain their permeability to increase by limiting the strain range to 0.2% or 0.3% even at points of local strain concentration. So it will be realistic for structural design to allow the permeability increase at local structural discontinuities in principle because such local increase should not give so much disturbance to the magnetic field. In ASME Code Section III, on the other hand, the stress limit of $S_n \leqq 2S_y (=3S_m)$ is prescribed to ensure that shake down occurs for the purpose of inhibition of low cycle fatigue failure. Although this stress limit is synonym-

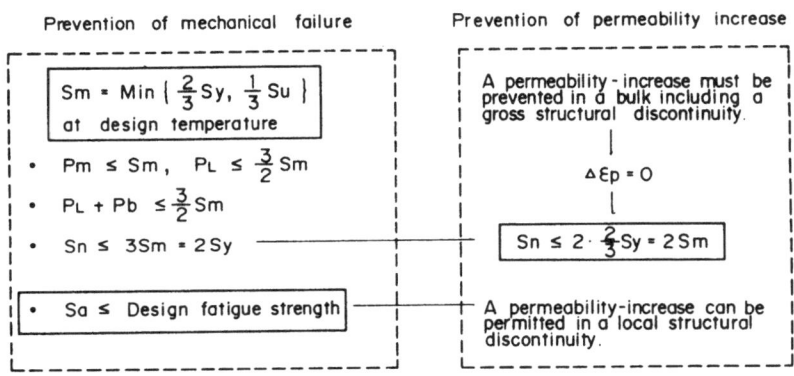

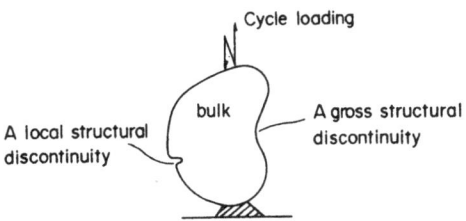

FIG. 18. Cryogenic fatigue design criterion on mechanical failure and permeability limit for type 304L and type 316L steels.

ous with a strain limit of $\Delta\varepsilon_t < 2\varepsilon_c$ where ε_c is defined as the strain corresponding to fictitious elastic stress S_y as illustrated in the insert in Table 3, magnetic permeability increases for strain amplitude of ε_c in both reversed and repeated strain cycling because the specimen does not show complete elastic behavior for ε_c. When ε_0 is defined as the upper strain amplitude in which $\Delta\varepsilon_p = 0$, ε_0 is almost the same for both strain cycling modes and $2\varepsilon_0$ is 0·3% at 4 K and 0·2% at 77 K for the two steels as described previously. Comparing ε_0 values with the mechanical properties of the steels in Table 3, ε_0 is nearly the same as, or little larger than, the proportional limit strain ε_A, and it also coincides with ε_B corresponding to the stress of 2/3 $S_y(= S_m)$. So the stress limit $S_n \leqq 2S_m (\Delta\varepsilon_t \leqq 2\varepsilon_B)$ gives the condition of inhibition of permeability increase of the gross structures in strain cycling, keeping the structure to elastic behavior. The basic design concept for protection against mechanical failure and magnetic permeability increase is summarized in Fig. 18.

Generally, the chemical compositions of steel vary depending on the material suppliers or material charges and this has a delicate effect on the stability of austenitic structure. The design stress limitations above were obtained for typical austenitic stainless steels of type 304L and type 316L from only one charge each; further research should be required to establish the general guideline of structural design of cryogenic materials.

6. CONCLUSIONS

The results of a study on general cryogenic structural materials of type 304L and type 316L to develop the cryogenic fatigue design concept taking the magnetic permeability limitation into consideration may be summarized as follows:

(1) The design fatigue curve in current ASME Code Section III possesses sufficient margin for mechanical failure of the test materials, type 304L and type 316L, and other austenitic stainless steels at 4 K and 77 K.

(2) It is reasonable to adopt the mechanical properties at design temperature for the determination of allowable design stress S_m in order to make good use of the advantage of high strength at cryogenic temperatures, presupposing the extended applications of ASME Code Section III to cryogenic structural design.

(3) The permeability increase is suppressed in strain cycling for strain ranges below 0·3% and 0·2% at 4 K and 77 K respectively, which correspond to the upper values to keep the plastic strain range equal to zero. Consequently, it is clarified that the permeability increase is inhibited by the stress limit $S_n = 2S_m$ for the gross structures, thereby forcing the plastic strain range to zero.

REFERENCES

1. SUZUKI, K. *et al.*, *J. Soc. Mater. Sci., Japan*, **34**, No. 385 (1985), 1206.
2. SUZUKI, K. *et al.*, *Proceedings of US–Japan Workshop on Low Temperature Structural Materials and Standard*, 1986, Sendai.
3. FICKETT, F. R. and REED, R. P., NBSIR78-884, 1978, p. 15.
4. VOYER, R. *et al.*, *Advanced Engineering in Cryogenics*, 1978.
5. LARBALESTIER, D. *et al.*, *Cryogenics* (1973), 160.
6. FUKUSHIMA, E. *et al.*, *J. Japan Inst. Metals*, **36-3** (1972), 195.
7. FUKUSHIMA, E, *et al.*, *J. Japan Soc. Strength and Fracture of Metals*, **10-1** (1975), 20.
8. HENNESSY, D. *et al.*, *Metallurgical Transactions* A, **7A** (1976), 415.
9. BAUDRY, G. *et al.*, *Mater. Sci. Engng*, **28** (1977), 229.

18

Fatigue Behavior of High-manganese Steel at Cryogenic Temperature*

T. MIZOGUCHI, S. YAMAMOTO, J. TSUKUDA** and M. AKAMATSU**

Applied Mechanics Center, Kobe Steel Ltd., Wakinohama-cho 1-3-18, Chuo-ku, Kobe, Japan.
***Mechanical Engineering Center, Kobe Steel, Kobe, Japan.*

ABSTRACT

The results of a research project on fatigue properties of high-manganese steels for cryogenic use under the sponsorship of Science and Technology Agency of Japan are summarized. The temperature dependence of cyclic deformation, low cycle fatigue life, notch sensitivity and near threshold fatigue crack growth rate are discussed.

1. INTRODUCTION

Structural materials for high field superconducting magnet devices such as fusion reactors will be exposed to huge magnetic forces and thermal stresses that will be repeated for each start-up and shut-down operation. Therefore it is necessary to determine the fatigue behavior of these materials over a wide range of temperature from ambient to liquid helium not only from the standpoint of materials developments but also from that of fatigue design and quality assurance of structures.

Most of the previous work on fatigue at liquid helium temperature has been restricted to the short-life region mainly because of the high running costs of the tests [1–6], and there is a lack of important design data such as $S-N$ curves in high-cycle regions or near-threshold crack growth rates at test frequencies sufficiently slow to avoid a temperature rise in the specimens. In order to provide these data, the authors developed a testing apparatus with a recondensing type helium re-

*Work performed through Special Coordination Funds of the Science and Technology Agency of the Japanese Government.

273

frigeration system which can conduct fatigue tests for more than hundreds of hours without an additional supply of liquid helium [12].

As pointed out by some previous investigators, the high-cycle fatigue strength of smooth specimens made of FCC or HCP alloys usually increases as the temperature decreases [1, 7]. However, the effect of notches on cryogenic fatigue, one of the essential factors in fatigue performance analysis of structures, has not been determined.

Much attention has been focussed recently on the discussion of fatigue crack growth at cryogenic temperatures from the standpoints of correlations with conventional mechanical properties, and also from metallurgical and microstructural aspects such as stability of the austenite phase [6, 8, 9, 11, 13] but most of these discussions are restricted to higher ΔK or higher growth rate regions. Less data for near-threshold fatigue crack growth rate have been obtained.

The recent development of large superconducting magnets has created new needs for alloys which are cheaper, stronger, and have a more stable austenite phase compared to the usual cryogenic steels. In order to satisfy the needs, high-manganese austenitic steels have been developed in Japan. However, no systematic work has been done on fatigue behavior of high-manganese steels.

With this background, emphasis will be placed on the cyclic deformation behavior, strain controlled low-cycle fatigue life, notch sensitivity, and near-threshold fatigue crack growth with respect to crack closure of a high-manganese steel at cryogenic temperature.

2. MATERIALS AND EXPERIMENTAL PROCEDURES

2.1. Materials Tested

Table 1 shows the chemical compositions of a high-manganese steel of nominal composition 22Mn–13Cr–5Ni and an austenitic stainless steel 304LN which were investigated in this study.

In the case of 22Mn steel, the ingot was hot-rolled to a 70-mm-thick

TABLE 1
Alloy compositions (wt%)

Material	C	Si	Mn	P	S	Ni	Cr	N	Nb
22Mn–13Cr–5Ni	0·04	0·34	21·82	0·013	0·004	4·94	12·84	0·212	
SUS 304 LN	0·03	0·61	1·70	0·021	0·001	11·52	19·06	0·14	0·046

TABLE 2
Mechanical properties of alloys

Material	Temp. (K)	Yield stress (MPa)	Tensile strength (MPa)	Elongation (%)	Reduction of area (%)	VE (J)	K_I (MPa$\sqrt{m}$)
22Mn Steel DWT	300	358	678	63	79	343	
	77	897	1437	61	54	152	
	4	1240	1659	37	49		189
22Mn Steel ST	300	298	627	70	78	419	
	77	785	1331	51	40	189	
	4	1145	1558	39			228
SUS 304LN ST	300	303	630	56	78		
	4	937	1585	56	61		232

plate. One part of the plate was water quenched directly after rolling. This process is referred to as Direct Water Toughening (DWT) hereinafter. The other part was solution treated followed by water quenching. This heat treatment is referred to as ST.

The mechanical properties for each material are given in Table 2.

2.2. Specimens

The specimen configuration used for strain controlled low-cycle fatigue tests is shown in Fig. 1a. Figures 1b and 1c show the smooth and notched specimens for the fully reversed high-cycle axial fatigue tests. Four kinds of specimens containing a circumferential V groove notch with different root radius ρ and stress concentration factor K_t as shown in Fig. 1c were prepared. The most mild notch had a ρ of 1·5 mm and K_t

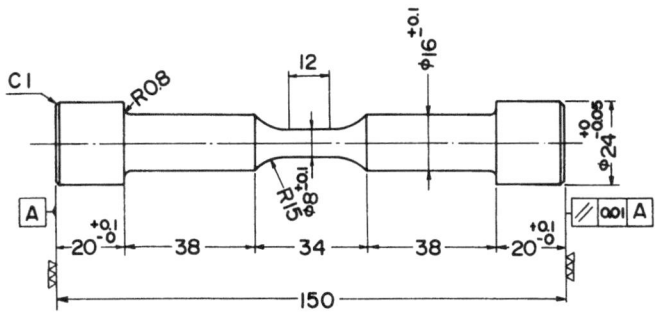

FIG. 1(a). Low cycle fatigue specimen.

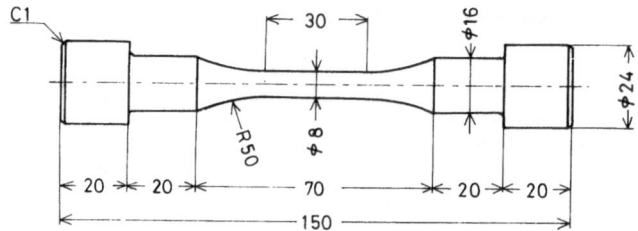

FIG. 1(b). Smooth specimen for high cycle fatigue.

Details of A

$\rho = 1.5$	$K_t = 1.35$
0.3	2.91
0.1	4.98
0.05	7.08

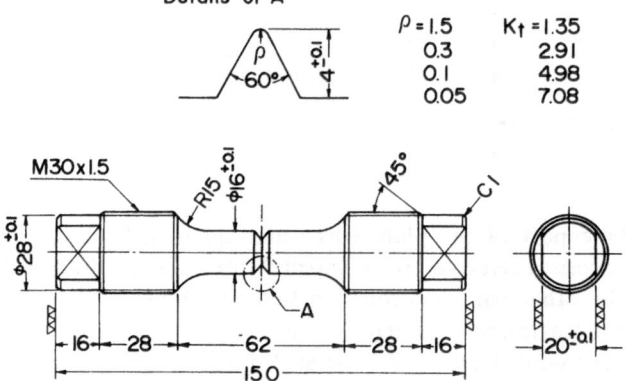

FIG. 1(c). Notched specimen for high cycle fatigue.

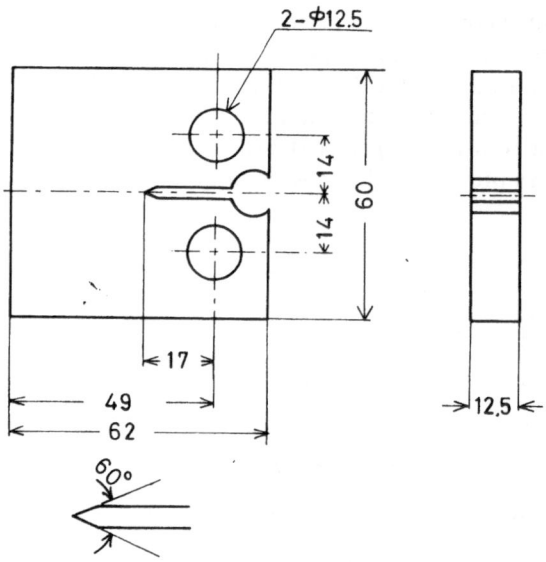

FIG. 1(d). CT specimen for fatigue crack growth test.

of 1·35. The most severe one had a ρ of 0·05 mm and K_t of 7·08. Figure 1d shows the CT specimen for fatigue crack propagation tests.

2.3. Experimental Test System

The fatigue tests were carried out by using a newly developed testing apparatus [12] the overall scheme of which is shown in Fig. 2. Liquid helium in the closed loop of a helium refrigeration system passes through recondensors which are attached to each cryostat of the three testing machines. Vaporized helium in the cryostats can be liquefied by the heat

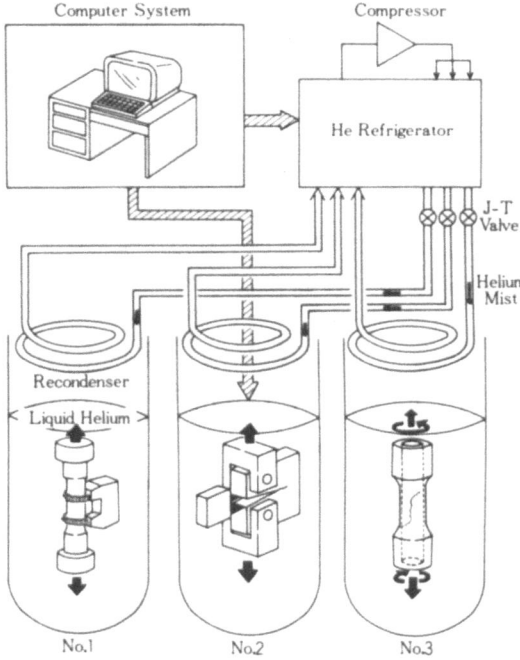

Fɪɢ. 2. Scheme of fatigue testing system.

exchanger on the surface of the recondensors. This system is controlled by a computer so that the warming and recooling operations of each cryostat do not disturb the refrigerated status of the rest of the system.

The loading mechanism and the cryostat with a specimen mounted in the load train is illustrated in Fig. 3. This apparatus made it possible to carry out fatigue tests for more than hundreds of hours without an additional supply of liquid helium.

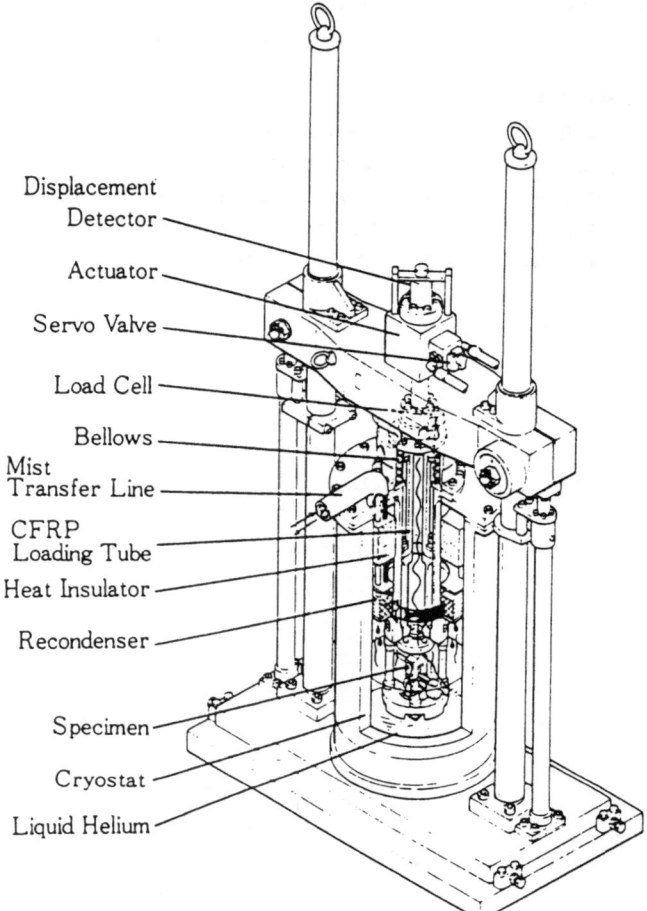

Displacement
Detector

Actuator

Servo Valve

Load Cell

Bellows

Mist
Transfer Line

CFRP
Loading Tube

Heat Insulator

Recondenser

Specimen

Cryostat

Liquid Helium

FIG. 3. Loading mechanism and cryostat.

2.4. Procedures

In the strain controlled low-cycle fatigue tests, cycling was performed under conditions of constant total strain amplitude at a strain rate of $5 \times 10^{-3} \, s^{-1}$ with a completely reversed tension–compression triangular wave.

Longitudinal strain was measured by a strain gauge type extensometer attached to the specimen by using two springs.

Tension–compression high-cycle fatigue tests up to 2×10^6 cycles at liquid helium temperature and up to 1×10^7 cycles at ambient temperature were carried out for smooth and notched specimens.

For smooth specimens, the test frequency was varied from specimen to specimen according to the applied stress level in order to avoid a temperature rise. The maximum frequency was set at 5 Hz. For notched specimens, all tests were performed at a frequency of 5 Hz.

A computerized crack length measuring and ΔK controlling system was used in the fatigue crack growth rate (FCGR) tests. The crack length was defined by the compliance computed from the load and displacement data sampled at a rate of 100 points per cycle. The method was found to be accurate within ± 0.4 mm in crack length.

Two test methods, continuously decreasing ΔK and constant load range were applied to get wide range FCGR data from near-threshold to a higher ΔK region. During the continuously decreasing ΔK test, the load-shedding schedule was controlled in accordance with the relationship

$$\frac{1}{\Delta K} \frac{\mathrm{d}\Delta K}{\mathrm{d}a} = -0.2 \qquad 1/\mathrm{mm}$$

where a is the instantaneous crack length.

In order to determine the effects of test frequency and stress ratio R on FCGR, three frequencies of 1, 5 and 20 Hz and two R values of 0.1 and 0.7 were applied.

Crack closure at the crack tip during the fatigue crack growth test was measured to evaluate the influence of temperature, stress ratio, and residual stress of specimens of FCGR.

3. TEST RESULTS AND DISCUSSION

3.1. Cyclic Deformation and Low-cycle Fatigue Life

Stress–strain hysteresis loops obtained in a certain range of strain rates were serrated with small variations in stress at liquid helium temperature. Figure 4a shows the strain rate dependence of the serrations for the 22Mn DWT steel. The serrations were not observed for strain rate below 1.7×10^{-4} s^{-1} and above 3.5×10^{-3} s^{-1} at a total strain amplitude of 1.35%.

The cause for serrations may be presumed to be unstable localized slip events due to localized heat generation which is induced as a result of the decrease in the thermal conductivity and the specific heat at cryogenic

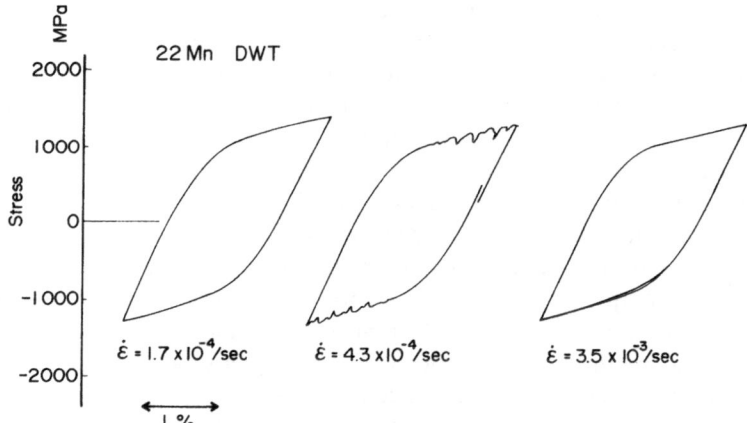

F IG. 4(a). Stress–strain hysteresis loop and serration: 22 Mn steel.

temperature. At a strain rate below 1.7×10^{-4} s^{-1}, the material may be thermally in equilibrium, and as a result, localized heat generation and slip may not take place.

In case of SUS 304LN, the results differ somewhat from those of 22Mn steel as shown in Fig. 4b. The upper limit of the strain rate where the serrations are observed is also 3.5×10^{-3} s^{-1}. However, serrations are observed even though the strain rate is reduced to less than 5.4×10^{-5} s^{-1}. Below this strain rate, the experimental data are not

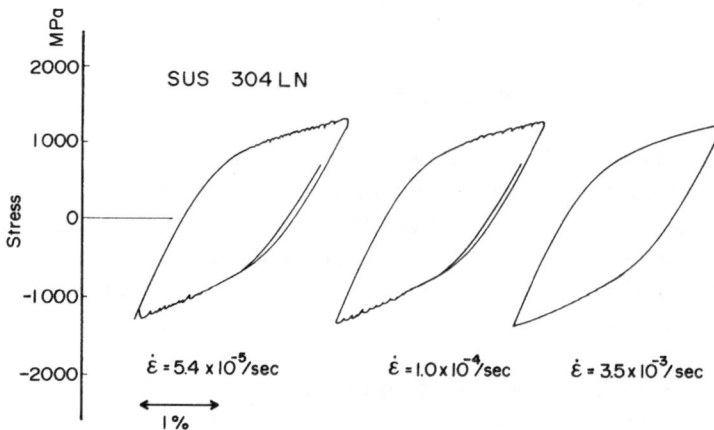

F IG. 4(b). Stress–strain hysteresis loop and serration: SUS 304LN.

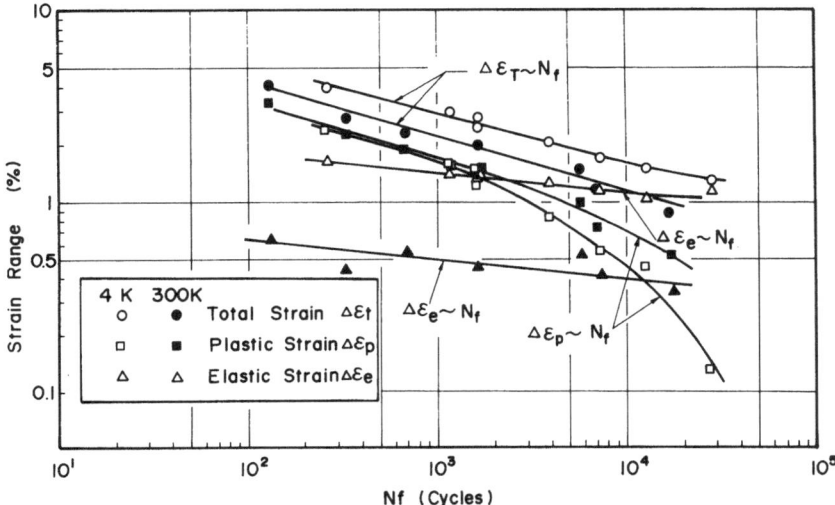

FIG. 5. Effect of temperature on low-cycle fatigue life.

available. The height of stress variations in the serrated stress–strain curves is smaller and the number of them is larger than those of 22Mn steel.

A slight cyclic hardening was observed in the first 10 cycles of a 304LN specimen at a strain amplitude and strain rate of 1.35×10^{-2} and $5 \times 10^{-3} \, s^{-1}$ respectively at liquid helium temperature. No initial cyclic hardening but slight softening was observed in 22Mn steels.

Figure 5 indicates the effect of temperature on low-cycle fatigue life of 22Mn DWT steel. In the relation of total strain range versus life, specimens at liquid helium temperature show longer life as compared to those at ambient temperature. However, no difference is observed in plastic strain range versus life curve between two temperatures in the range up to 2×10^3 cycles. It must be pointed out that the elastic strain range decreases at liquid helium temperature. This means that below 2×10^3 cycles, the fatigue life is controlled by plastic strain range and the difference in total strain versus life comes from the difference in the cyclic stress–strain relationship at ambient and liquid helium temperature. In the life range above 2×10^3, the plastic strain range versus life curve at liquid helium temperature has a tendency to bend downward.

Figure 6 compares the total strain range controlled low-cycle fatigue lives of three steels investigated. The shaded region of Fig. 6 summarizes

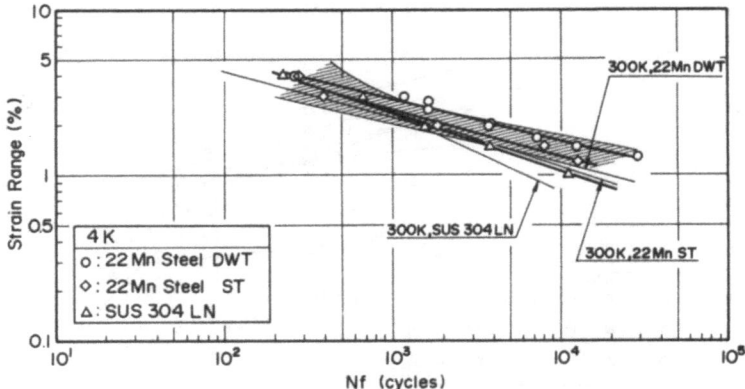

FIG. 6. Comparison of low-cycle fatigue lives.

the data of austenitic steels at 4·2 K by other investigators. Among three steels, 22Mn DWT indicates longest life at temperatures of both air and liquid helium. The total strain-life curve of this steel also locates near the upper limit of the scatter band of other austenitic steels.

3.2. High-cycle Fatigue Strength and Notch Factor

Figure 7 shows a comparison of S–N curves of smooth and notched

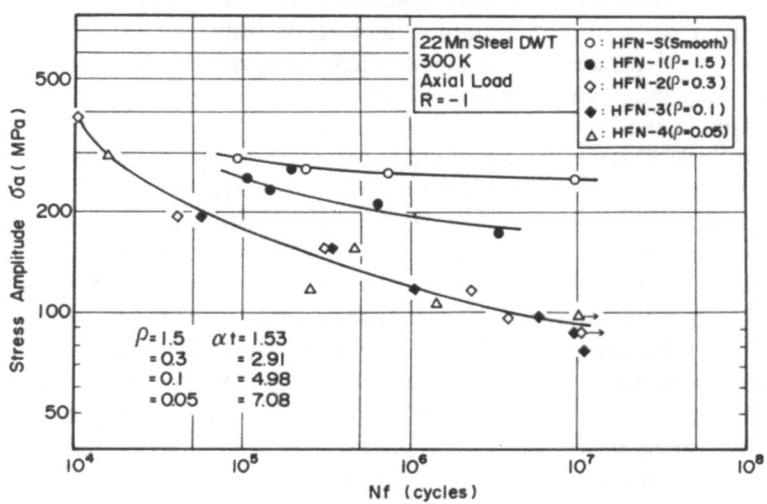

FIG. 7. S–N curves at room temperature.

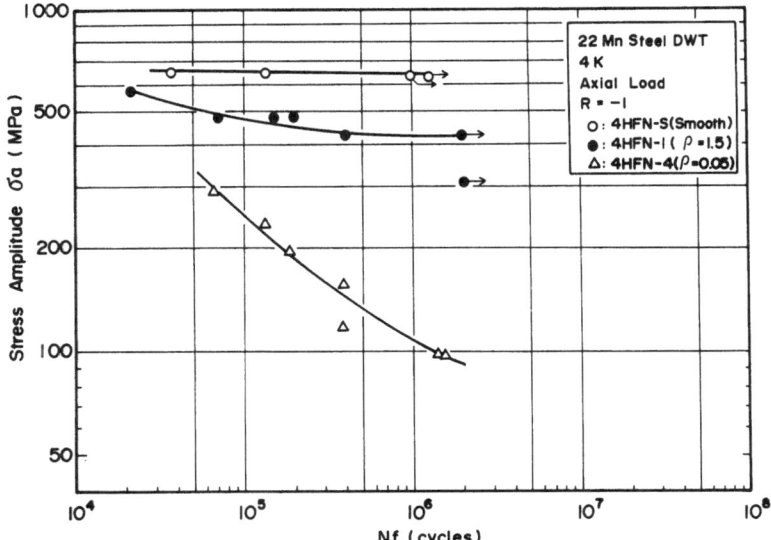

F IG. 8. *S–N* curves at liquid helium temperature.

specimens of 22 Mn DWT steel at room temperature. The fatigue strength of smooth specimens at 10^6 cycles was estimated to be 255 MPa. At room temperature, increasing the stress concentration factor K_t from 1 to 2·91 decreases fatigue strength. However, for K_t greater than 2·91, the experimental data are plotted in the same master curve independent of K_t or notch root radius.

Figure 8 shows the S–N curves at liquid helium temperature. In case of smooth specimens, decreasing the temperature increases the fatigue strength at 10^6 cycles to 637 MPa, whereas the specimens with a severe notch do not show an increase in the fatigue strength. Figure 9 indicates the relationship between the fatigue strength at 10^6 cycles and the temperature for each notch configuration.

Figure 10 demonstrates the connection between K_t versus K_f (fatigue notch factor) relations and temperature. At 300 K, the K_f value does not exceed 2·2. But, at cryogenic temperatures, K_f takes higher values.

Fatigue crack initiation due to the cyclic plastic deformation may be a dominant factor governing the fatigue strength of smooth specimens. Therefore the low temperature that increases the plastic deformation resistance leads to high fatigue strength of smooth specimens. On the other hand, fatigue crack growth may play an important role in the

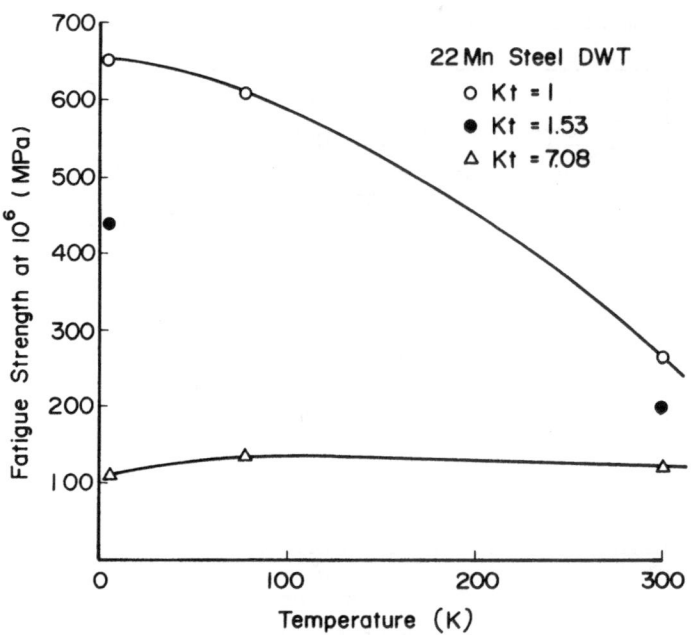

FIG. 9. Effect of temperature on fatigue strength at 10^6 cycles.

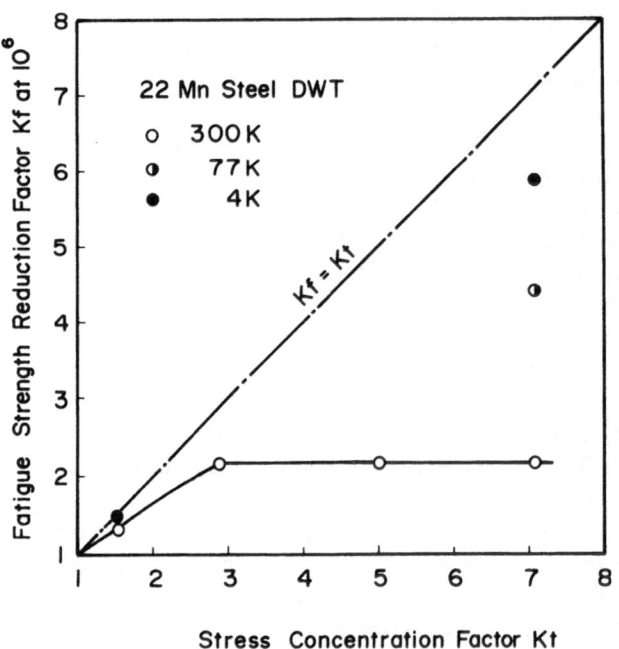

FIG. 10. Effect of temperature on notch sensitivity.

specimen with a severe notch. The temperature dependance of FCGR of 22Mn steel is not as pronounced as shown later. This may be a reason for the weak temperature dependency of the high-cycle fatigue strength of notched specimens.

3.3. Fatigue Crack Growth Behavior

The fatigue crack growth rates (FCGR) at a stress ratio of 0·1 from near-threshold to higher ΔK region of 22Mn DWT steel at room and liquid helium temperature are illustrated in Fig. 11. The shaded area of the upper right hand side indicates the general data trend of austenitic stainless steels summarized by Tobler and Yi-Wen Cheng [9]. FCGR curves of 22Mn DWT are located in the middle of the aggregate of the data of other austenitic steels.

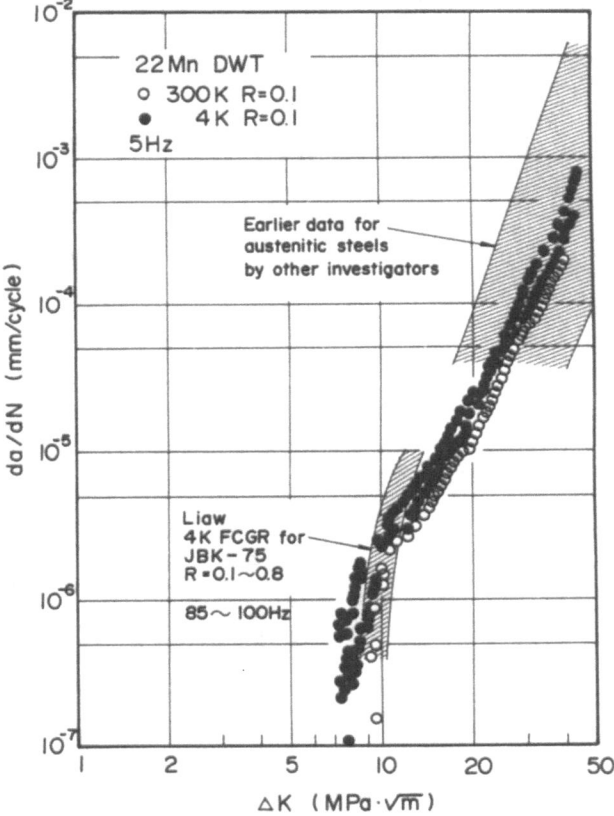

FIG. 11. Effect of temperature on FCGR.

It seems that decreasing the temperature increases FCGR slightly both at the near threshold condition and the higher crack growth rates. This result seems to be opposed to that of JBK-75 steel presented by Liaw *et al.* [11]. But it must be kept in mind that FCGR curves of 22Mn DWT scatter considerably.

FCGR tests at liquid helium temperature under the conditions of 5 Hz in frequency and 0·1 in stress ratio were repeated three times. The results for ΔK_{th} vary from specimen to specimen. The minimum ΔK_{th} is 7·5 MPa$\sqrt{m}$ and the maximum is 8·9 MPa$\sqrt{m}$. In the case of room temperature, the minimum ΔK_{th} in two data is 4·5 MPa$\sqrt{m}$ and the maximum is 9·3 MPa$\sqrt{m}$. The reason why such scatter takes place may

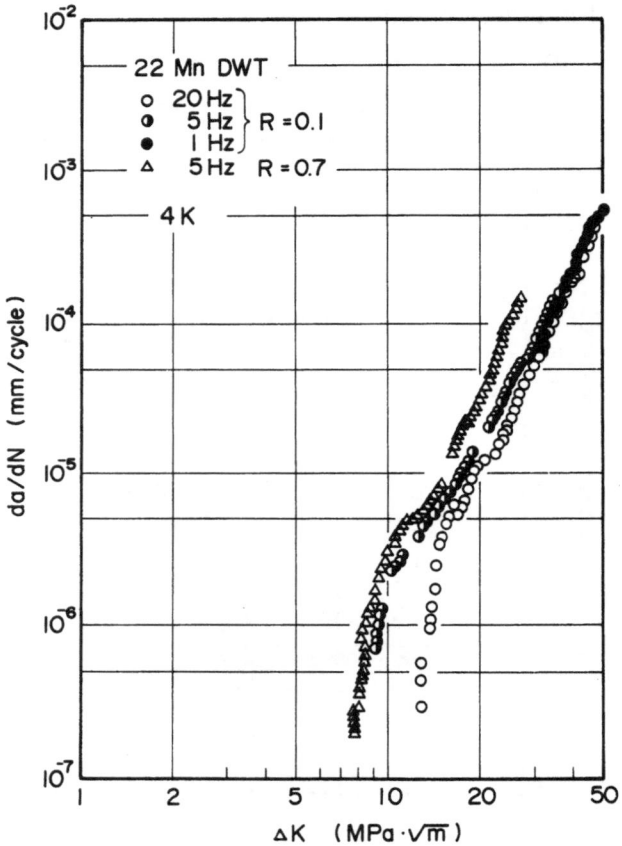

FIG. 12. Effect of test frequency on FCGR.

be due to the residual stress which is formed during the rapid cooling process of heat treatment.

Liaw *et al.* [11] carried out FCGR tests and indicated that at 4·2 K, the near-threshold FCGR properties were insensitive to load ratio compared to the case of room temperature as shown in the lower shaded area of Fig. 11. However, the tests were conducted at a frequency range from 85 to 100 Hz. It may be necessary to take account of heat generation at the crack tip. It may be very difficult to measure the localized temperature rise at the crack tip. An experiment on the effect of test frequency on FCGR is assumed to be one of the ways to evaluate the heat generation.

The da/dN versus ΔK curves at three test frequencies of 1, 5 and 20 Hz at liquid helium temperatures are illustrated in Fig. 12. At the higher crack growth rate, the effect of test frequency is found to be minor.

Figure 12 also indicates the effect of stress ratio R on FCGR. As in the result presented by Liaw *et al.* [11], the stress ratio does not have a significant effect. But it is necessary to pay attention again to the problem of scatter.

In order to evaluate the temperature dependence of the near-threshold FCGR eliminating the effects of residual stress, the effective ΔK was defined based on the measurement of crack closure. The curve in Fig. 13

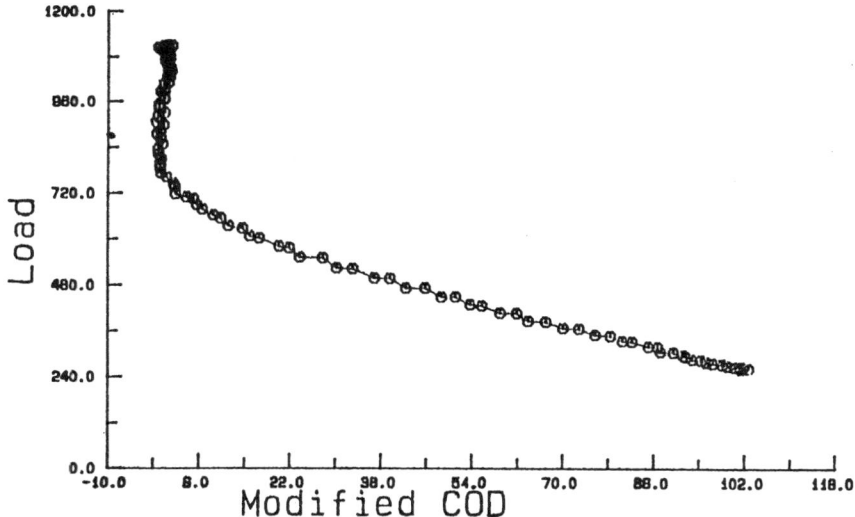

FIG. 13. Example of load–displacement curve.

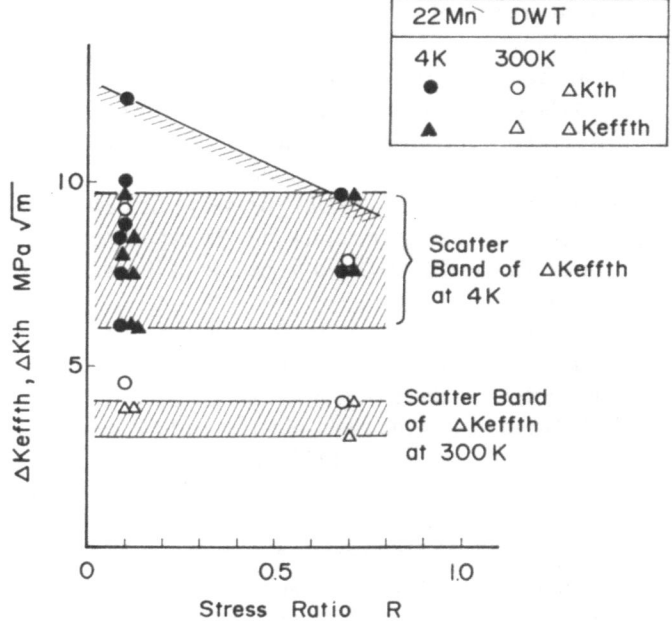

FIG. 14. Effect of temperature on effective threshold.

is an example of the relationship between load and subtracted displacement which is given by subtracting the product of unloading elastic compliance and the load from the real displacement. The vertical lines of the curve in Fig. 13 correspond to the crack opening level.

The crack opening ratio U was defined as the ratio of the length of this vertical line to the total height of the curve. The effective threshold ΔK_{th} was determined as

$$\Delta K_{effth} = U \Delta K_{th}$$

The effects of the temperature and the stress ratio on the threshold ΔK_{th} and the effective threshold ΔK_{effth} are summarized in Fig. 14.

The values of ΔK_{th} for all stress ratios fall in a wide scatter band irrespective of the temperature, whereas the data of ΔK_{effth} are divided into two groups of narrower scatter bands, depending on the temperature. The values of ΔK_{effth} at liquid helium temperature are greater than those at room temperature as shown in Fig. 14.

4. CONCLUSIONS

The fatigue behavior of 22Mn steels at cryogenic temperature is summarized as follows.

1. There exists a lower limit of strain rate above which serrations occur during strain controlled low-cycle fatigue of 22Mn steel at liquid helium temperature.
2. No difference was found in plastic strain range versus low-cycle fatigue life relations between 4 K and 300 K for lives less than 2×10^3 cycles.
3. The direct water toughened 22Mn steel showed the longest life in total strain controlled low-cycle fatigue tests as compared to other austenitic steels.
4. Decreasing the temperature increases the fatigue notch factor of specimens with severe notches.
5. The near-threshold FCGR data fall in a wide scatter band irrespective of temperature. However, the effective threshold values are divided into two narrower scatter bands depending on the temperature. The scatter band for 4 K is located above that for 300 K.
6. The effect of test frequency on FCGR is minor in the region of higher growth rates.

REFERENCES

1. SCHWARTZBERG, F. R. *et al., Adv. Cryog. Engng,* **10** (1965), 1–14.
2. NACHTIGAL, A. J., ASTM STP 579, 1975, pp. 378–96.
3. CHAMBERLAIN, D. W., *Adv. Cryog. Engng,* **9** (1964), 131–8.
4. WILKOV, M. *et al., Mater. Res. Stand.,* ASTM, **4** (1964), 6.
5. FOWLKES, C. W. and TOBLER, R. L., *Engng Fract. Mech.,* **8** (1976), 487–500.
6. READ, D. T. and TOBLER, R. L., *Adv. Cryog. Engng,* **28** (1982), 17–28.
7. GIGEON, D. N. *et al., Adv. Cryog. Engng,* **7** (1962), 503–8.
8. TOBLER, R. L. and REED, R. P., *Adv. Cryog. Engng,* **22** (1976), 35–46.
9. TOBLER, R. L. and YI-WEN CHENG, ASTM.STP 857, 1985, pp. 5–30.
10. SHIBATA, K. *et al.,* ASTM STP 857, 1985, pp. 31–46.
11. LIAW, P. K. *et al.,* ASTM STP 857, 1985, pp. 173–190.
12. AKAMATSU, M. *et al.,* presented at the CEC-85, 1985.
13. OGAWA, R. and MORRIS, J. W., ASTM STP 857, 1985, pp. 47–59.

4. CONCLUSIONS

REFERENCES

19

Fatigue Crack Growth at Elevated Temperature in Ferritic Steels

A. J. MᶜEᴠɪʟʏ, K. Mɪɴᴀᴋᴀᴡᴀ* and H. Nᴀᴋᴀᴍᴜʀᴀ**

Metallurgy Department U-136, University of Connecticut, 97N. Eagleville Rd., Storrs, CT 06268, USA.

*NKK Technical Research Center, Minamiwatarida-cho 1-1, Kawasaki, Japan.

**Dept. of Mechanical Engineering Science, Tokyo Institute of Technology, Ohokayama 2-12-1, Meguro-ku, Tokyo, Japan.

ABSTRACT

At elevated temperatures in air, both creep as well as oxidation have the potential for influencing the fatigue crack growth process. Both the mechanism of crack growth as well as the crack closure level can be affected by these variables. In order to evaluate the importance of creep and the environment, fatigue crack growth data and the associated crack closure levels have been obtained at room temperature and at 538°C in air and in vacuum, for a series of ferritic steels. It was found in vacuum over the range of frequencies of 0·3–30 Hz that the rate of crack growth was independent of frequency, an indication that the basic mechanical process of fatigue crack growth is athermal in nature for the test conditions employed. At 538°C in air however the rate of crack growth is frequency dependent, a dependency which relates to the thermally activated process of oxidation. These results as well as a discussion of the modelling process for the rate of fatigue crack growth will be presented. The importance of the chromium content on the fatigue crack growth behavior in air at 538°C will also be discussed.

1. INTRODUCTION

As design temperatures increase for various structural components there is a need to have available a knowledge of material behavior under both static and dynamic loading in the environment of interest in order for reliable and economic performance. At temperatures above 0·4 of the absolute melting point both creep and thermal fatigue are already well recognized as being life limiting for certain applications. Fatigue, particularly strain controlled, low-cycle fatigue with consideration of hold times and stress relaxation may also be an important factor in initiating cracks. Once the cracks are initiated their propagation either under steady load, i.e. creep crack growth, or under cyclic load, i.e. fatigue crack growth, become the important considerations. This paper will concentrate on the latter of these processes for ferritic steels containing chromium in the range 2·25–9 weight %, and molybdenum in the range of from 1 to 2 weight %. Such steels can be economic alternatives to the austenitic stainless steels provided that they possess sufficient resistance to oxidation or to other environmental effects.

2. MATERIALS AND EXPERIMENTAL PROCEDURES

The compositions, processing histories, and mechanical properties of the ferritic steels tested are listed in Tables 1, 2 and 3, respectively.

The specimens used in the fatigue crack growth studies were of the

TABLE 1
Chemical compositions

Alloy	Cr	Si	Mo	V	C	Mn	Nb	Fe
Mod. 9Cr–1Mo	8·47	0·19	0·88	0·21	0·04	0·37	0·07	bal.
9Cr–2Mo	9·58	0·36	2·13		0·06	0·53		
2·25Cr–1Mo	2·17	0·18	0·99		0·12	0·44		

TABLE 2
Plate processing histories

Mod. 9Cr–1Mo	
	Normalized at 1177°C for 1 h, air cooled
	Normalized at 1038°C for 1 h, air cooled
	Tempered at 760°C for 1 h, air cooled
9Cr–2Mo	Normalized at 750°C, air cooled
	Tempered at 750°C, air cooled
2·25 Cr–1 Mo	Normalized at 954°C for 1 h, air cooled
	Tempered at 727°C for 2 h, air cooled

TABLE 3
Mechanical properties

	Young's modulus (E)	$\sigma_y (MPa)$	$\sigma_u(MPa)$	$Cl\%$	$RA\%$	Hardness (R_B)
Mod. 9Cr–1Mo	$10^5 MPa$					
20°C	2·0	531	668	26	73	96
538°C	1·57	371	389			
9Cr–2Mo						
20°C	2·0	479	647	24	75	91·5
538°C	1.57	347	438	27	83	
2·25Cr–1Mo						
20°C	2·0	390	550	29	76	89
538°C	1·68	341	484			

standard ASTM compact type, 6 mm in thickness. The specimens were fatigue precracked at room temperature and then tested in air or in a 5×10^{-5} torr vacuum at 538°C (approximately 1/2 of the melting point). Crack closure measurements were made with an LVDT extensometer system which determined displacements at the mouth of the specimen.

3. RESULTS AND DISCUSSION

The fatigue crack growth rate in vacuum at frequencies of the order of 30 Hz as a function of ΔK is shown in Fig. 1 [1]. Several interesting points are to be noted. First of all, although the alloys differ with respect to mechanical properties there is little if any influence of alloy type on the fatigue crack growth rate. Further there is no influence of stress ratio R on the growth rate. Finally in these tests crack closure was virtually absent, so that the lack of any dependence of the growth rate on R can be attributed to the lack of closure. The similarity in behavior of the different alloys suggests that the dislocation arrays developed at the crack tips are similar in the different alloys, and that the rate of growth is dependent on this circumstance rather than the initial properties. The absence of closure may be due to the elimination of any potential roughness effects by creep during the unloading portion of the cycle.

Figure 2 indicates the effect of the air environment on the crack growth rate for the 9Cr–2Mo alloy. It is noted that the threshold is higher in air than in vacuum, but at higher ΔK levels the rate of growth

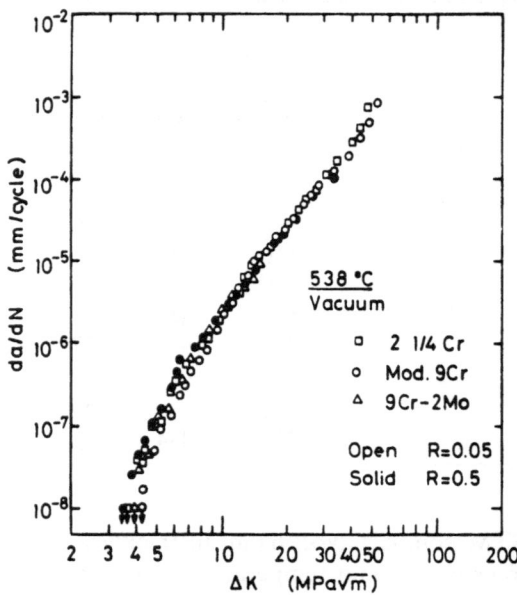

FIG. 1. Fatigue crack growth rate (da/dN), versus stress intensity range (ΔK), for Mod. 9Cr–1Mo, 9Cr–2Mo and 2·25Cr–1Mo steels in vacuum at 538°C.

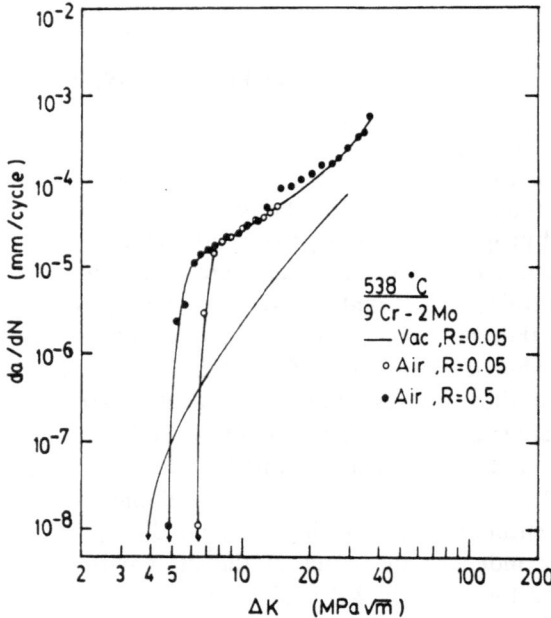

FIG. 2. Fatigue crack growth rate (da/dN), versus stress intensity range (ΔK), for 9Cr–2Mo steel in air at 538°C.

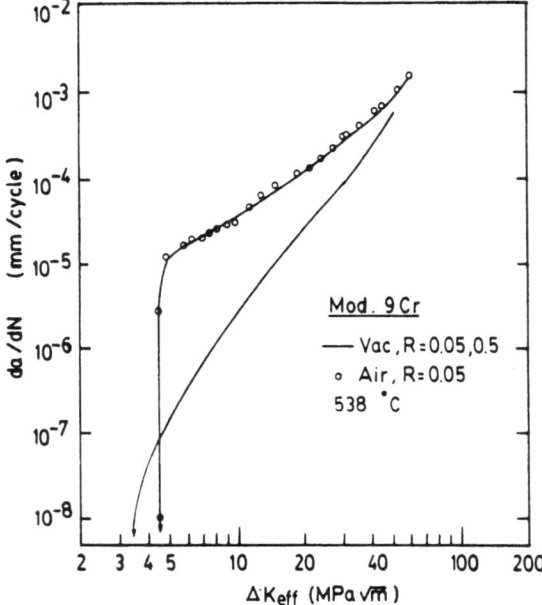

FIG. 3. Fatigue crack growth rate (da/dN), versus effective stress intensity range (ΔK_{eff}), for Mod. 9Cr–1Mo steel in air at 538°C.

in air is higher than in vacuum. There is also an R-dependency, at least in the near-threshold region. If crack closure is taken into account the results shown in Fig. 3 are obtained. In this case the ΔK_{eff} value is similar to the results for $R = 0.5$ in air, for in that case closure was minimal. Results for the modified 9Cr–1Mo alloy were identical.

The increase in threshold in air as compared to vacuum can be attributed to the behavior of the oxide film formed at the crack tip. Auxiliary measurements at room temperature have shown that strains of the order of 4% are needed to rupture the oxide film on the 9Cr alloys. Since the crack in air cannot propagate until tip strains of sufficient magnitude to rupture the oxide have been developed it is postulated that the increase in threshold over that in air is due to the strengthening effect of the oxide [2]. Further it is noted that as soon as the threshold level in air is exceeded the rate of growth increases to the order of 10^{-5} mm per cycle, which is in the range of the oxide thickness. Growth per cycle includes both the rupture of the oxide film as well as the advance in the alloy itself as in vacuum. At the highest growth rates the difference between air and vacuum results diminishes as the contribution from

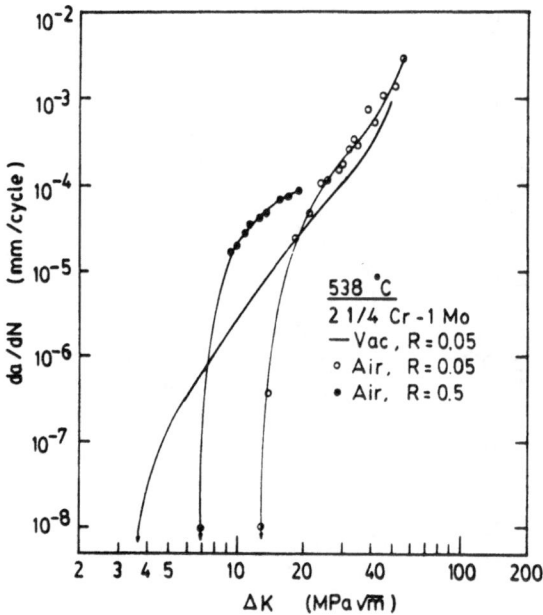

FIG 4. Fatigue crack growth rate (da/dN), versus stress intensity range (ΔK), for 2·25Cr–
1Mo steel in air at 538°C.

mechanical effects increases, with the contribution from oxidation remaining constant with ΔK.

The fatigue crack growth behavior of the 2·25 Cr–1Mo alloy at 538°C
at 30 Hz is shown in Fig. 4. For this alloy an even greater increase in
threshold is noted as compared to the threshold in vacuum. However in
this case the oxide is quite weak. The main effect of oxidation is to
increase the closure level due to the development of a much thicker oxide
than in the case of the 9Cr steels. In fact since this oxide continues to
thicken with time it is possible to increase the threshold level further
simply by holding at a temperature near threshold for extended time
periods. Therefore in such an alloy the effects of oxidation are time
dependent and correlations based on ΔK alone are not sufficient to
define the crack growth behavior in air. For an alloy such as the 9Cr–
2Mo where the oxide that forms at 538°C is more protective the time
dependency will be much less pronounced.

In addition to crack growth behavior in base metals, fatigue crack
growth behavior in weldments is also of interest. Fatigue cracks have

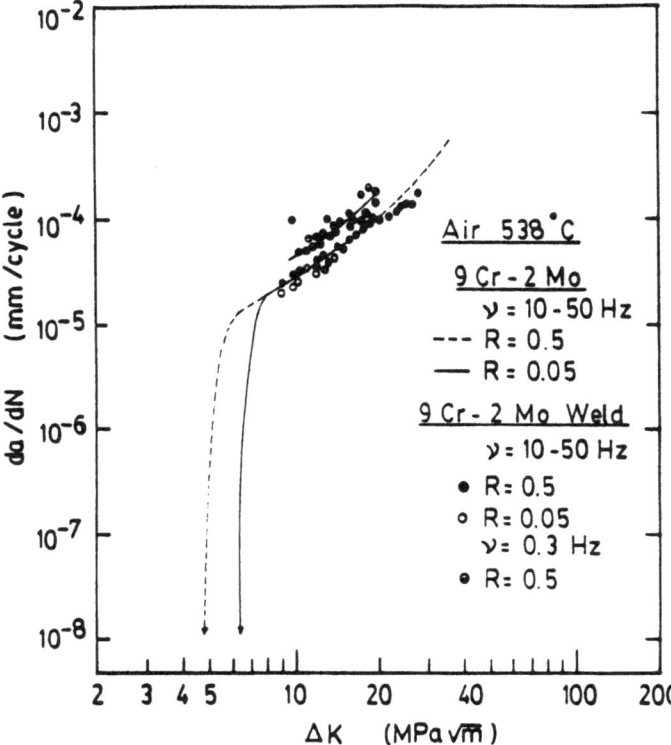

FIG. 5. Fatigue crack growth rate as function of ΔK in weldment of 9Cr–2Mo alloy in air at 538°C.

been grown in the weld metal of butt-welded 9Cr–2Mo specimens in air at 538°C. The results obtained are compared with base metal data in Fig. 5. At a test frequency of 30 Hz there is no difference between weld metal and base metal behavior. When the frequency is reduced to 0·3 Hz the rate increases. Is this increase the result of additional oxidation or is it due to additional creep? To answer this question additional tests of these weldments have recently been carried out in vacuum at 538°C as a function of frequency [3]. The crack growth rates obtained were identical to those for the base metal, again reflecting an independence of the specific ferritic alloy type, and were independent of frequency at least down to 0·3 Hz. This latter finding indicates that creep is not a factor under these test conditions affecting the rate of fatigue crack growth in air as a function of frequency, rather it is oxidation that is primarily responsible.

The absence of frequency effects in the vacuum tests also indicates that the fatigue crack growth process is an athermal process. Correlations of fatigue crack growth at different temperatures have been obtained simply on the basis of the dependency of the elastic modulus on temperature as shown in Fig. 6. In air however the correlation is not possible. We note also that at room temperature where the oxide is thinner than 538°C the threshold is lower than at 538°C. In this case the oxide is easily ruptured and crack growth is facilitated even in the near threshold region.

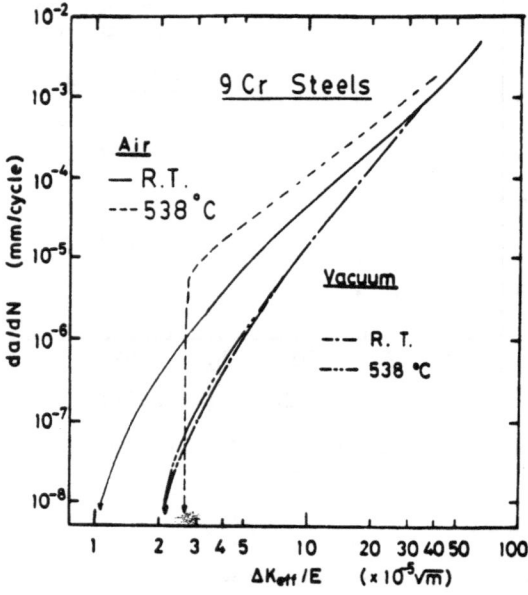

FIG. 6. Fatigue crack growth rate (da/dN) plotted with respect to $\Delta K_{eff}/E$ in order to consider the effect of temperature in vacuum and air.

In recent studies of the 9Cr–2Mo weld metal behavior carried out in vacuum at 538°C at 0.3 Hz we have noted that interesting transients can be observed [3]. It is recalled that at this temperature closure is not observed. However on decreasing the ΔK level by 9% in a decreasing ΔK test, the rate of growth initially decreased by an order of magnitude before gradually returning to the rate expected in Fig. 1 for that K level. If the K level is increased there is a corresponding transient increase in the growth rate before settling back into the expected rate.

In Fig. 1 it is noted that the rate of crack growth is independent of R. However if at a ΔK of 15 MPa$\sqrt{m}$ the R ratio is changed from 0·5 to 0·05 the crack growth rate can be reduced to the threshold level [3]. Examination of crack grown at the lower R ratio indicated that it was tightly closed not only at the surface but at a depth of 25% of the thickness. The crack behaves as if the closure level at $R = 0·5$ were equal to the minimum level. If at $R = 0·05$ the minimum level is reduced there will be a transitional period during which the crack tip region undergoes a shake down to reflect the new conditions of cycling. On increase of R at the same K level an acceleration is predicted since the closure level will initially be below the minimum level. Work is continuing to show more about these transitional effects. It is clear that a phenomenon different from either roughness induced closure or plasticity induced closure is involved, and that the delay is not associated with plane stress associated effects as in the case of an overload.

4. SOME ADDITIONAL CONSIDERATIONS

As has been discussed, at 538°C creep played a negligible part in affecting the rate of fatigue crack propagation down to a frequency of 0·3 Hz. It is known however that these alloys can fail in creep in this temperature range. We have recently investigated the effect of off-loading the 9Cr steels at various intervals during constant load creep tests to determine the nature of any effects induced by this type of cyclic loading [4]. The creep tests lasted a time of the order of two weeks and failure occurred by a necking instability process. Little effect on the lifetime to failure was noted if the specimens were unloaded after every hour under load, but if the time under load was reduced to less than 10 min then the lifetime increased. When failure occurred it still did so because of necking instability. The main influence of off-loading was to reduce the creep rate during the short time hold tests.

We have also examined the effect of cycling at a frequency of 30 Hz under constant load amplitude (rather than constant strain range) cyclic conditions [4]. At room temperature it is usual practice to establish a modified Goodman diagram to indicate the effect of mean stress on fatigue for constant load amplitude conditions. At 538°C, however, it developed that the specimens preferred to ratchet rather than fail in fatigue. Examination of the dislocation arrays developed in the tests at elevated temperature indicated a random array of dislocations, i.e. no indications for strain localization were observed. In the absence of such localization as is developed at a crack tip it is not possible to nucleate fatigue cracks.

Such results indicate that the circumstances leading to true creep fatigue interactions may be limited, and there is a need to define the nature and circumstances more precisely. Our results at positive mean stresses indicate that cycling can modify dislocation densities, solid solution effects, and precipitation kinetics. These modifications influence the creep ratchetting process, but do not lead to fatigue failures. It is also noted that in a recent investigation of fatigue crack growth in 2·25Cr–1Mo steel which had been exposed to hydrogen at elevated temperature (hydrogen attack), that although the grain boundaries contained a high density of voids which greatly reduced fracture toughness, the fatigue crack growth rate in the near threshold region was little affected by the presence of these voids, thereby suggesting a minimal creep–fatigue interaction [5].

5. CONCLUSIONS

(1) At 538°C at cyclic frequencies as low as 0·3 Hz creep is not a factor affecting the rate of fatigue crack growth in ferritic alloys.

(2) At 538°C the fatigue crack growth rate in vacuum is insensitive to alloy composition and original mechanical properties of the ferritic steels investigated. It is concluded that the dislocation arrays developed at the crack tips are similar, and that these arrays determine the rate of crack growth.

(3) Oxidation at 538°C can lead to an increase of the threshold level and an increase in crack propagation rate above threshold.

(4) A new type of transitional retardation–acceleration effect has been identified. This effect is a through-thickness effect and can develop on change of mean stress or K level.

(5) In the creep range it is difficult to obtain fatigue failures in unnotched, axially loaded specimens tested at positive mean stresses. There is an absence of strain localization, and failure occurs as the result of ratchetting.

ACKNOWLEDGEMENT

The support of the US Department of Energy, Office of Basic Energy Sciences, Division of Materials Sciences, Grant DE-FG02-84ER45109 is gratefully acknowledged.

REFERENCES

1. NAKAMURA, H., MURALI, K., MINAKAWA, K. and MCEVILY, A., in *Proc. Conf. Microstructure and Mechanical Behavior of Materials.* Xi'an, PRC, to be published by EMAS, Warwick, England, 1985.
2. MCEVILY, A., in *Proc. Symp. Modeling Environmental Effects on Crack Initiation and Propagation,* to be published by AIME, 1985.
3. MCEVILY, A. and YANG, Z., University of Connecticut, to be published.
4. BUNCH, J. and MCEVILY, A., University of Connecticut, to be published.
5. PENDSE, R. D. and RITCHIE, R. O., in *Proc. Symp. Modeling Environmental Effects on Crack Initiation and Propagation,* to be published by AIME, 1985.

20

Fatigue Crack Growth Resistance of Structural Materials in Vacuum

M. JONO

Department of Mechanical Engineering, Osaka University, Yamadaoka 2-1, Suita, Osaka, Japan.

ABSTRACT

Fatigue crack growth tests were carried out in vacuum and in air under ΔK-increasing and ΔK-decreasing conditions on several kinds of structural materials, and crack growth rates and crack closure behavior were investigated by using the unloading elastic compliance method.

Fatigue crack growth resistance in terms of ΔK_{eff} as well as ΔK was generally increased in vacuum. The degree of increase, however, was found to depend on material and on crack growth rate regime. For aluminum alloys, due to the ease of deformation in vacuum, it was found that the crack could advance at a much lower ΔK_{eff} regime than the threshold condition of ΔK_{eff} for crack growth in air, which implied the unconservative situation for use of aluminum alloys at high R ratio in vacuum. The variable environment test results were also discussed.

1. INTRODUCTION

Fatigue crack growth resistance has been increasingly recognized to be an important characteristic of structural materials in connection with the damage tolerant design concept, and much effort has been made to compile fatigue crack growth resistance data on many kinds of materials under conventional laboratory environments, corrosive environments and high and low temperature conditions [1, 2]. A vacuum environment is also thought to have an effect on the fatigue crack growth characteristics of materials [3–6]. It is also widely accepted now that the crack closure plays an important role in fatigue crack growth, and much attention has been focused on clarifying the closure mechanism. At present, three types of mechanisms have been suggested; namely plasticity induced crack closure [7], oxide induced crack closure [8–10] and roughness induced crack closure [11]. Both oxide induced crack closure and roughness induced crack closure are relevant to fatigue crack growth characteristics in the low stress intensity regime, and oxide induced crack closure, in particular, is thought to be strongly environment-sensitive.

In this study, in order to investigate the effect of environment on fatigue crack growth and crack closure, crack growth tests were carried out in vacuum on several kinds of structural materials, and compared with those in air, under ΔK-increasing and ΔK-decreasing conditions in the relatively low stress intensity regime. Air-to-vacuum-to-air variable environment tests were also performed under ΔK constant conditions.

2. MATERIALS AND TESTING PROCEDURE

The materials used as examples of conventional structural materials are a medium carbon steel, JIS S35C, and an aluminum alloy, JIS A5083-O. A P/M (Powder Metallurgy) Al–Si–Fe alloy is also investigated. Concerning the last material, although various improved properties such as high tensile strength, excellent wear resistance and low thermal expansion coefficient due to the high content of Si and Fe are reported [12] little is known as to the fatigue strength or fatigue crack growth characteristics [13, 14]. Chemical composition and mechanical properties of materials used are listed in Tables 1 and 2, respectively. A photograph of the microstructure of the P/M alloy is represented in Fig. 1, where the layered structure can be observed along the extrusion

TABLE 1
Chemical composition of materials used (%)

	C	Si	Mn	P	S	Ni	Cr	Mo	Al	V	Ti
S35C	0·34	0·26	0·74	0·015	0·004	0·03	0·04	0·009	0·02	0·003	0·011

	Si	Fe	Cu	Mn	Mg	Cr	Ti	Al			
A5083–O	0·09	0·21	0·01	0·46	4·20	0·17	0·03	rest			
P/M Alloy	12·0	5·0	4·0	0·3	0·5	—	—	rest			

TABLE 2
Mechanical properties of materials used

Material		Yield point 0·2% proof stress $\sigma_y(MPa)$	Tensile strength $\sigma_B (MPa)$	Elongation $\delta(\%)$	Reduction of area $\psi(\%)$
S35C		339	586	33·3	59·9
A5083–O		140	301	24·0	42·1
P/M Alloy	C	246	326	0·83	2·2
	L	231	368	3·30	5·4

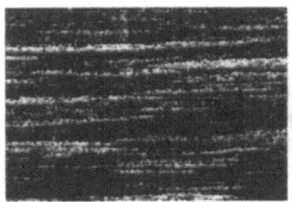 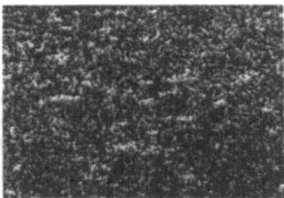

FIG. 1. Microstructure of P/M alloy. (Left) longitudinal direction; (right) circumferential direction.

direction. The specimens used were side-grooved small CT specimens, the configuration of which is shown in Fig. 2, where the root radius of side groove is changed depending on materials to obtain near plane strain condition through the specimen thickness and to make the crack front straight so that the crack opening point can be detected clearly [15]. The specimen is cut in L–T direction for S35C and A5083-O while

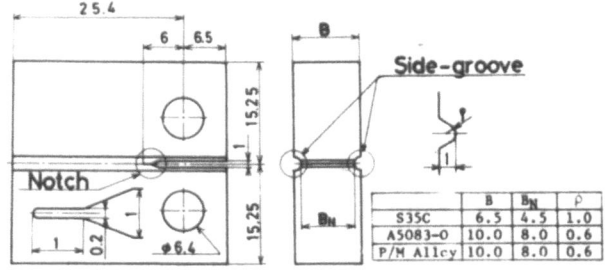

FIG. 2. Specimen configuration (all dimensions in mm).

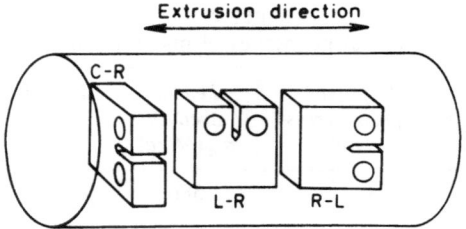

FIG. 3. Specimen direction of P/M alloy.

three specimen directions are chosen for P/M alloy as shown in Fig. 3 in order to investigate the effect of anisotropy due to extrusion on fatigue crack growth resistance. Fracture toughness values, K_{IC}, for three directions of P/M alloy are given in Table 3.

TABLE 3
Fracture toughness values, K_{IC}, of P/M alloy

Direction	L–R	R–L	C–R
$K_{IC}(MPa\sqrt{m})$	11·8	9·0	9·7

Fatigue crack growth tests were conducted on a closed loop servo-hydraulic testing system at a cycling frequency of 20 Hz in air and in vacuum (4 mPa) under both ΔK-increasing and ΔK-decreasing conditions at an R ratio of 0·1. The K-increasing rate was chosen in the range of $(1/K)(dK/da)=0·25\sim0·3$ mm^{-1}, and the decreasing rate $-0·12\sim-0·2$ mm^{-1}, which is slightly high as compared with ASTM's recommendation. In addition, for S35C ΔK-constant tests were carried out in air-to-vacuum-to-air environment. Crack length and crack opening points were monitored throughout the tests by the mini-computer aided unloading elastic compliance method using a back face strain gauge [16].

3. EXPERIMENTAL RESULTS OF CRACK GROWTH RATE AND CRACK CLOSURE

3.1. The Case of the Medium Carbon Steel, S35C

Figure 4 shows the relationship between crack growth rate, da/dn, and the stress intensity range, ΔK, for the medium carbon steel, S35C. Open

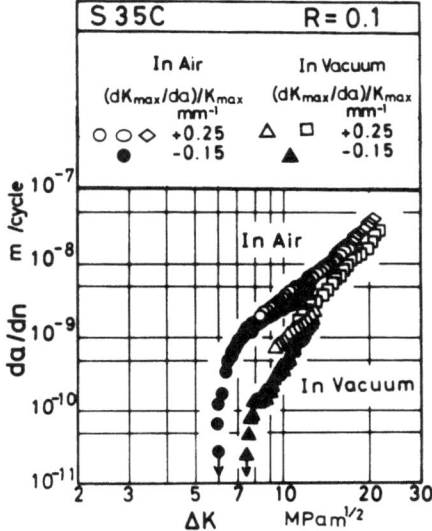

FIG. 4. Relationship between da/dn and ΔK on S35C.

symbols and solid symbols indicate the results under ΔK-increasing and ΔK-decreasing conditions, respectively. As to the growth rates in air, there was found little difference between ΔK-increasing and ΔK-decreasing test results, probably because the ΔK-decreasing test was started from a relatively low stress intensity level. In vacuum, however, ΔK-decreasing test results were lower than under ΔK-increasing tests. Although it was found, in general, that crack growth rates in vacuum were lower than in air under both ΔK-increasing and ΔK-decreasing tests, the difference of growth rates between in air and in vacuum was found to depend on testing procedure. In ΔK-increasing tests, growth rates in air were about 2–3 times higher than in vacuum over the range tested, but the difference in ΔK-decreasing tests increased with decrease of ΔK. The threshold value for crack growth, ΔK_{th}, was found to be $6.0\,\mathrm{MPa}\sqrt{m}$ in air, being lower than that value in vacuum of about $7.4\,\mathrm{MPa}\sqrt{m}$. Thus superior fatigue crack growth resistance in vacuum is similar to that reported for other materials [3–6] and the reason may be that the rewelding of newly slipped surfaces at the crack tip can occur in vacuum environment [17].

In Figs. 5a and 5b, the variation of crack opening stress intensity, K_{op}, is plotted against the maximum stress intensity, K_{max}, in air and in

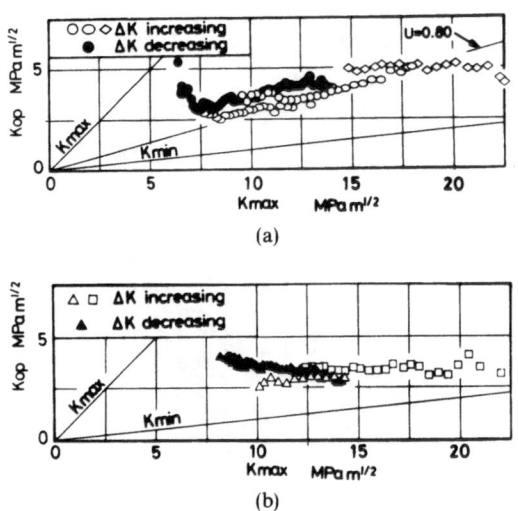

FIG. 5. Behavior of K_{op} on S35C: (a) in air; (b) in vacuum.

vacuum, respectively. K_{op} in ΔK-increasing tests in air increased approximately in proportion to K_{max}, and took on an almost constant value at the high K_{max} regime. A similar tendency was observed for K_{op} in vacuum in ΔK-increasing tests, except that the crack opening stress intensity values were slightly lower than those in air. Under ΔK-decreasing conditions, however, a different behavior of K_{op} was found depending on the test environment. In air, at first K_{op} was almost constant and then gradually decreased with decrease of K_{max}, until it increased abruptly at $K_{max} \simeq 7\cdot5\,\mathrm{MPa}\sqrt{m}$, corresponding to the growth rate of $10^{-9}\mathrm{m/cycle}$, and reached the threshold condition. On the other hand, K_{op} in vacuum was found to continuously increase with decrease of K_{max} and reached the threshold condition without showing the abrupt increase of K_{op}, which resulted in the large value of $(\Delta K_{eff})_{th}$.

The crack growth rates are plotted in Fig. 6 against the effective stress intensity range, $\Delta K_{eff}(=K_{max}-K_{op})$, where the difference of crack growth rates between ΔK-increasing and ΔK-decreasing conditions, which was observed in terms of ΔK as shown in Fig. 4, was eliminated in both air and vacuum environment. Moreover, each environmental test result could be represented approximately by a single straight line except near threshold regime, which means that the governing parameter of fatigue crack growth in vacuum is also the effective stress intensity range, ΔK_{eff}, as well demonstrated in air [18]. Fatigue crack growth resistance in

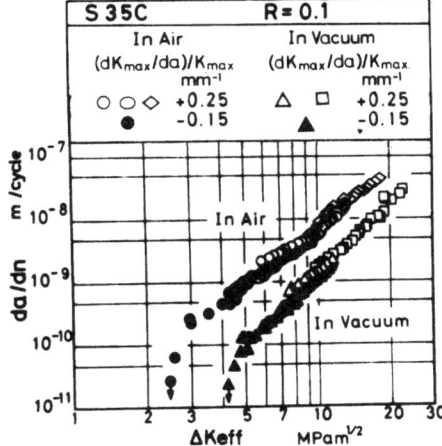

FIG. 6. Relationship between da/dn and ΔK_{eff} on S35C.

terms of ΔK_{eff} was found superior in vacuum to that in air, especially in the low ΔK_{eff} regime.

3.2. The Case of the Aluminium Alloy, A5083-O

Figure 7 shows the relationship between da/dn and ΔK for the aluminum alloy, A5083-O. In the case of the vacuum environment test, as was observed for S35C, the fatigue crack growth rates under ΔK-decreasing conditions, represented by solid triangular symbols, were found to be lower than those obtained under ΔK-increasing conditions, and gave a relatively high threshold value compared with that in air. The fatigue crack growth rate curve in air exhibits a trilinear form in the region above the growth rate of 2×10^{-9} m/cycle (so-called region II of fatigue crack growth) due to the transition of the crack growth mechanism [18], while the curve in vacuum under ΔK-increasing conditions may be expressed by a simple relationship in terms of ΔK.

The behavior of crack opening stress intensity, K_{op}, in air and in vacuum is shown in Figs. 8a and 8b, respectively. Crack opening points in air in the ΔK-increasing test were found to take higher values than those in vacuum at the low K_{max} level and likely to merge in the high K_{max} regime. In contrast to the behavior of K_{op} under ΔK-decreasing conditions in air which showed the decrease of K_{op} with decrease of K_{max}, K_{op} in vacuum under a ΔK-decreasing test monotonously in-

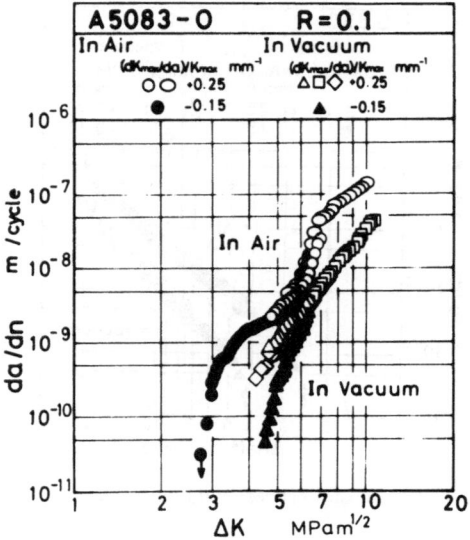

FIG. 7. Relationship between da/dn and ΔK on A5083-O.

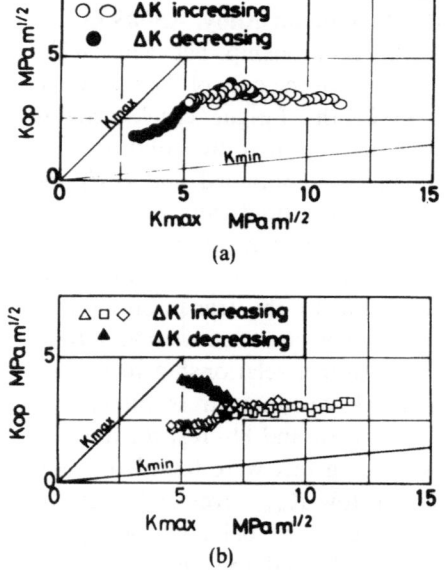

FIG. 8. Behavior of K_{op} on A5083-O: (a) in air; (b) in vacuum.

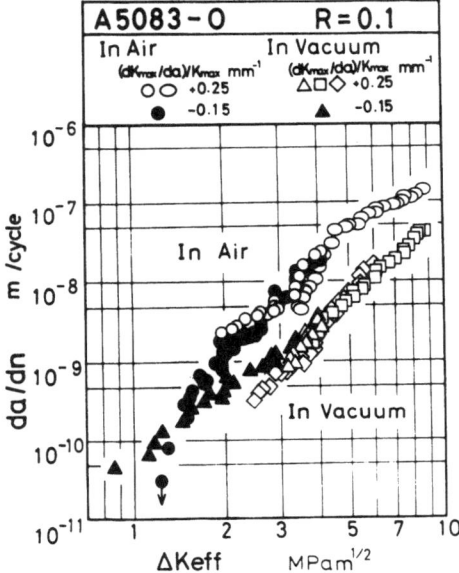

FIG 9. Relationship between da/dn and ΔK_{eff} on A5083-0.

creased with decrease of K_{max}, and, within the limit of this experiment, the crack was found to grow even at the very low stress intensity level, so that the threshold condition might be given by the condition of $\Delta K_{\text{eff}} = 0$.

In Fig. 9, crack growth rates were plotted against the effective stress intensity range. It was found that the load history effect could be eliminated by taking into consideration the crack closure behavior, although slightly lower growth rates were observed in vacuum under ΔK-increasing conditions than under ΔK-decreasing conditions. The crack growth resistance in vacuum was found to be about five times higher than that in air in the regime above the ΔK_{eff} of $4\,\mathrm{MPa}\sqrt{\mathrm{m}}$, while in the lower stress intensity regime growth rates under both environments seemed to approach each other. Further, in the regime below a growth rate of 10^{-10} m/cycle, the growth rates in air abruptly decreased to the threshold condition, whereas the cracks in vacuum continued to advance down to the condition of $\Delta K_{\text{eff}} = 0$, and showed therefore an inferior crack growth resistance. It may be concluded from above mentioned test results that the fatigue crack growth resistance of A5083-O is dependent on the growth rate regime.

3.3. The Case of the P/M Al–Si–Fe Alloy

Figure 10 shows the test results for L–R specimens of the P/M alloy. Relationships between growth rate, da/dn, and ΔK and da/dn and ΔK_{eff} are shown on the right and left hand side, respectively. With respect to the growth rates as a function of ΔK, da/dn in vacuum was found lower than that in air over the entire test region, and the threshold condition,

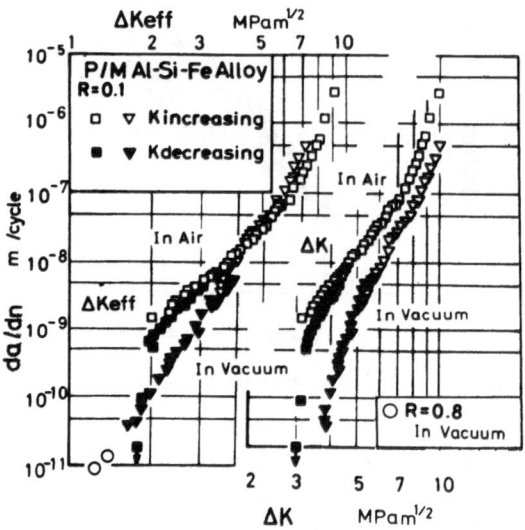

FIG. 10. Fatigue crack growth rate on P/M alloy (L–R direction).

ΔK_{th}, was about $4\,\mathrm{MPa}\sqrt{\mathrm{m}}$ in vacuum being higher compared with ΔK_{th} in air of $3\,\mathrm{MPa}\sqrt{\mathrm{m}}$. However, the slope of the crack growth curve was found to become rather steep at a growth rate of about $10^{-7}\,\mathrm{m}/$ cycle, where $K_{max} = \Delta K/(1-R)$ approached the K_{IC} value of the material. K_{IC} was relatively low for this material as shown in Table 3, resulting in a narrow range of stable crack growth governed by the Paris law. On the other hand, the relative crack growth resistance against ΔK_{eff} in vacuum compared with in air was found dependent on the crack growth rate regime; although the growth rates under both environments were nearly the same in the relatively high growth rate regime, da/dn in vacuum became lower than that in air in the region below a growth rate of $10^{-8}\,\mathrm{m/cycle}$. Moreover, at the extremely low growth rate regime, growth rates in vacuum seemed to monotonously decrease with decreasing

ΔK_{eff} crossing the growth curve in air which showed threshold condition against ΔK_{eff}, as was observed for A5083-0 aluminum alloy. This fact is important enough to be noted, because when aluminum alloys are used for structural components subjected to loads with high R ratio, where $\Delta K = \Delta K_{\text{eff}}$, in vacuum, there seems to exist no threshold condition and a crack can advance even at the very low ΔK values, as shown for example with $R = 0.8$ in Fig. 10 by open circles.

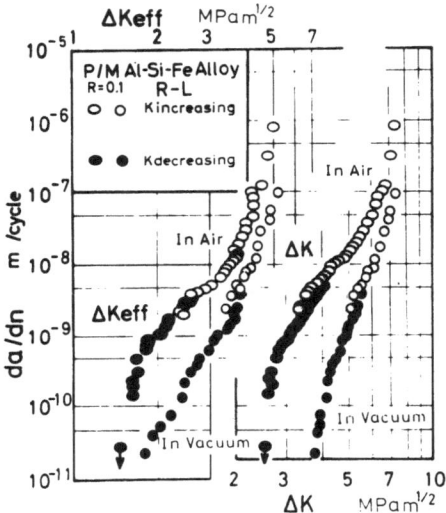

FIG. 11. Fatigue crack growth rate on P/M alloy (R–L direction).

Figures 11 and 12 show the test results on R–L and C–R specimens, respectively. Similar tendencies were observed in general, but the difference of crack growth rates between in air and in vacuum was found large compared with that for L–R specimens.

In Fig. 13 is shown the average growth rate curves for three specimen directions. In air the L–R direction showed superior growth resistance in terms of ΔK and ΔK_{eff}, while in vacuum although crack growth rate on C–R specimens was found lowest in the low growth rate regime, it was replaced by L–R specimens in the region above the growth rate of 5×10^{-8} m/cycle. Thus the crack growth resistance in vacuum may be concluded to depend on the growth rate regime.

Figures 14–16 indicate the crack closure behavior in air and in vacuum under ΔK-increasing and ΔK-decreasing conditions on L–R, R–

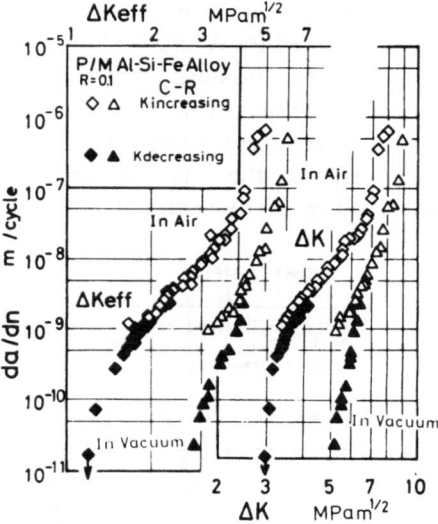

FIG. 12. Fatigue crack growth rate on P/M alloy (C–R direction).

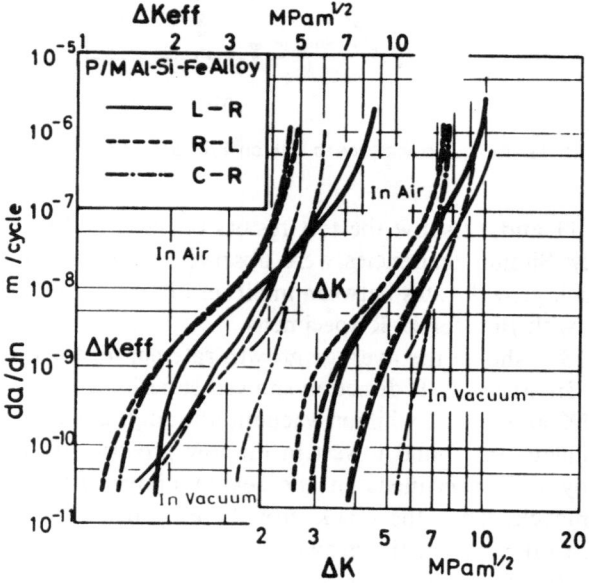

FIG. 13. Average crack growth rate curve for three specimen directions.

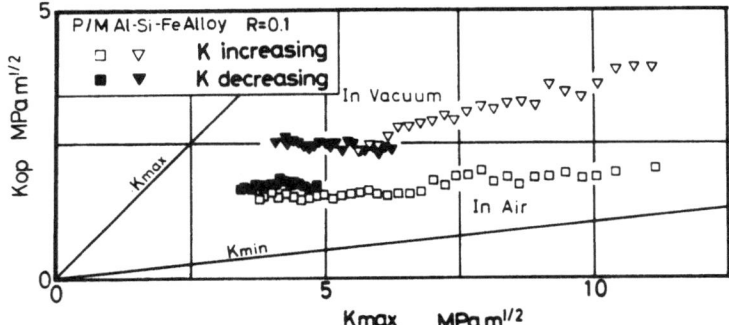

FIG. 14. Behavior of K_{op} on P/M alloy (L–R direction).

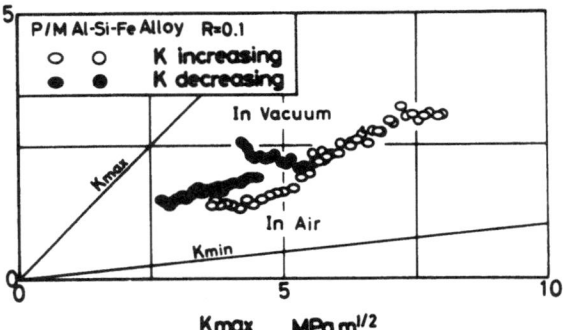

FIG. 15. Behavior of K_{op} on P/M alloy (R–L direction).

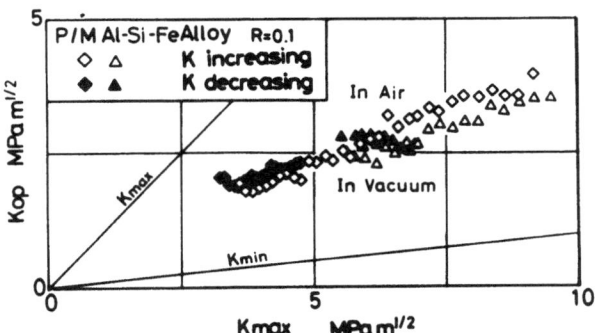

FIG. 16. Behavior of K_{op} on P/M alloy (C–R direction).

L and C–R specimens, respectively. In L–R specimens the crack opening point, K_{op}, under ΔK-increasing conditions in vacuum increased in proportion to K_{max} and was higher than that in air which gradually increased with K_{max} and took almost constant values in the high K_{max} regime. On the other hand, K_{op} under ΔK-decreasing conditions was found to increase with decreasing K_{max} in vacuum. With respect to R–L specimens, K_{op} under ΔK-increasing conditions, which increased with increasing K_{max}, took similar values both in vacuum and in air. However, different behavior of K_{op} under ΔK-decreasing conditions was observed in different environments; an increase of K_{op} with decrease of K_{max} was found in vacuum, while it decreased in air. Similar behavior was also observed for the C–R specimens, except that K_{op} in air was higher than that in vacuum under ΔK-increasing conditions.

4. FRACTOGRAPHIC FEATURES OF FRACTURE SURFACE

Figures 17a and 17b show the macroscopic fractographs of fracture surfaces both in air and in vacuum of S35C and A5083-O, respectively. In

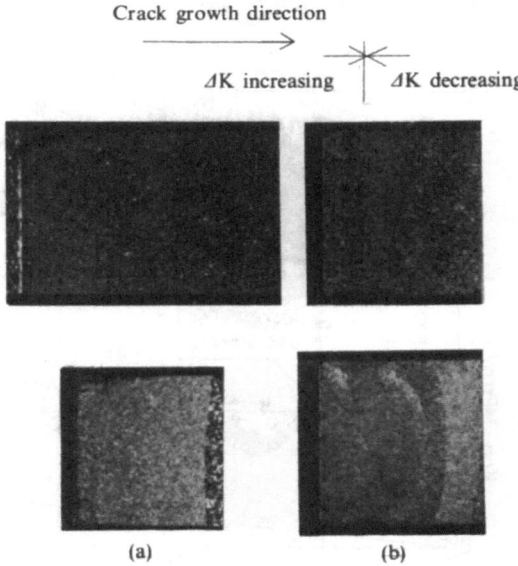

FIG. 17. Macroscopic fractographs of fracture surface. (a) S35C: (top) in air, (bottom) in vacuum; (b) A5083-O: (top) in air, (bottom) in vacuum.

the case of S35C, fracture surfaces manifest a uniform appearance irrespective of crack growth rates, within the range of this experiment, but different as a function of environment. The appearance in vacuum was smooth and glossy white, while dark and rough in air. On the other hand, the fracture surface of A5083-O in air reveals different features depending on growth rate and load history. The appearance varies from dark to rather white during ΔK-increasing tests, and further turns gray under ΔK-decreasing conditions. The variation of surface appearance in ΔK-increasing tests was due to the variation of the crack growth mechanism, as pointed out in the previous work [18], and the glossy white portion corresponded to the transition region of the growth rate curve shown in Fig. 7, where the intergranular fracture was found to be the main feature. The dark colored fracture surface was found to be rougher than that of S35C. The surface feature of A5083-O in vacuum

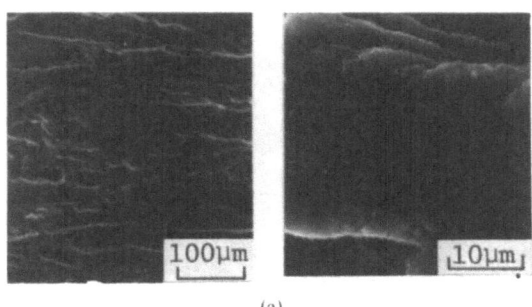

(a)

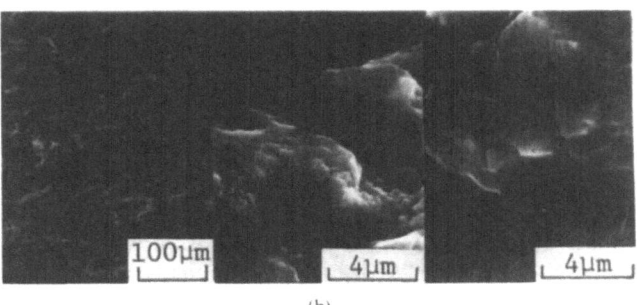

(b)

FIG. 18. Microscopic fractographs of fracture surface of A5083-O. (a) In vacuum; $\Delta K_{\text{eff}} = 3.70$ MPa· m$^{1/2}$, da/dn = 2.19×10^{-9} m/cycle; (b) in air; (left and middle) $\Delta K_{\text{eff}} = 2.17$ MPa· m$^{1/2}$, da/dn = 2.78×10^{-9} m/cycle; (right) $\Delta K_{\text{eff}} = 3.36$ MPa· m$^{1/2}$, da/dn = 8.32×10^{-9} m/cycle. Crack growth direction: left→right.

exhibits the glossy white appearance as was seen on S35C in vacuum. The crack fronts were usually found to be straight due to the effect of constraint of deformation near the specimen surface by the side-grooves. However, the crack front of A5083-O in vacuum was found to be curved, which implied the plane strain condition could not be satisfied in that case, probably because of the easily deformable characteristics of the material in vacuum, so that the crack advanced faster interior than near the surface as in the case of the specimen without side-grooves.

Figure 18 shows the microscopic fractographs of A5083-O in vacuum under a ΔK-increasing test (Fig. 18a) compared with those in air with similar growth rate (left hand side of Fig. 18b) and with similar ΔK_{eff} values (right hand side of Fig. 18b). When compared at a similar growth rate, the surface appearance in air was found rougher than that in vacuum. And deformation features in vacuum were similar or smoother in appearance to that in air for the identical ΔK_{eff} value.

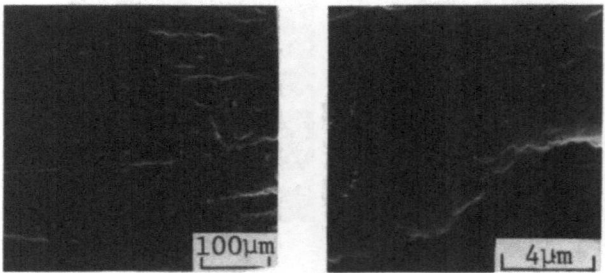

FIG. 19. Fractographs in vacuum at low growth rate regime of A5083-O. $\Delta K_{eff} = 1.13$ MPa·m$^{1/2}$, da/d$n = 6.4 \times 10^{-11}$ m/cycle. Crack growth direction: left→right.

Microfractographs of A5083-O in vacuum at a crack growth rate of 6.4×10^{-11} m/cycle are shown in Fig. 19, where the flat and deformed feature of the fracture surface was still observed even at such a very low growth rate.

Figure 20 shows macroscopic fractographs of fracture surfaces for three directions of P/M alloy. A discernible difference was not observed between the appearances in air and in vacuum in each direction, although fracture surfaces in vacuum generally revealed glossy white features, while dark and oxided ones were observed in air. Råther smooth appearances were observed on the fracture surfaces corresponding to the relatively low crack growth rate regime. However, rough

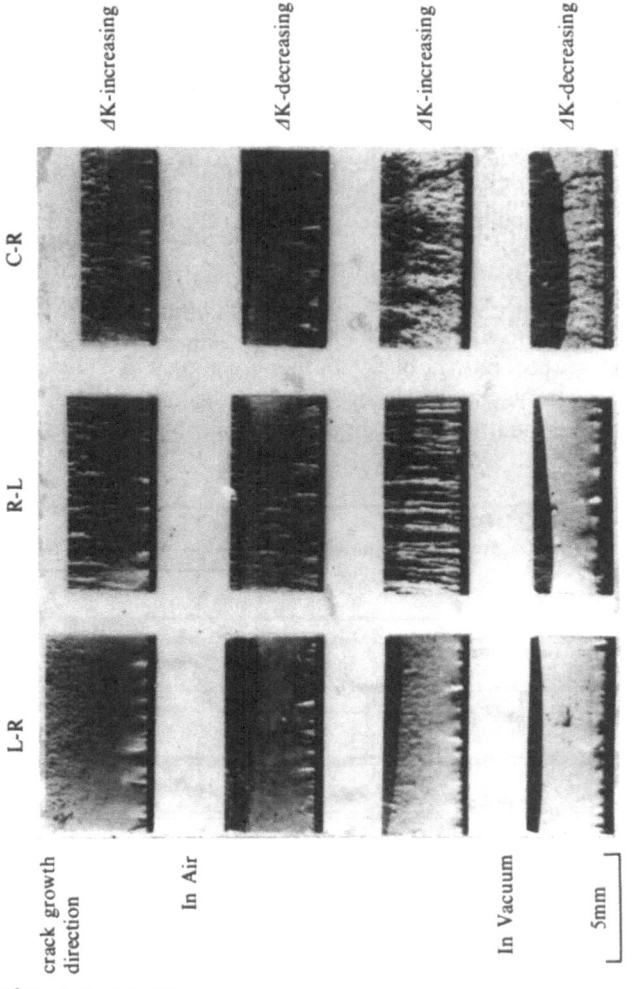

FIG. 20. Macroscopic fractographs of fracture surface of P/M alloy.

surfaces with stripe patterns along the extrusion direction were found in
the high growth rate regime, especially on the specimens of R–L and C–
R directions.

5. CRACK GROWTH UNDER VARIABLE
ENVIRONMENT TEST

Figure 21 shows the crack growth rates and crack opening points on
S35C under air-to-vacuum-to-air variable environment tests at three
different ΔK levels. The open symbols represent the results in air and
solid symbols in vacuum. The crack growth rate was the averaged one
over the crack growth increment of about 30 μm. When the specimen
chamber was exhausted by a vacuum pump, the crack growth rates were
found to gradually decrease with increasing vacuum, from the high value
corresponding to the air test to the low value corresponding to the
vacuum test, and to return back to the high growth rate to air immediately after the vacuum was broken. On the other hand, the crack
opening points seemed not to change corresponding to the environment
change.

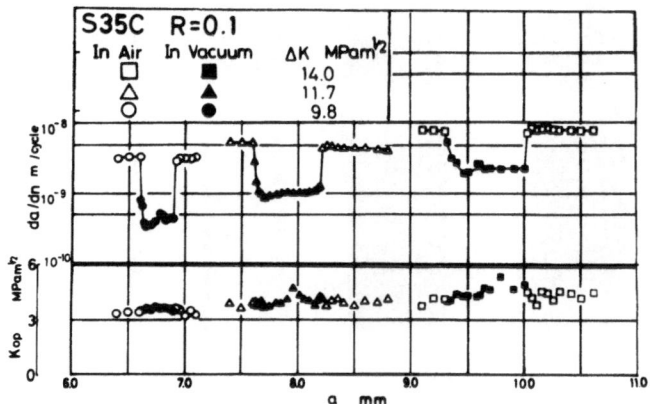

FIG. 21. Variation of da/dn and K_{op} under variable environment test.

In Fig. 22 is plotted the crack growth rates against the effective stress
intensity range, ΔK_{eff}. It was found that the growth rates under variable
environment tests well coincided with those of each environment, which
indicated the crack growth rate was hardly affected by environmental
history.

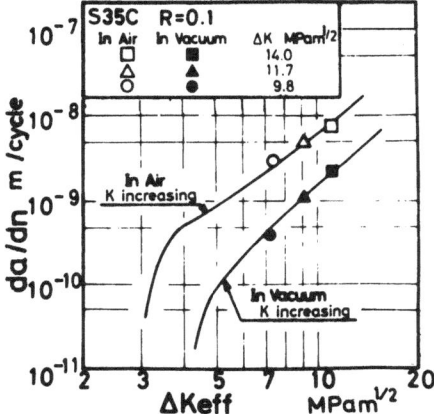

FIG. 22. Relationship between da/dn and ΔK_{eff} under variable environment test.

6. DISCUSSION

From the experimental results and fractographic observations performed in this study, the effects of vacuum environment on fatigue crack
growth are considered as follows.

In the case of I/M (Ingot Metallurgy) alloys, the vacuum environment
made the material at the crack tip plastically deform as already reported
by Endo and Komai [4] and Rangnathan and Petit [19]. However, the
extent of deformability was found to be dependent on materials, and Fig.
10 indicated that this effect was remarkable for A5083-O aluminum alloy.
The crack opening points are expected to be affected by the above
mentioned deformation behavior, i.e. the ease of deformation may cause
much crack closure if it is induced by plasticity, to result in an increase of
the crack opening point. On the other hand, the ease of deformation
brought about a deformed flat fracture surface, which might reduce
closure contribution resulting from roughness induced crack closure.
Crack opening points in vacuum are thought to be determined by the
competition between the above mentioned two effects, and lower crack
opening points in vacuum than in air observed in the experiments, as
shown in Figs. 5 and 8, indicated that, along with the lack of oxide
induced crack closure, the roughness induced crack closure was dominant in the low crack growth rate regime for those materials.

However, the situation is thought to be different for the P/M Al–Si–Fe alloy. As already shown in Figs. 14–16, crack opening points of P/M alloy under ΔK-increasing conditions in air were found to become higher in the order of L–R, R–L, C–R directions corresponding to the roughness of the fracture surface. Therefore, roughness induced crack closure may be concluded to be the main cause of crack closure of this material in air. On the other hand, the fact that crack opening points took similar high values in vacuum irrespective of specimen directions, then irrespective of macroscopic roughness of fracture surface, suggested that the crack closure in vacuum was strongly associated with the rewelding of the crack surface.

The second effect of vacuum environment appears in the fact that the crack growth resistance in vacuum was generally increased probably because of the rewelding of the crack tip [16] and of increase of fracture ductility [18]. And no transition of crack growth mechanism was found in vacuum for A5083-O within the range of this experiment, which resulted in the simple relationship between da/dn and ΔK or ΔK_{eff} without showing the trilinear form of crack growth curve as observed in air. And further, it should be noted that, because the alternative slip at the crack tip could occur in vacuum in the very low ΔK_{eff} regime, the crack could advance down to the condition of $\Delta K_{eff} = 0$, which implied an inferior growth resistance in vacuum in the very low growth rate regime.

Thirdly, the characteristic of crack growth behavior in vacuum to be pointed out was the increase of K_{op} with decrease of K_{max} as shown in Figs. 3a, 6b and 14–16. Although the reason for this peculiar behavior is not clear at present and should be clarified in future study, this behavior led to lower crack growth rates under ΔK-decreasing conditions than those under ΔK-increasing conditions, so that it should be avoided in using the growth rate curve in terms of ΔK obtained under ΔK-decreasing conditions as the design curve for the machine and/or structure operated in vacuum.

7. CONCLUSIONS

Fatigue crack growth tests were carried out both in air and in vacuum under ΔK-increasing and ΔK-decreasing conditions on three kinds of structural materials and crack closure behaviors were investigated. Fractographic observation and air-to-vacuum variable environment tests were also conducted. The main results obtained in this study are as follows:

(1) Fractography indicated that the material ahead of the crack tip deformed more easily in vacuum than in air in the case of I/M alloys.

(2) The ease of deformation of material in vacuum along with the lack of oxidation caused a smooth fracture surface within the range of the present experiments, which resulted in making the crack opening point lower in vacuum than in air.

(3) However, the fact that the crack opening point of the P/M alloy in vacuum was found to take high values irrespective of the roughness of the fracture surface suggested rewelding might cause the crack closure in vacuum.

(4) The crack opening stress intensity under ΔK-decreasing conditions in vacuum was found to increase with decrease of ΔK. This caused a different crack growth rate between ΔK-increasing and ΔK-decreasing tests at the identical ΔK value.

(5) Such differences in crack growth rate, da/dn, in terms of the stress intensity range, ΔK, as mentioned above, were eliminated when da/dn was expressed in terms of the effective stress intensity range, ΔK_{eff}, in air and vacuum. This result implies that the governing parameter of crack growth in vacuum, as well as in air, is ΔK_{eff}.

(6) Fatigue crack growth resistance of material in terms of ΔK_{eff} was generally increased in vacuum. The degree of increase was found to depend on the material, specimen directions and on crack growth rate regime. For aluminum alloys, however, due to the ease of deformation in vacuum, the crack could advance in a much lower ΔK_{eff} regime than the threshold condition for crack growth in air.

(7) The variable environment test results showed that fatigue crack growth resistance was scarcely affected by environmental history, and the crack growth rate was found to coincide with that of constant amplitude tests in each environment.

REFERENCES

1. *Damage Tolerant Design Handbook*, Battelle, Metals and Ceramics Information Center, 1972.
2. *Data Book on Fatigue Crack Growth Rates of Metallic Materials*, Society of Materials Science, Japan, Vol. 1 and 2, 1983, p. 2.

3. COOKE, R. J., IRVING, P. E., BOOTH, G. S. and BEEVERS, C. J., *Engng Fract. Mech.*, **7** (1975), 69.
4. ENDO, K. and KOMAI, K., *J. Soc. Mater. Sci., Japan*, **26** (1977), 143.
5. KOMAI, K., MATOBA, H. and KIKUCHI, J., *J. Soc. Mater. Sci.*, Japan, **33** (1984), 566.
6. DAVIDSON, D. L. and LANKFORD, J., *Met. Trans.*, **15A** (1984), 1931.
7. ELBER, W., ASTM STP 486, 1971, p. 280.
8. ENDO, K., KOMAI, K. and OHNISHI, K., *Memoirs Fac. Engng Kyoto University*, (1969), 25.
9. STEWART, A. T., *Engng Fract. Mech.*, **13** (1980), 463.
10. RITCHIE, R. O., SURESH, S. and MOSS, C. M., *Trans. ASME, J. Engng. Mater. Technol.*, **102** (1980), 293.
11. MINAKAWA, K. and MCEVILY, A. J., *Scripta Met.*, **15** (1981), 633.
12. DIXON, C. F. and SKELLY, M. M., *Int. J. Powd. Met.*, **1** (1965), 28.
13. MINAKAWA, K. and MCEVILY, A. J., *Fatigue Threshold*, EMAS, UK, Vol. 1, 1982, p. 373.
14. VASUDEVAN, A. K. and SURESH, S., *Met. Trans.*, **13A** (1982), 2271.
15. KIKUKAWA, M., JONO, M., KONDO, Y., YAMAKI, T. and YAMADA, K., *J. Soc. Mater. Sci., Japan*, **29** (1980), 155.
16. KIKUKAWA, M., JONO, M. and TANAKA, K., *Proc. ICM-II*, Special Vol., 1976, p. 254.
17. KIKUKAWA, M., JONO, M. and ADACHI, M., ASTM STP 675, 1979, p. 234.
18. JONO, M. and SONG, J., *Fatigue 84*, **II** (1984), 717.
19. RANGNATHAN, N. and PETIT, J., ASTM STP 811, 1983, p. 464.

21

Role of Closure Mechanisms on Fatigue Crack Growth in Steels under Service Conditions

HIDEO KOBAYASHI and HARUO NAKAMURA

Department of Mechanical Engineering Science, Tokyo Institute of Technology, Ohokayama 2-12-1, Meguro-ku, Tokyo, Japan.

ABSTRACT

The crack closure mechanisms which are encountered under expected service conditions were examined. The accuracy of the back face strain technique under some crack closure mechanisms were also examined by comparing the da/dN versus ΔK_{eff} curves with closure free da/dN versus ΔK curves at a high stress ratio. Several differences in closure effects for three kinds of oxide, fretting oxide debris, oxidation films and corrosion products, are emphasized, and their roles and mechanisms in crack closure are clarified.

1. INTRODUCTION

It is well known that opening stress intensity, K_{op}, values determined by using a back face strain unloading compliance technique [1] show good agreement with real K_{op} when only plasticity induced crack closure [2] prevails. At elevated temperatures or in corrosive environments, the mechanisms of crack closure may be different from those originally proposed by Elber [2]. Conventionally, these effects on fatigue crack growth rates, da/dN, have been examined by comparing da/dN versus ΔK_{eff} curves with da/dN versus ΔK curves, where ΔK_{eff} and ΔK are the effective stress intensity range and the stress intensity range. In quantitative discussions of the effect of crack closure, it is essential that the experimentally determined K_{op} values should agree with the real K_{op} values. Under certain crack closure mechanisms such as oxide induced crack closure near threshold, however, it has been pointed out that this method leads to an incorrect estimation of the K_{op} value [3], which may sometimes lead us to the wrong conclusion on the effect of crack closure.

In this study, the crack closure mechanisms which are encountered under expected service conditions were examined. The accuracy of the back face strain technique as influenced by several crack closure mechanisms were also examined by comparing the da/dN versus ΔK_{eff} curves with closure free da/dN versus ΔK curves at a high stress ratio, R.

Several differences in closure effects for three kinds of oxide, fretting oxide debris [4, 5], oxidation films [6–8] and corrosion products [9, 10], are emphasized, and their roles and mechanisms in crack closure are clarified.

2. MATERIAL AND EXPERIMENTAL PROCEDURE

The materials used in this study were a low alloy steel, ASTM A508-3, and high strength steels, AISI 4340, tempered at 573 K and 773 K. The chemical compositions and the mechanical properties are shown in Tables 1 and 2.

Specimens used were of the ASTM compact type with the width, $W = 51.0$ mm and the thickness, $B = 12.5$ mm. In all of the tests the data analysis procedure for ASTM E647 [11] and ASTM E24·04 working document [12] were adhered to.

Crack closure was measured by a back face strain (BFS) technique. The linear portion of the load, P, versus BFS, ε, curve was subtracted by using a differential electric circuit. The crack opening load was de-

TABLE 1

Chemical compositions (weight %)

(a) A508-3

C	Si	Mn	P	S	Ni	Cr	Cu	Mo	V	Al
0·18	0·21	1·43	0·004	0·004	0·67	0·14	0·04	0·57	0·01	0·028

(b) 4340

C	Si	Mn	P	S	Ni	Cr	Mo	Cu
0·40	0·25	0·71	<0·003	0·003	1·81	0·82	0·25	0·01

TABLE 2
Mechanical properties

(a) A508-3

Temperature (°C)	σ_{ys} (MPa)	σ_B (MPa)	δ (%)	ψ (%)	E (GPa)
R.T.	448	548	30·9	74·8	209
150	407	539	—	—	203
288	402	543	21·2	75·0	192

(b) 4340

Heat treatment		σ_{ys} (MPa)	σ_B (MPa)	δ (%)	ψ (%)
γ-temp.	Tempering temp.				
850°C (1123 K) × 1 h	300°C (573 K) × 1 h	1 430	1 700	15·4	43·0
	500°C (773 K) × 1 h	1 110	1 360	22·5	48·5

termined as a deviation point from the linear portion of the curve during each load cycle.

The following three test programs were conducted:

(1) tests in an air environment at a room temperature (RT);
(2) tests in an air environment at elevated temperatures from 353 to 623 K;
(3) tests in synthetic sea water at a temperature of 298 K.

Details of experiments and results are given in Refs. 3, 7, 8 and 10.

2.1. Room Temperature Test

For A508-3, two types of tests, stress ratio R-constant tests ($R = 0.06$, 0·3 and 0·7) and maximum stress intensity K_{max}-constant tests ($K_{max} = 10.8$, 15·5 and 31·0 MPa$\sqrt{m}$), were conducted. The test was started by stress intensity range ΔK-decreasing conditions until the threshold stress intensity range, ΔK_{th}, was achieved. Then the test was continued under ΔK-increasing conditions. The R-constant, ΔK-decreasing tests involve the stepping down of K_{max} as well as minimum stress intensity, K_{min}, for a given R value. On the other hand, the K_{max}-constant, ΔK-decreasing tests involve the stepping up of K_{min} for a given K_{max} value. In these cases, the crack length was measured by using a travelling microscope.

For 4340, room temperature tests were performed to compare them with those tested in the synthetic sea water.

2.2. Elevated Temperature Tests

For A508-3, the R-constant, ΔK-decreasing tests were conducted at a frequency of 25 Hz in a furnace. The temperature range tested was from 353 K (80°C) to 623 K (350°C). For some specimens, the tests were interrupted at ΔK_{th} to observe the fracture surface appearance and the section profile near the crack tip. For other specimens, the tests were continued by increasing ΔK. In these cases, the crack length was measured by using a DC potential drop technique.

2.3. Synthetic Sea Water Tests

For 4340, the R-constant, ΔK-decreasing tests were conducted at a frequency of 5 Hz, in synthetic sea water (temperature = 298 K (25°C), pH = 8·2). In these cases, the crack length was measured by using compliance data obtained by the BFS technique.

For comparison, stress corrosion cracking (SCC) tests were conducted to obtain the threshold stress intensity, K_{Iscc}, using WOL type specimens.

3. CRACK CLOSURE INDUCED BY FRETTING OXIDE

Figure 1 shows relations between da/dN and ΔK in the R-constant tests for A508-3. For low R values ($R = 0·06$ and $0·3$), da/dN deviates from the Paris relation, and decreases abruptly as ΔK decreases below $\Delta K = 10 \, \mathrm{MPa}\sqrt{m}$. For $R = 0·7$, the Paris relation is followed above $da/dN = 10^{-10}$ m/cycle.

The specimens were sectioned after the tests and observation was carried out by scanning electron microscopy. The fretting oxide debris existed between the crack surfaces in the near threshold region for a low R value. For a high R value, however, it did not exist except as a narrow region corresponding to ΔK_{th}. Near the crack tip, transgranular and intergranular shear modes of crack growth are dominant and the crack contour is rough, which may promote partial contact of crack surfaces [13]. As a result, the fretting oxide debris is produced not uniformly but partially between the crack surfaces. That is to say, the fracture surface roughness may be a trigger for the production of the fretting oxide debris.

If ΔK_{eff} instead of ΔK is taken as a parameter, the crack growth rates converge to a narrow band as shown in Fig. 2, where $\Delta K_{eff} = K_{max} - K_{op}$ for $K_{op} \geqq K_{min}$ or $\Delta K_{eff} = K_{max} - K_{min}$ for $K_{op} \leqq K_{min}$. It is interesting to

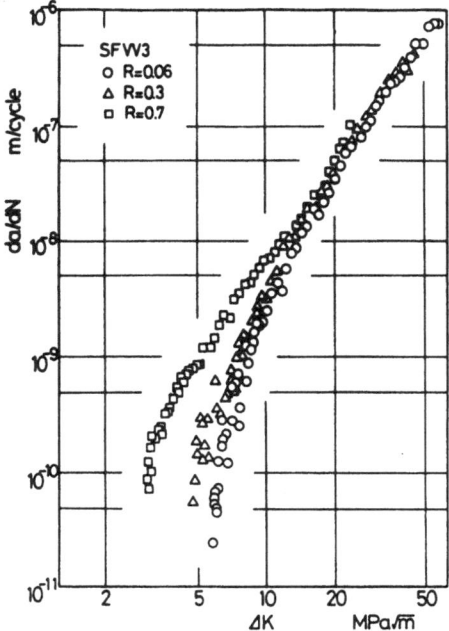

FIG 1. da/dN–ΔK relation in R-const. tests (A508-3, in air, at R.T.).

note that, for $da/dN \leqq 5 \times 10^{-9}$ m/cycle, da/dN deviates from the Paris relation and decreases with decreasing ΔK_{eff}, especially at $R = 0.06$ and 0·3, where fretting oxide induced crack closure is dominant. So, it may be concluded that the BFS method can not detect sufficiently the real K_{op} value in those regions.

For 4340 tempered at 573 K and 773 K, the general trends are the same as Figs. 1 and 2 for A508-3, except the following.

(1) ΔK_{th} of the 573 K tempered material is lower than that of the 773 K tempered material.

(2) Near threshold, the da/dN versus ΔK_{eff} curve for the 573 K tempered material for a given R tends to be located at the lower portion of that for the 773 K tempered material as ΔK decreases as shown in Fig. 3. So, the linear relationship between $\log(da/dN)$ and $\log(\Delta K_{eff})$ can stand for a wider range of ΔK in the former case, which suggests that the real K_{op} value is successfully detected by the BFS method.

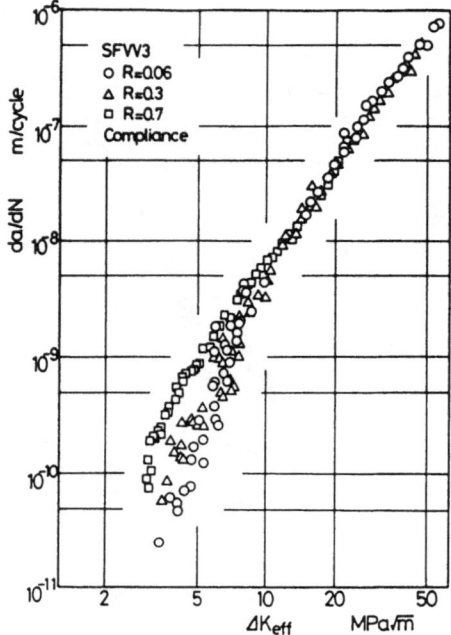

FIG 2. da/dN–ΔK_{eff} relation in R-const. tests (A508-3, in air, at R.T.).

The yield strength of the 573 K tempered material is higher than that of the 773 K tempered material. So, the above mentioned phenomena indicate that the higher the strength of the material the lower the amount of fretting oxide debris produced.

The K_{max}-constant tests for A508-3 gave a slightly lower ΔK_{th} compared with that for R = 0·7, which may be due to a possible influence of oxide induced crack closure on the latter.

4. CRACK CLOSURE INDUCED BY OXIDATION FILMS

The results of ΔK decreasing and increasing tests at higher temperatures are typically shown in Fig. 4 (373 K) and Fig. 5 (623 K).

In intermediate growth rate region ($\simeq 10^{-8} - 10^{-6}$ m/cycle), da/dN versus ΔK curves fall on one line regardless of the test temperature and R. So, the well-known linear relation between $\log(da/dN)$ and $\log(\Delta K)$ is

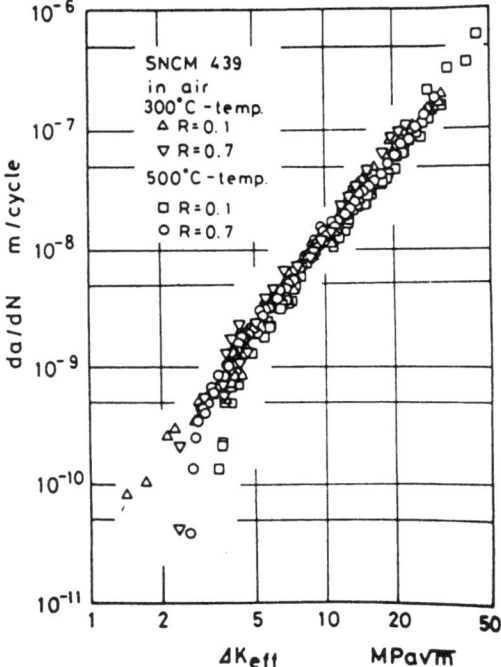

FIG. 3. da/dN–ΔK_{eff} relation in air (4340 tempered at 573 K and 733 K).

obeyed. On the other hand, in the near-threshold region ($\leqq 10^{-8}$ m/cycle), the test temperature and R have a significant influence on the relations, as summarized below.

(1) At a given temperature, the da/dN curve tends to deviate downwards from the extrapolation line of the intermediate da/dN curve as ΔK decreases. The phenomenon becomes more remarkable with decreasing R.

(2) At 623 K, especially, it deviates abruptly near ΔK_{th} (da/dN $\simeq 10^{-9}$ m/cycle). By changing the cyclic frequency, f, it was confirmed that the change of $f=1$ to 80 Hz has no influence on the curve above da/dN $\simeq 2 \times 10^{-9}$ m/cycle, and below it, da/dN decreases as f decreases. So, the deviation point is a function of f.

(3) At low R ($R=0.06$), ΔK_{th} decreases with increasing test temperature, T, up to $T=373$ K. A further increase in T increases ΔK_{th}.

(4) At high R ($R=0.7$), the relation between da/dN and ΔK remains unchanged up to $T=423$ K. Hereafter, da/dN shifts upwards with

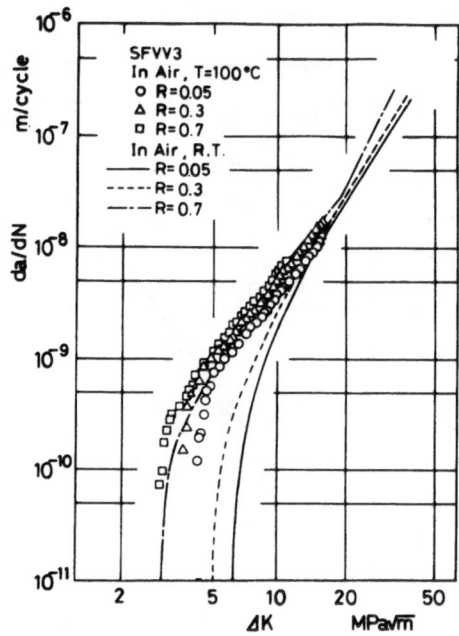

FIG 4. da/d$N-\Delta K$ relation at $T = 373$ K (A508-3, in air).

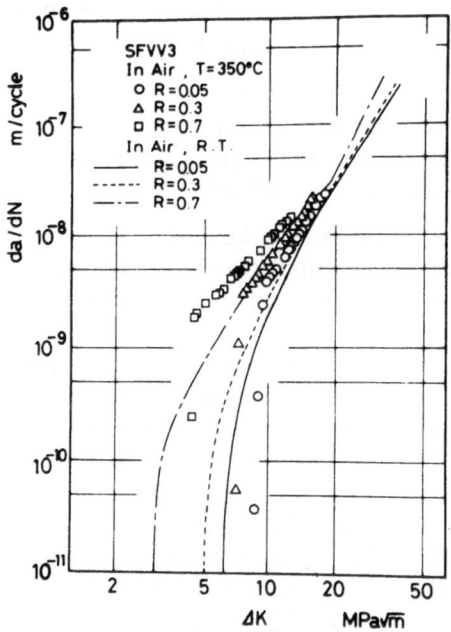

FIG. 5. da/d$N-\Delta K$ relation at $T = 623$ K (A508-3, in air).

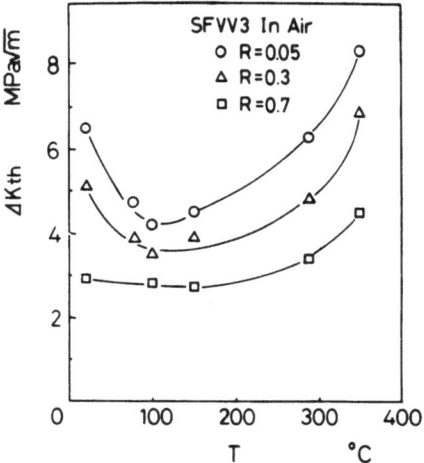

FIG. 6. $\Delta K_{th}-T$ relation (A508-3, in air).

increasing T. However, it drops rapidly near $da/dN = 10^{-9}$ m/ cycle at $T = 623$ K. So, ΔK_{th} at 623 K is higher than that at room temperature.

The dependence of T and R on ΔK_{th} is given in Fig. 6. At room temperature, there is a marked influence of R on ΔK_{th}. The R dependency decreases with increasing T up to $T = 373$ K, and it takes a minimum value there. Then it increases with increasing T.

These trends have been interpreted as occurrence of oxide induced crack closure triggered by roughness induced crack closure [13]. In such a case, the conventional approach is to introduce the effective stress intensity, ΔK_{eff}, instead of ΔK. The results obtained by the BFS method are typically shown in Figs. 7 (373 K) and 8 (623 K).

Usually, Young's modulus, E, decreases as T increases. And it is known that E is also a controlling parameter at elevated temperature at intermediate growth rates. For A508-3, E at room temperature and 623K are 209 and 192 GPa, respectively. So, the results at $T = 623$ K are modified by using $\Delta K_{eff}/E$ as a parameter. From these results, the following can be pointed out.

(5) The result at $R = 0.7$ for each test temperature gives an upper bound of the relation.

(6) At $T = 373$ and 423 K, all data fall on one curve.

(7) At $T = 623$ K, the data at $R = 0.7$ are located at the upper portion of the curve obtained at room temperature at $R = 0.7$ except near

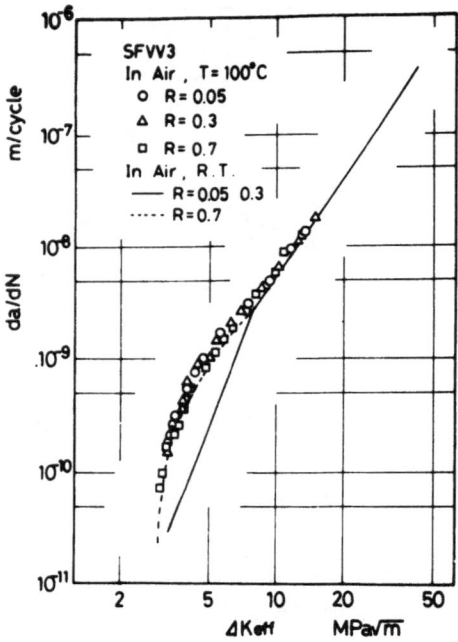

FIG. 7. da/dN − ΔK_{eff} relation at $T = 373$ K (A508-3, in air).

threshold. Near threshold, they drop suddenly to the lower portion of the curve.

The macroscopic observation of the fracture surface reveals that there existed noticeable fretting oxide debris near the threshold region tested at $R = 0.06$ at room temperature, and a thick oxidation film on the whole fracture surface tested at $T = 623$ K. The former can be easily distinguishable from the latter by their color. To evaluate the thickness, the sectioning technique was applied to each specimen. After that, the section profile was observed by using a scanning electron microscope. Thickness of fretting oxide debris decreased with increasing T up to $T = 423$ K. On the other hand, above $T = 561$ K, thickness of oxidation film increased with increasing T. These trends coincide with the results of ΔK_{th} as shown in Fig. 6.

The phenomenon is found to be coincident with Fe_3O_4 instead of Fe_2O_3 becoming a noticeable oxide constituent at higher temperature and the rate controlling factor in fatigue crack growth changing from

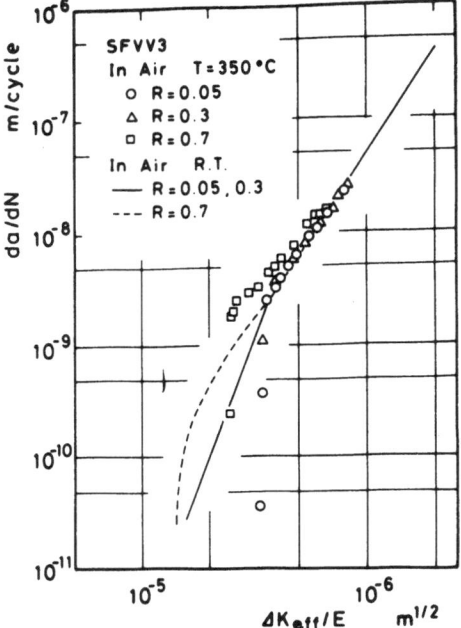

FIG. 8. da/dN-$\Delta K_{eff}/E$ relation at $T = 623$ K (A508-3, in air).

fretting to oxidation as shown in Fig. 9. So, the higher the testing temperature up to 373 K, the greater the wear protection resulting in the thinner fretting oxide debris (Fe_2O_3) on the fracture surface. Above 561 K, as the testing temperature increases, the thickness of oxidation film (Fe_3O_4) on the fracture surface increases which causes a higher value of ΔK_{th} compared with that at 373 K.

According to recent analytical work on fatigue crack, the experimentally detected K_{op} value, by using the unloading compliance method, leads to an incorrect estimation of the real K_{op} when asperities are attached on the crack surface [14]. Briefly, the following are the results obtained.

(a) When asperities are attached near crack tip, the experimentally detected K_{op} value leads to an underestimation of the real K_{op} value.

(b) When they are attached remote from the crack tip, the experimentally detected K_{op} value leads to an overestimation of the real K_{op} value. Even if thickness of the asperities is in the order of the maximum residual plastic stretch, δ_R, during steady state fatigue crack growth, this pheno-

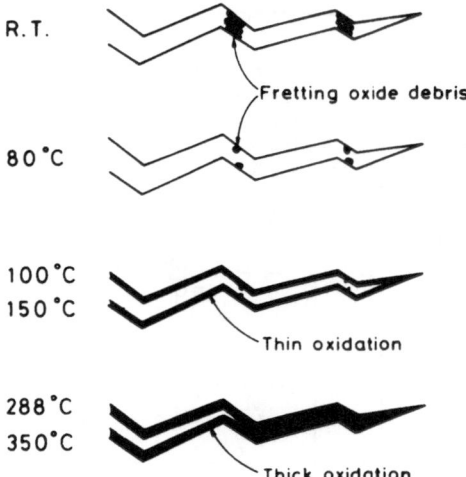

F IG. 9. Schematic diagram showing change from fretting oxide to oxidation.

menon does occur. Here, δ_R is given by the following equation,

$$\delta_R = C \, \delta_{max} \tag{1}$$

$$\delta_{max} = K^2_{max}/E\sigma_{ys} \tag{2}$$

where δ_{max} is the crack tip opening displacement for an ideal crack given in eqn. (2), σ_{ys} is the yield stress and the constant C increases with increasing R from 0·86 (for $R=0$) to 1·0 (for $R=1·0$), hence C is nearly 1·0. It should be noted that δ_{max} is different from the actual crack tip opening displacement, $\Delta\delta_{max}$, for the fatigue crack. For example, $\Delta\delta_{max} = \delta_{max} - \delta_R = 0·14 \times \delta_{max}$ at $R=0$.

From the results of (a) and (b), above mentioned items (1)–(6) can be interpreted as follows.

It requires many loading cycles and high compressive stress for the fretting oxide debris to be produced, which suggests the fretting oxide debris becomes thicker as R decreases due to high compressive stress at $K = K_{min}$, and as da/dN reaches threshold due to increase of surface contact cycles at a given location. In such a case, K_{op} is underestimated as stated in the result (a), because the fretting oxide debris is attached only near the crack tip. Figures 2 and 8 clearly indicate such trend.

It is well known that the harder the matrix results in increase of the resistance to fretting wear. The thickness of oxidation film increases as T increases. Up to $T = 423$ K, as the oxidation films are harder than matrix, they protect the fracture surfaces from fretting wear, which reduces the thickness of fretting oxide debris. This is confirmed from the result in Fig. 7.

Above $T = 423$ K, further increase in thickness of oxidation film occurs. The thickness of oxidation film increases as the fracture surfaces are exposed longer to the environment. Accordingly, it decreases at reaching the crack tip. In this case, K_{op} is overestimated as stated in the result (b). However, as da/dN decreases, the overestimated as stated in the result (b). However, as da/dN decreases, the oxidation film can be produced even near the crack tip, which ressults in the underestimation of K_{op} as stated in the result (a). The typical example is shown in Fig. 8.

5. CRACK CLOSURE INDUCED BY CORROSION PRODUCTS

The K_{Iscc} values were obtained as $15 \cdot 2$ MPa$\sqrt{}$m (tempered at 573 K) and $69 \cdot 6$ MPa$\sqrt{}$m (tempered at 773 K). In the following section, the fatigue crack growth characteristics especially below these values are considered.

The da/dN versus ΔK curves for the 573 K tempered material are shown in Figs. 10 ($R = 0 \cdot 1$) and 11 ($R = 0 \cdot 7$). In Fig. 10, da/dN in

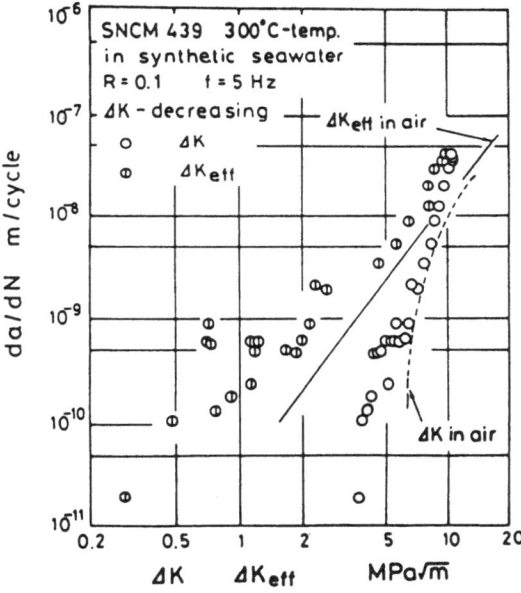

FIG. 10. $da/dN - \Delta K$, ΔK_{eff} relations for $R = 0 \cdot 1$ in synthetic seawater (4340 tempered at 573 K).

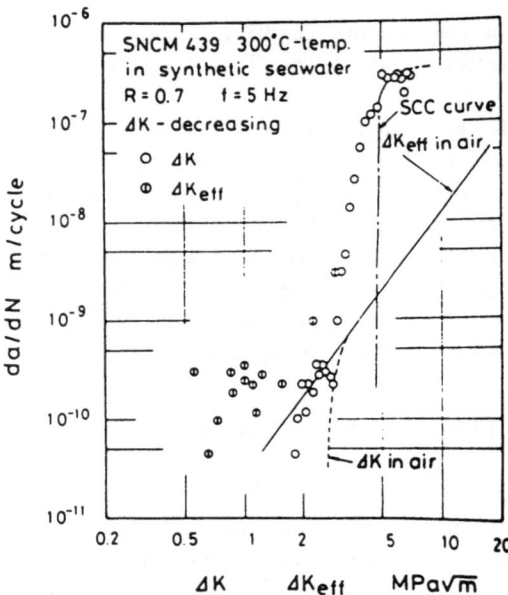

Fig. 11. da/dN − ΔK, ΔK_eff relations for R = 0·7 in synthetic seawater (4340 tempered at 573 K).

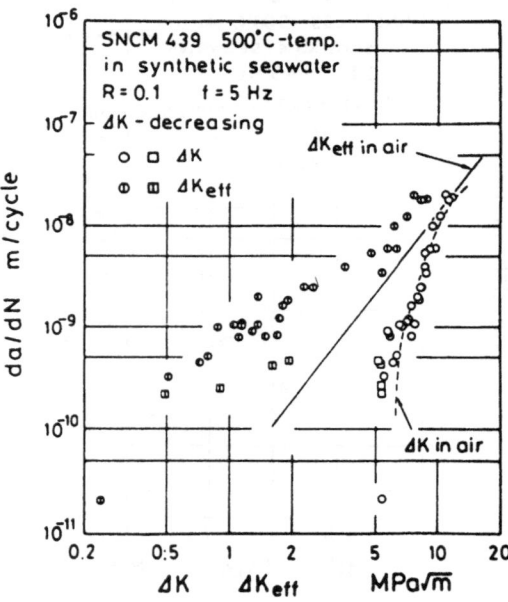

Fig. 12. da/dN − ΔK, ΔK_eff relations for R = 0·1 in synthetic seawater (4340 tempered at 773 K).

synthetic sea water is remarkably accelerated as compared to the rate in air due to the interaction effect between SCC and the fatigue crack growth for $\Delta K \geqq 9$ MPa$\sqrt{\text{m}}$ ($K_{max} \geqq 10$ MPa$\sqrt{\text{m}}$). For $\Delta K \leqq 9$ MPa$\sqrt{\text{m}}$, the former is slightly accelerated as compared to the latter. However, the former does not exceed the $da/dN - \Delta K_{eff}$ curve in air.

In Fig. 11, the same trend can be observed at $\Delta K \geqq 3$ MPa$\sqrt{\text{m}}$ ($K_{max} \geqq 10$ MPa$\sqrt{\text{m}}$). Again, the da/dN versus ΔK curve does not exceed the da/dN versus ΔK_{eff} curve in air at $R = 0.7$ as in Fig. 10.

The results for the 773 K tempered material are shown in Figs. 12 ($R = 0.1$) and 13 ($R = 0.7$). In both cases, da/dN decreases with decreasing ΔK following the da/dN versus ΔK curve in air. Near threshold, they show a slightly accelerated trend compared to those in air. However, da/dN does not exceed the $da/dN-\Delta K_{eff}$ curve in air as for the 573 K tempered material.

In Figs. 10–13, da/dN versus ΔK_{eff} curves in synthetic sea water are also shown. In every case, those curves show the accelerated trend compared to the da/dN versus ΔK_{eff} curve in air.

These results have been interpreted as the effect of the stress induced

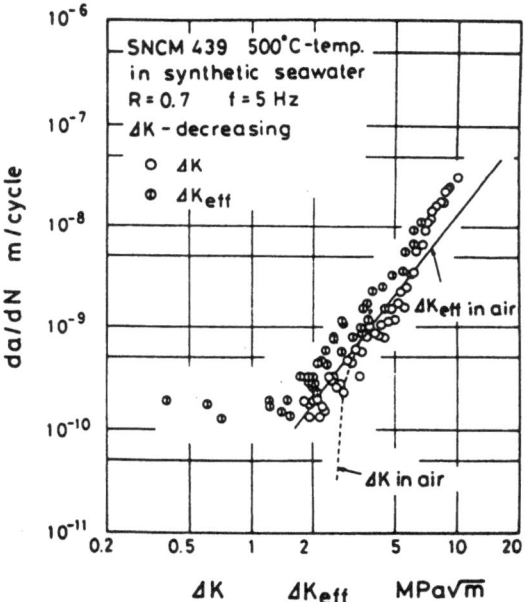

FIG. 13. $da/dN - \Delta K$, ΔK_{eff} relations for $R = 0.7$ in synthetic seawater (4340 tempered at 773 K).

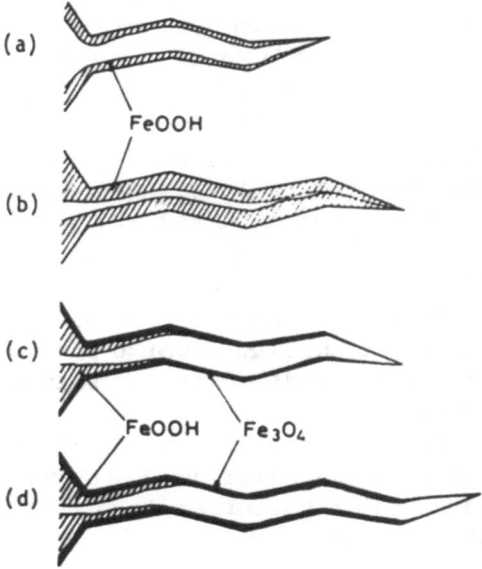

FIG. 14. Schematic diagram showing corrosion products attached on crack surfaces.

corrosion dissolution at the crack tip [9]. To examine such effect, fractographic observation was performed. For the 573 K tempered material, the fracture mechanism was intergranular for both R's when $K_{max} \geqq 10 \ MPa\sqrt{m}$. For the 773 K tempered material, it appeared when $K_{max} \geqq 20 \ MPa\sqrt{m}$ only for $R = 0.7$. It should be noted an indication of dissolution was not observed for either case. Therefore, the acceleration in these regions should be attributed to the interaction effect between hydrogen induced cracking and the fatigue crack growth. In other regions, the fracture mechanism was transgranular. Although, for the 573 K tempered material, intergranular fracture was observed near threshold as in air.

Examination of the fracture surfaces reveals that corrosion products attached to the crack surfaces were FeOOH and Fe_3O_4. The latter covers the whole fracture surfaces except near the crack tip (between the crack tip and 0.5 mm behind). On the other hand, the former is found to exist only near the initial notch tip which is well behind the crack tip, and is considerably thick compared with the latter. Generally, when the dissolved oxygen level becomes lower, Fe_3O_4 is produced. As the crack grows from the initial notch tip, the corrosion products attached near the notch tip prevent dissolved oxygen transportation to the region near the crack tip. This promotes the production of Fe_3O_4 near the crack tip. These aspects are schematically shown in Fig. 14. It is interesting to note that FeOOH is formed only in the near-threshold region. Thus, dissolution occurred remote from the crack tip which reduces the thickness of the plastic residual stretch, the surface roughness and the fretting oxide debris on the fracture surface. And this causes acceleration of da/dN compared to data in air as shown in Figs. 10–13.

It may therefore be concluded that FeOOH has no influence on the actual crack growth rate and the relation between da/dN and ΔK near threshold. However, the thicker the FeOOH products, the K_{op} value determined by using the back face strain method becomes higher which leads to the overestimation of the real K_{op} value. Because, when asperities are attached remote from the crack tip as shown in Fig. 15, the experimentally detected K_{op} value leads to the overestimation of the real K_{op} value as mentioned earlier. Even if the thickness of the asperities is in the order of the maximum residual plastic stretch, δ_R, during steady state fatigue crack growth, this phenomenon does occur.

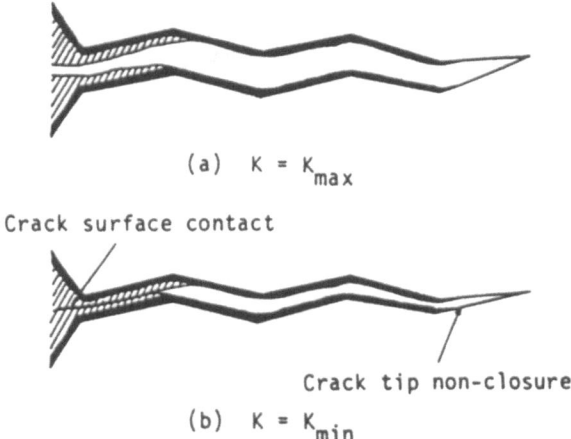

(a) $K = K_{max}$

Crack surface contact

Crack tip non-closure

(b) $K = K_{min}$

FIG. 15. Schematic diagram showing crack surface contact near initial notch tip.

Accordingly, at least for those particular materials, it may be suggested that da/dN versus ΔK_{eff} curve in air can give a conservative estimate of the da/dN versus ΔK curve in synthetic sea water.

6. CONCLUSION

In this study, the crack closure mechanisms which are encountered under expected service conditions were examined. The accuracy of the back face strain technique under some crack closure mechanisms were also examined by comparing the da/dN versus ΔK_{eff} curves with closure free da/dN versus ΔK curves at a high stress ratio, R.

Several differences in effects between three kinds of oxide, fretting oxide debris, oxidation films and corrosion products, were emphasized, and their roles and mechanisms in crack closure were clarified. It is suggested that the experimentally detected K_{op} values do not give correct information regarding the real K_{op} values under some conditions. Especially in corrosive environments, the experimentally detected K_{op} value leads to an overestimation of the real K_{op}. On the other hand, in the near-threshold region in an air environment, the K_{op} value leads to an underestimation.

REFERENCES

1. KIKUKAWA, M., JONO, M. and TANAKA, K. *Proc. Second Int. Conf. on Mechanical Behavior of Materials*, ASM special volume, 1976, pp. 254–77.
2. ELBER, W., ASTM STP 486, 1971, pp. 230–42.
3. KOBAYASHI, H., OGAWA, T., NAKAMURA, H. and NAKAZAWA, H. *Advances in Fracture Research, Proc. 6th ICF Inter. Conf. on Fracture*, Vol. 4, Oxford, Pergamon Press, 1984, pp. 2481–8.
4. STEWART, A. T., *Engng Fract. Mech.*, **13** (1980), 463–78.
5. SURESH, S., ZAMISKI, G. F. and RITCHIE, R. O., *Metall. Trans.*, **12A** (1981), 1435–43.
6. LIAW, P. K., SAXENA, A., SWAMINATHAN, V. P. and SHIH, T. T., *Metall. Trans.*, **14A** (1983), 1631–40.
7. KOBAYASHI, H., TSUJI, U., PARK, K. and NAKAZAWA, H., *Trans. Japan Soc. Mech. Engrs (JSME). Ser. A*, **50** (1984), 1003–9 (in Japanese).
8. KOBAYASHI, H., PARK, K., TSUJI, U., KANAZAWA, H. and NAKAZAWA, H. *Ibid.* **51** (1985), 799–804 (in Japanese).
9. ENDO, K., KOMAI, K. and SHIKIDA, T., ASTM STP 801, 1983, pp. 81–95.
10. KOBAYASHI, H. ISHIZAKI, S. and GAO, H., *Trans. Japan Soc. Mech. Engrs (JSME). Ser. A*, **52** (1986), 1778–85 (in Japanese).
11. Standard Test Method for Constant-Load-Amplitude Fatigue Crack Growth Rates above 10^{-8} m/cycle, ASTM Standards, E647–83.
12. Proposed ASTM Test Methods for Measurement of Fatigue Crack Growth Rates, ASTM STP 738, 1981, pp. 340–56.
13. MINAKAWA, K. and McEVILY, A. J., *Scripta Met.*, **15** (1981), 633–6.
14. NAKAMURA, H. and KOBAYASHI, H., appeared in ASTM STP, Fatigue Crack Closure, 1987.

22

Fatigue Strength of Shot Peened Carburized Steel

ATSUTOMO KOMINE, MASAO KIKUCHI, TOHRU YAMAGUCHI and YASUTADA KIBAYASHI

Technical Research Center, Komatsu Ltd, 1200 Manda, Hiratsuka, Kanagawa, 254 Japan.

ABSTRACT

The common problem of carburized steels is the reduction of the fatigue strength by the existence of surface structure anomalies and the sensitivity of the mechanically induced notches. In order to improve the surface origin fatigue fracture resistance, shot peening has been applied. The significance of the distribution of residual stresses and hardness within the shot peened surface layer of the specimen is discussed. Approaches to developing extremely high fatigue strength are presented.

1. INTRODUCTION

Many treatments have been applied in the past to improve the fatigue strength of mechanical parts. Table 1 shows the comparison of the rotating bending fatigue strength of typical treatments for high fatigue durability. Although thermomechanical treatment was found to be the most effective method to improve fatigue strength, its application is restricted to the very simple shape parts because of the difficulty of the process [1]. In contrast to thermomechanical treatment, gas carburizing is quite an effective heat treatment to obtain high fatigue durability and is widely accepted in industry. The problem of carburized steels is the reduction of the fatigue strength by the existence of the internal oxides precipitated along the grain boundaries and the accompanying non-martensitic microstructure near the surface [2–4].

In this study, the influence of the surface structure anomalies on the

343

TABLE 1
Comparison of the rotating bending fatigue strength

MPa

	0	500	1000	1500
Ausforming			▬	
Quench and temper		▬▬▬▬		
Maraging		▬▬		
Tufftride		▬		
Induction hardening		▬▬▬		
Carburizing		▬▬▬		

fatigue behavior of carburized steel was evaluated through the rotating bending fatigue tests. Shot peening has been applied to improve the surface-originated fatigue fracture resistance. The optimization of the shot peening condition and heat treatment process to obtain high fatigue strength is discussed.

2. MATERIAL AND EXPERIMENTAL PROCEDURE

The material used in this study was Cr–Mo steel (JIS SCM415). The chemical composition is shown in Table 2. The geometries of the round bar type rotating bending fatigue specimen are shown in Fig. 1. After machining, specimens were carburized in the condition shown in Fig. 2.

Shot peening was conducted using a rotating blowing-type peening machine. A specimen without surface structure anomalies was prepared by removing the surface layer about 50 μm in depth after carburizing. Fatigue tests were carried out by a rotating bending fatigue test machine

TABLE 2
Chemical composition

C	Si	Mn	P	S	Cr	Mo
0·16	0·26	0·74	0·012	0·013	1·01	0·18

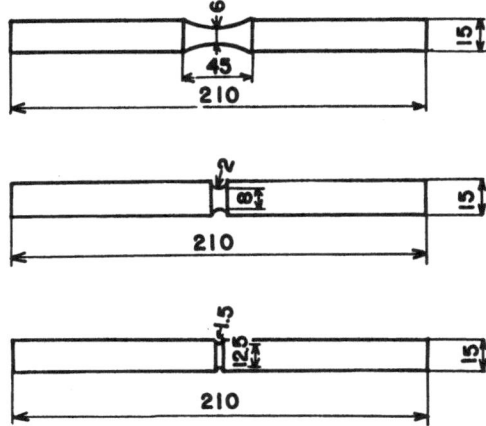

FIG. 1. Dimensions of specimen.

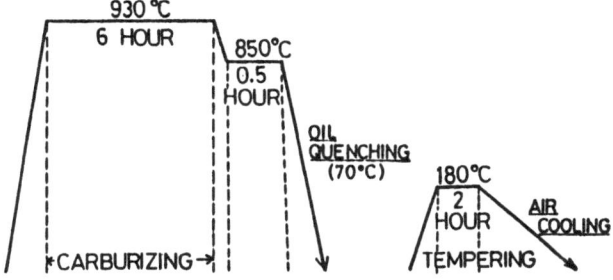

FIG. 2. Condition of heat treatment.

of 98 N m in load capacity and 3000 rpm in cycle speed in a laboratory atmosphere. The Vickers' hardness distribution near the surface was measured on the taper section of the specimen. The residual stress was measured using X-ray stress measuring equipment.

3. RESULTS AND DISCUSSION

3.1. Fatigue Strength of Carburized Steel

Because of the existence of the surface structure anomalies (Fig. 3), the hardness near the surface of the carburized specimen decreases as shown in Fig. 4. The residual stress also decreases near the surface. Figure 5

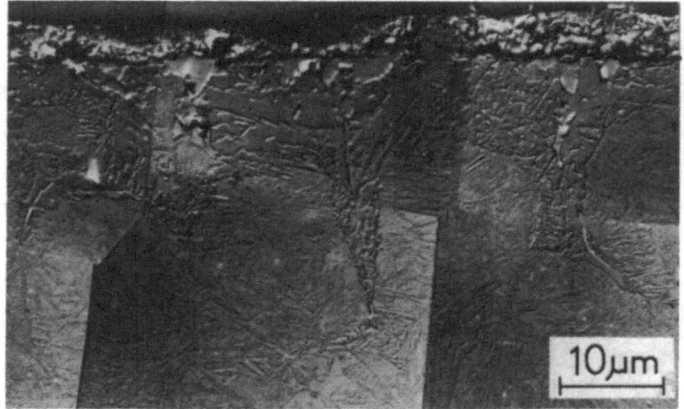

FIG. 3. Microstructure near the surface of fatigue test specimen after heat treatment.

shows the difference of the fatigue strength of the carburized smooth specimen with and without surface structure anomalies. The fatigue strength of the specimen without surface structure anomalies is higher than that of the as-carburized specimen. The notch sensitivity of the carburized specimen was also examined. Figure 6 shows the S–N diagram of the notched carburized specimen. The notch reduces the fatigue strength severely.

3.2. Hardness, Residual Stress and Retained Austenite
Figure 7 shows the difference in the hardness, compressive residual stress

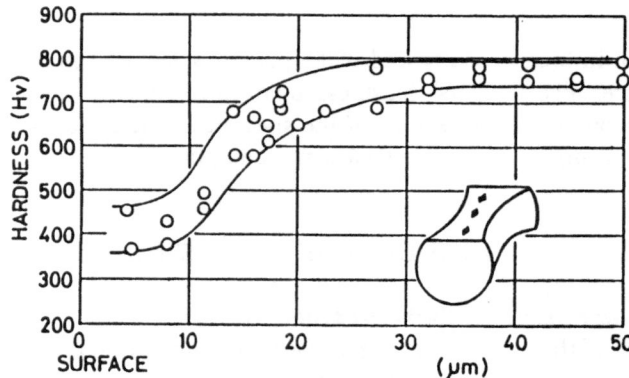

FIG. 4. Hardness distribution near the surface of fatigue specimen after heat treatment.

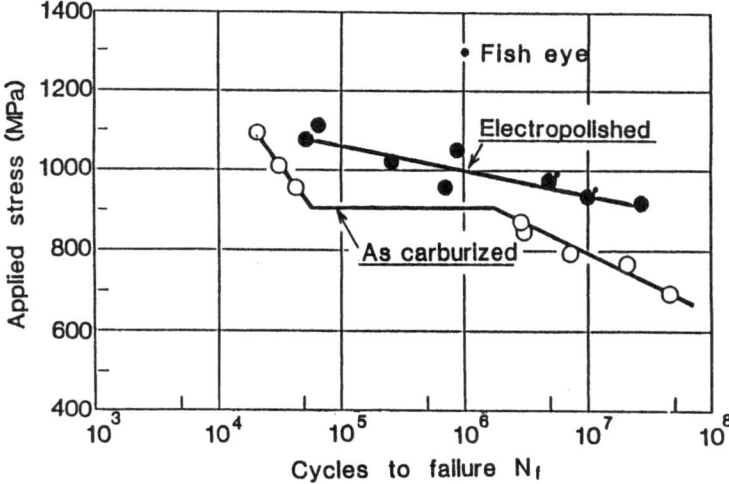

FIG. 5. *S–N* diagram on carburized smooth specimen both with and without surface structure anomalies.

and retained austenite distribution near the surface between the as-carburized specimen and the shot peened carburized specimen. Shot peening conditions were 0·6 mm diameter shot particles, 46 m/s shot speed and 10 min peening time. The hardness and compressive residual

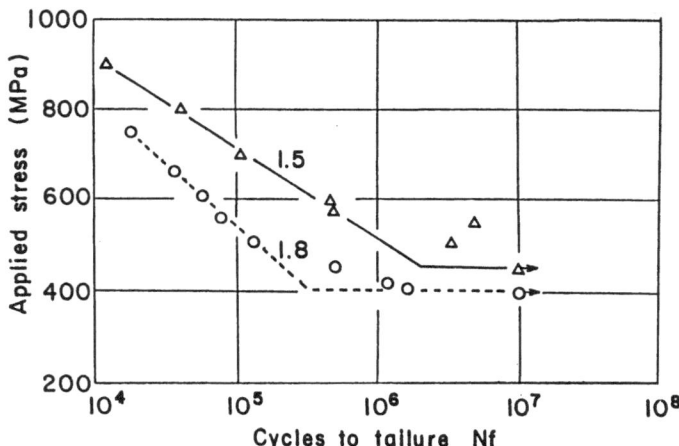

FIG. 6. *S–N* diagram on carburized notched specimen (stress concentration factors 1·5 and 1·8).

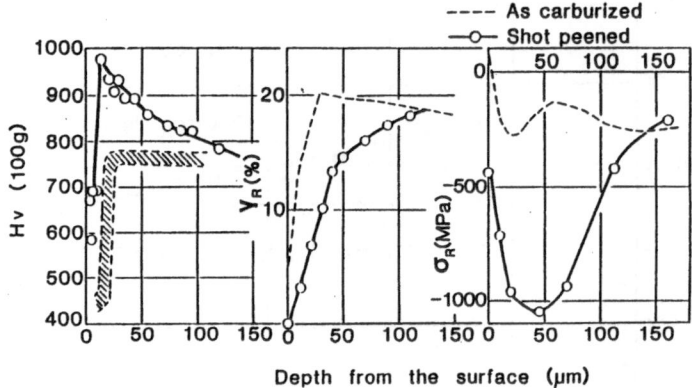

FIG. 7. Hardness, residual stress and retained austenite distribution near the surface of the specimen.

stress increased by shot peening and the amount of the retained austenite decreased. The change of these properties were observed to a depth of about 130 μm. In other words, the affected depth of shot peening was 130 μm.

In the case of 0·8 mm diameter shot particles with the same speed and the same time of those of 0·6 mm diameter shot particles, the affected zone depth of the shot peening was increased to 200 μm.

3.3. Fatigue Strength of Shot Peened Smooth Specimen

Substantial gain in fatigue strength of carburized steel [6] can be seen from the results of the increase of the hardness and compressive residual stress. Figure 8 shows the results of the rotating bending fatigue test on shot peened specimens treated with three sizes of shots, 0·3, 0·6, 0·8 mm in diameter. Shot speed and shot peening time were 46 m/s and 10 min. The fatigue strength of the shot peened specimens was about 50% higher than that of the as-carburized specimen and increased with increasing shot size. The fractographic difference observed on these specimens is the position of the fracture origin. In high-cycle fractured specimens with 0·3 mm shot, the origin of fracture was at the surface of the specimen, for others with 0·6 and 0·8 mm shot the origins were subsurface.

Figure 9 shows the difference in the fatigue strength with different peening times. The shot size used is 0·6 mm and the shot speed is 46 m/s. Fatigue strength of the shot peened specimen increased with increasing peening time and saturated over 5 min of the peening time.

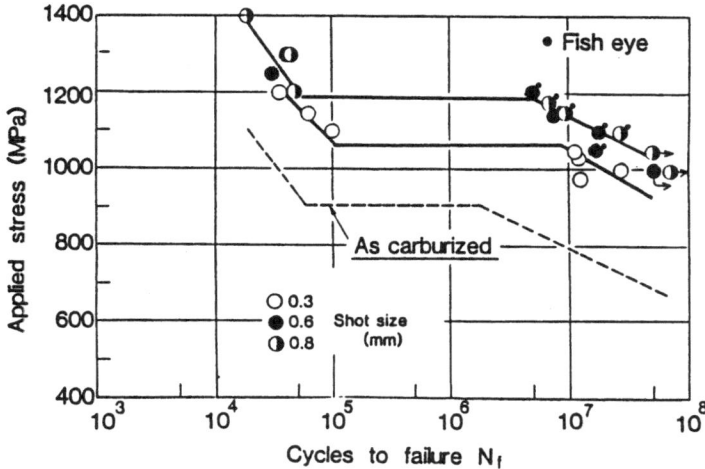

FIG. 8. *S–N* diagram for the smooth specimen shot peened with three sizes of shot.

It was evident that the fatigue strength of the specimen without surface structure anomalies was higher than that with surface structure anomalies. Figure 10 shows the results of the rotating bending fatigue tests on the shot peened specimen both with and without surface structure anomalies. The fatigue strength of both specimens increased and became comparable after shot peening.

3.4. Fatigue Strength of Shot Peened Notched Specimen
Figure 11 shows the *S–N* diagram on the shot peened notched

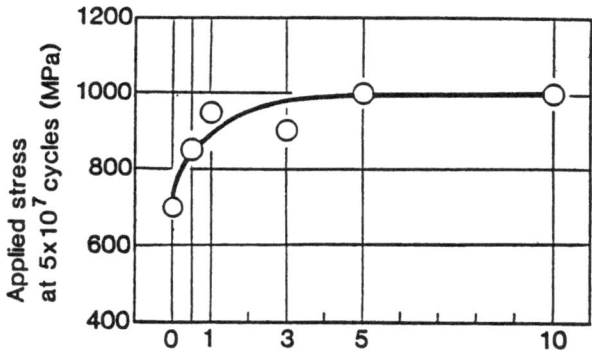

FIG. 9. The effect of shot peening time on the fatigue strength of smooth specimens.

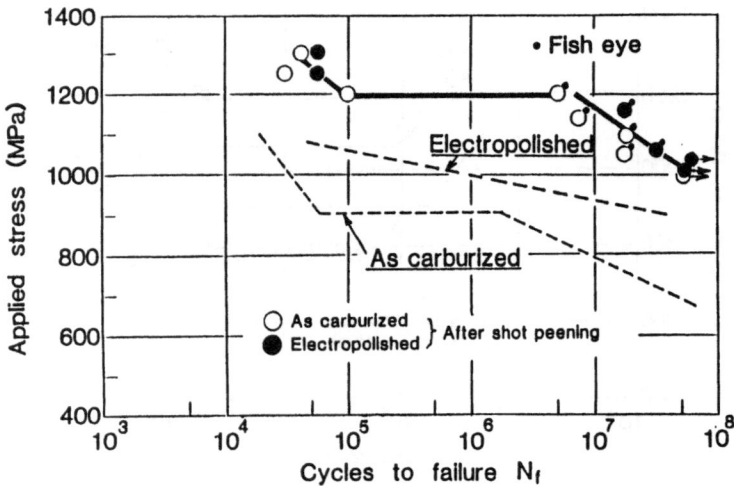

FIG. 10. S–N diagram for shot peened smooth specimens both with and without surface structure anomalies.

specimens (stress concentration factors 1·5 and 1·8). The conditions of shot peening of these specimens was 0·8 mm shot, 46 m/s shot speed and 10 min peening time. Fatigue strength increased with shot peening as for the smooth specimen. The 10^7 cycle fatigue limit of the shot peened

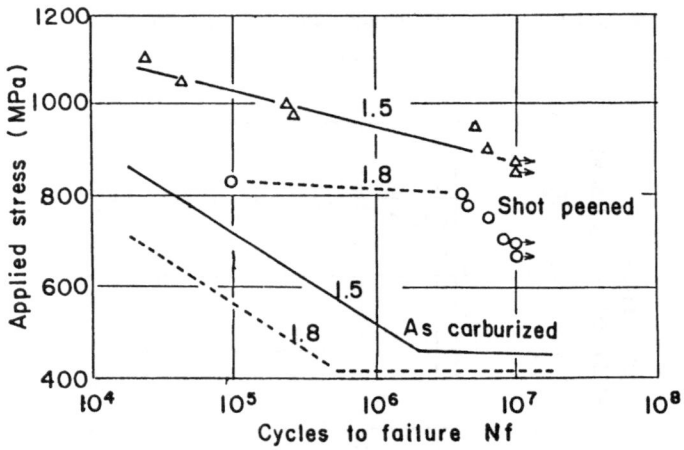

FIG. 11. S–N diagram for as-carburized and shot peened notched specimen (stress concentration factors 1·5 and 1·8).

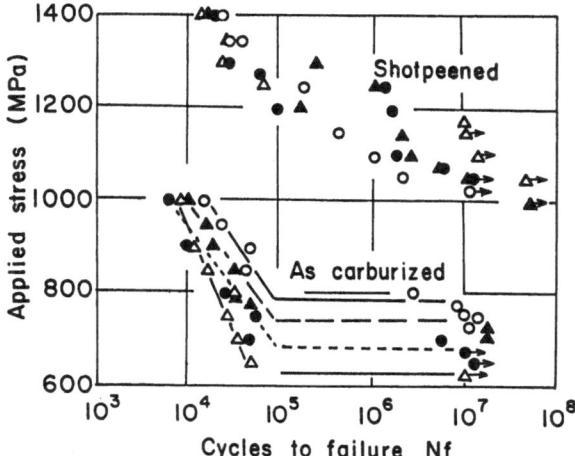

FIG. 12. *S–N* diagram for as-carburized and shot peened smooth specimen with varying surface roughness (surface roughness before shot peening; ○ 1·3 s, ▲ 3 s, ● 10 s, △ 30 s).

specimen with stress concentration factor 1·5 is 90% higher than that of the as-carburized specimen. In the case of stress concentration factor 1·8, the increase is 75%. The increase of fatigue strengths of notched specimens is higher than those of smooth specimens. The difference in increase between notched and smooth specimens is due to the difference in the affected depth by shot peening.

In the practical case of machine parts, the surface usually has severe machine marks. The fatigue strength of the carburized specimen decreases with increasing surface roughness. Figure 12 shows the results of rotating bending fatigue tests on a shot peened specimen with varying surface roughness. The conditions of shot peening were 0·8 mm shot size, 46 m/s shot speed and 10 min peening time. The fatigue strength of the specimens with surface roughness ranging from 1·3 s to 30 s increased and became comparable after shot peening. This means that shot peening is practically quite effective to improve the fatigue strength of carburized machine parts.

4. CONCLUSION

(1) Surface structure anomalies of the carburized specimens reduced fatigue strength.

(2) Shot peening increased fatigue strength of carburized smooth specimens as much as 50%.
(3) Fatigue strengths of carburized specimens both with and without surface structure anomalies became comparable to each other after shot peening.
(4) The increase rate of the fatigue strength of the notched carburized specimens by shot peening is higher than that of smooth specimens.
(5) Fatigue strength of the specimen with surface roughness ranging from 1·3 s to 30 s increased and became comparable after shot peening.

REFERENCES

1. SHYNE, J. C., *Trans, ASM.,* **52** (1960), 346.
2. NAKANISHI, E., UEDA, H. and KAJIURA, T., *J. Soc. Mater. Sci. Japan,* **26**, 280 (1977), 68.
3. NAITO, T., UEDA, H. and KIKUCHI, M., *Metall. Trans. A,* **15A** (1984), 1431.
4. CHATTERJEE-FISHER, R., *Metall. Trans. A,* **9A** (1978), 1553.
5. STRAUB, J. C., SAE paper 730800 1973.
6. KIKUCHI, M., UEDA, H., HANAI, K. and NAITO, T., *Advances in Surface Treatments,* Vol. 3, Oxford, Pergamon Press, (in press).

23

Mode I Propagation of Delamination Fatigue Cracks in CFRP

MASAKI HOJO, CLAES-GÖRAN GUSTAFSON*, KEISUKE TANAKA** and RYUICHI HAYASHI

Industrial Products Research Institute, Higashi 1-1-4, Yatabe-machi, Tsukuba-gun, Ibaraki 305, Japan.

*Swedish Plastics and Rubber Institute, Roslagsvägen 101, Hus 15, S104-05, Stockholm, Sweden.
**Department of Engineering Science, Kyoto University, Kyoto 606, Japan.

ABSTRACT

A test method to obtain the fatigue crack growth data near the threshold region was first established by using unidirectional laminates made from Ciba Geigy prepregs 914C (T300/914) and Toray prepregs P305 (T300/No. 2500). Then the effect of the stress ratio on near-threshold growth of delamination fatigue cracks was investigated. A new type of loading apparatus for the double cantilever beam specimen was developed to measure the crack length by the compliance method. During load shedding under a stress ratio between 0·1 and 0·7, a normalized gradient of energy release rate was kept constant over the range of $-0·3$ to $-3\,mm^{-1}$. The relation between the crack propagation rate and the fracture mechanics parameters is independent of the load-shedding condition from the threshold region to the high rate region. The results of a load-increasing test agree well with those of a load-decreasing test. In the region of rates above about $5 \times 10^{-10}\,m/cycle$, the rate is expressed as a power function of the fracture mechanics parameters. Below this region, there exists the growth threshold. The influence of the stress ratio becomes smaller when the rate is correlated to the energy release rate range than when the rate is correlated to the stress intensity range or the maximum energy release rate. The relative contributions of the maximum stress and of the cyclic stress on the crack growth rate are discussed on the bases of fractographic observation and mechanism consideration.

1. INTRODUCTION

Interlaminar fracture or delamination is a predominant mode of fatigue failure in laminated composite structures subjected to cyclic loading [1, 2]. In cross-ply or angle-ply laminates, high interlaminar stresses are developed at the edge of the laminates owing to the difference in stiffness of individual plies [3]. Small interlaminar defects which are often introduced into composite structures during the molding process generally give rise to nucleation of delamination fatigue fracture, which is driven by high interlaminar stresses.

The fracture mechanics approach has been successfully used in damage tolerance design of metallic structures. This approach often adopts the growth behavior of fatigue cracks near the threshold for fail-safety assessment of structures. In composite laminates, fracture mechanics was recently used to analyze the growth behavior of delamination fatigue cracks [4–6]. However, most of these studies dealt only with crack growth at high rates. Thus, researches on fatigue crack growth in the low rate region [7, 8] are urgently required to assure the reliability of composite laminates used as main structural components.

In the present paper, a test method to obtain the fatigue growth data of delamination cracks near the threshold region was first developed. The loading mode was mode I opening. Then, the influence of the stress ratio on near-threshold crack growth was investigated, by using two different laminates of unidirectional CFRP made of different prepregs. The controlling fracture mechanics parameter was discussed in conjunction with the fatigue crack growth mechanism.

2. EXPERIMENTAL PROCEDURE

2.1. Materials

The laminates were made from prepregs of Ciba Geigy 914C and prepregs of Toray P305. Prepreg 914C (carbon fiber Toray T300/epoxy resin Ciba Geigy 914) has a 175°C curing temperature and is used in the aircraft industries. Prepreg P305 (carbon fiber Torey T300/epoxy Toray No. 2500) has a 120°C curing temperature and is used for general purposes. The constitution of laminates and the elastic moduli are summarized in Table 1 [9].

2.2. Specimens

Two different set-ups for tests of delamination fatigue crack pro-

TABLE 1
Materials and elastic constants

Prepreg Carbon fiber Epoxy	Ciba geigy 914C Toray T300 Ciba geigy 914	Toray P305 Toray T300 Toray no. 2500
Volume fraction of fiber	66·3%	60%
Constitution of laminate	$(0)_{60}$	$(0)_{32}$
Elastic constants (GPa)	$E_1 = 138, E_2 = 10, G_{12} = 3\cdot6$ $v_{12} = v_{23} = v_{13} = 0\cdot35$	$E_1 = 102, E_2 = 7\cdot8, G_{12} = 4\cdot1$ $v_{12} = 0\cdot34$

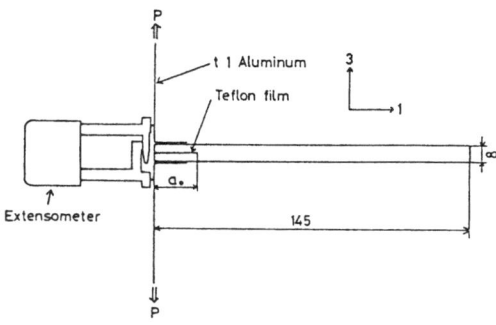

FIG. 1. DCB specimen with Al tabs (dimensions are in mm).

pagation were built as shown in Figs. 1 and 2. In both cases, double cantilever beam (DCB) specimens (width = 20 mm) were used, and the fiber direction was parallel to the crack growth direction (X_1 axis).

The DCB specimen shown in Fig. 1 was used by Wilkins *et al.* [4]. Aluminum tabs are bonded to the specimen for load transmission. While fairly long tabs are required to avoid the influence of bending, these long tabs often cause unnecessary vibration during testing. Moreover, tabs are easy to debond during fatigue testing. Thus we designed a new type of loading apparatus which was shown in Fig. 2. Two aluminum blocks are bonded on the specimen and the cyclic load is applied to the specimen through hinges attached to the aluminum blocks. A starter slit of about 20 mm in length is introduced into the specimen by inserting PTFE film (10–30 µm in thickness) during fabrication. Crack opening displacement on the load line was measured by an extensometer attached to the specimen as shown in Figs. 1 and 2.

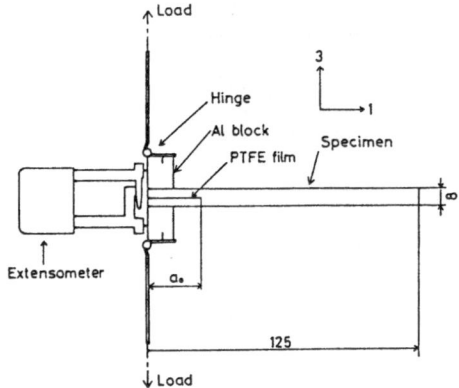

FIG. 2. DCB specimen with hinges (dimensions are in mm).

2.3. Fracture Mechanics Parameters

Two fracture mechanics parameters, i.e. the stress intensity factor and the energy release rate are calculated by the following procedure.

The energy release rate, G, is related to the change in compliance, C, as

$$G = 1/(2b) P^2 dC/da \qquad (1)$$

where P is the applied load, b is the width of the specimen, and a is the crack length. The relation between energy release rate and stress intensity factor is given by

$$G = HK^2 \qquad (2)$$

where coefficient H is a function of the elastic moduli (E_{ij}, v_{ij}) [10]. The H value is calculated as follows for each laminate under the plain strain condition: $H = 8.40 \times 10^{-2} \text{GPa}^{-1}$ for 914C laminates, and $H = 9.24 \times 10^{-2} \text{GPa}^{-1}$ for P305 laminates.

Figure 3 shows the relation between compliance and crack length obtained experimentally by using a specimen with Al tabs (Fig. 1). The specimen was used for three tests under different stress ratios. For each case of the stress ratio, the compliance is expressed as a power function of crack length.

$$C = Da^n \qquad (3)$$

where D and n are the coefficients computed using a least squares program. Jumps in C were introduced when the stress ratio was changed.

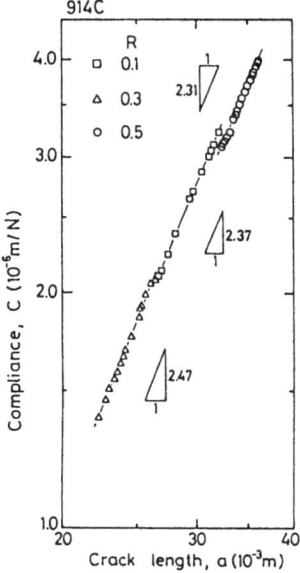

F IG. 3. Change of compliance for specimen with Al tabs.

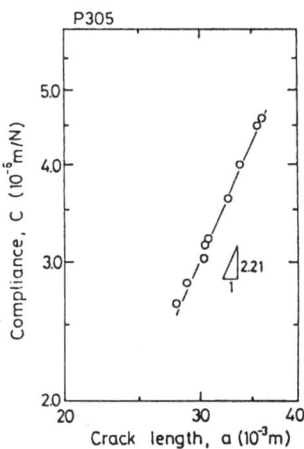

F IG. 4. Change of compliance for specimen with hinges.

These jumps are probably due to small deformations of long Al tabs during re-chucking.

Figure 4 shows the result obtained by using the newly designed specimen with hinges (Fig. 2). Although the stress ratio was changed between 0.2 and 0.7, C is a unique function of a.

The experimentally obtained compliance is larger than that computed by the elementary beam theory. This deviation comes from the deformation of the material ahead of the crack tip [9]. The values of G and K are calculated from eqns (1–3) in the following experiments. Since the coefficient of eqn. (3) is affected by the bonding condition of Al blocks and the thickness of the specimen, the coefficients were measured experimentally for each specimen with hinges. On the other hand, for the specimen with Al tabs, these coefficients were determined under each test condition to avoid the influence of stress ratio on C.

2.4. Crack Length Measurement

The length of a fatigue crack was measured on both sides of the specimen with traveling microscopes at magnifications of 400×. The surface of the specimen was polished and painted with white ink to make crack-length measurement easier.

The length of a crack was also computed from the measurement of compliance by using eqn. (3). The specimen compliance was determined from the load–displacement relation averaged over successive 10 cycles at an interval of 1000 cycles. Figure 5 shows the crack length calculated by the compliance method. The triangle marks show the crack length

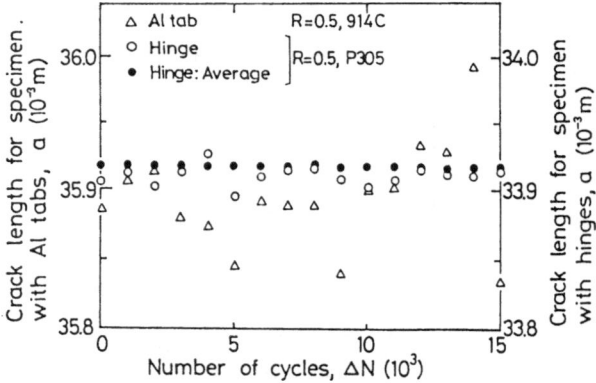

FIG. 5. Crack length calculated from compliance near the threshold.

obtained by using the specimen with Al tabs (Fig. 1). The resolution of crack length is about $\pm 50\,\mu$m, which is not enough for the tests near the threshold region. On the other hand, the crack length measurement by using the specimen with hinges (Fig. 2) has a resolution of $\pm 10\,\mu$m (see the open circles). The solid circles show the average of 50 points indicated with the open circles. This method gives a resolution of about $1\,\mu$m. The crack length determined from compliance yields the average crack length in the width direction of the specimen.

In the measurement of crack length in the present experiment, the compliance method was used for the specimen with hinges, while the direct measurement with microscopy was used for the specimen with Al tabs.

2.5. Fatigue Tests

The fatigue tests were carried out in a computer-controlled 10 kN servohydraulic fatigue testing machine. A load cell of 490 N capacity was attached to the testing machine. The stress ratio, R, of the minimum to the maximum stress was kept constant during each test. A normalized gradient of energy release rate, $(1/G)dG/da$, was controlled. The frequency of stress cycling was 5 Hz for 914C laminates under $R = 0.1$ and 0.3, and 10 Hz for the other tests. The tests were carried out in laboratory air ($23°C$, 50% RH).

3. EXPERIMENTAL RESULTS AND DISCUSSION

3.1. Effect of Load Shedding Condition

G-decreasing tests were conducted by controlling the normalized gradient of energy release rate, $(1/G)dG/da$, which is equal to twice as much as the normalized gradient of stress intensity factor, $(1/K)dK/da$. Tests of 914C laminates under $R = 0.2$ were carried out at normalized G-gradients of -0.32, -1.1 and -3.2 mm^{-1}.

Figure 6 shows the change in the energy release rate range, ΔG, with crack length, where $\Delta G = G_{max} - G_{min}$ (G_{max} and G_{min} are the maximum and minimum values of the energy release rate corresponding to the maximum and minimum stresses). In this figure, the slope of the straight line is the normalized gradient of G. In Fig. 7, the rate of fatigue crack propagation, da/dN, is plotted against ΔG. The load shedding condition does not influence the relation between da/dN and ΔG.

A G-decreasing test at a gradient of -0.77 mm^{-1} was followed by a G-

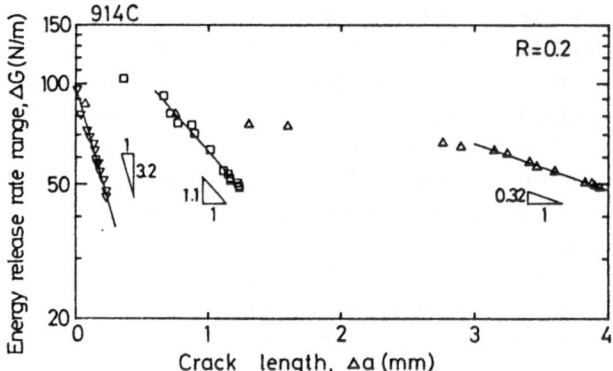

FIG. 6. Change in ΔG with a in 914C laminate (G-decreasing test).

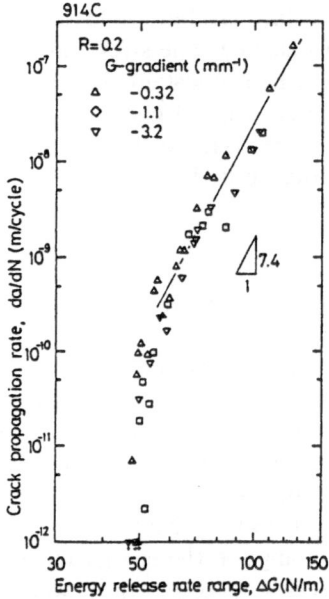

FIG. 7. Effect of G-gradient in 914C laminate (G-decreasing test).

increasing test at a gradient of 0·60 mm^{-1}. The variation of ΔG with a is shown in Fig. 8, and the corresponding da/dN–ΔG relation is shown in Fig. 9. The results of the G-decreasing test (the open circles) agree well with those of the G-increasing test (the solid circles) for the whole region.

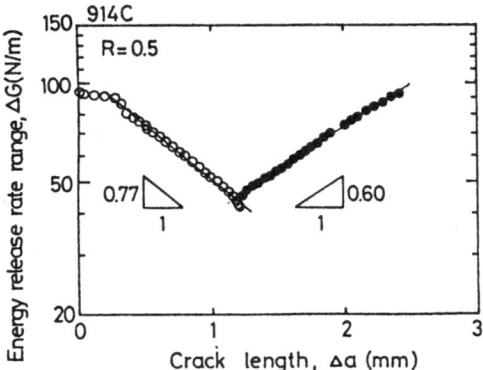

FIG. 8. Change in ΔG with a in 914C laminate (G-decreasing and increasing test).

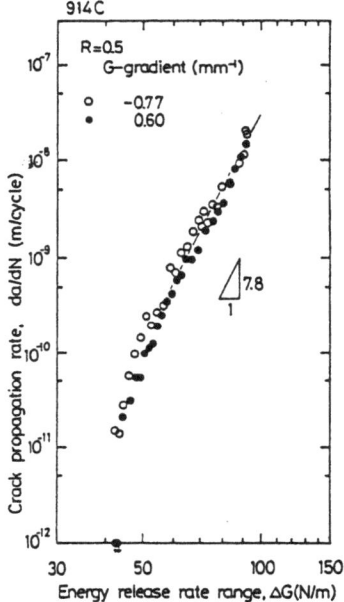

FIG. 9. Effect of G-gradient in 914C laminate (G-decreasing and increasing test).

Similar results for the case of P305 laminates are shown in Figs. 10 and 11. The relation between da/dN and ΔG is again independent of the G-gradient.

On the basis of the above results, it can be concluded that a normal-

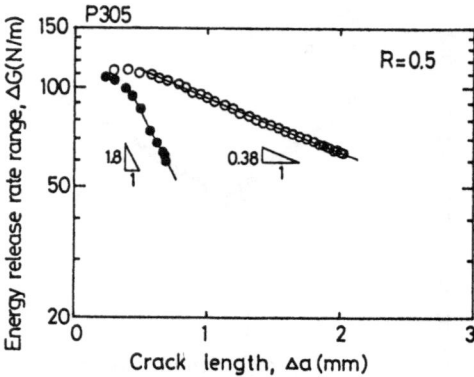

FIG. 10. Change in ΔG with a in P305 laminate (*G*-decreasing test).

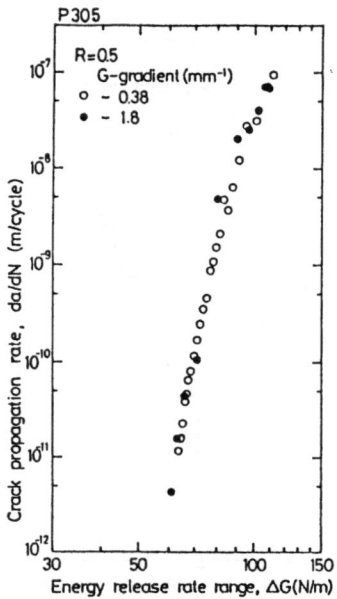

FIG. 11. Effect of *G*-gradient in P305 laminate (*G*-decreasing test).

ized *G*-gradient of about $-1\,\text{mm}^{-1}$ is slow enough to avoid the influence of the load shedding condition on the fatigue crack growth rate for the case of the present experimental materials. This allowable limit of the normalized *G*-gradient for CFRP in *G*-decreasing tests is rather high compared with the ASTM recommendation for metallic materials (-0.16 mm^{-1}) [11].

Bathias and Laksimi [12] reported that glass fiber/epoxy laminates had no retardation effect under temporary overloads, but showed instead a sudden instantaneous crack growth. They also reported that damage accumulation was approximately linear under programmed loading at two levels. These facts also suggest the least amount of load interaction.

3.2. Effect of Stress Ratio

The relation between da/dN and ΔK for 914C laminate is shown with the open marks in Fig. 12 and that for P305 laminate is shown in Fig. 13. For both materials, da/dN is expressed as a power function of ΔK in the region where da/dN is larger than about 5×10^{-10} m/cycle. Below this region, da/dN deviates to the lower rate from the power functions, and there exists the growth threshold of fatigue cracks. The figures attached to the lines in Figs. 12 and 13 indicate the exponents of the power functions. These exponents are much higher than those of metallic materials. In particular, the exponent increases with the increase of stress ratio in P305 laminate. When compared at the same ΔK, da/dN increases with increasing R value.

Only near the threshold region of 914C laminate under $R = 0.1$, was crack closure detected with the compliance method. Thus, the effective range of stress intensity factor, $\Delta K_{eff} = K_{max} - K_{op}$, was calculated, where K_{max} and K_{op} are the stress intensity factors corresponding to the maximum and crack opening stresses. The relation between da/dN and ΔK_{eff} is shown with the solid squares in Fig. 12. The amount of crack closure is small. The values of K_{op} or P_{op} are almost constant without respect to the rate or P_{max}. This crack closure may come from the roughness of the fracture surface.

In Figs. 14 and 15, the growth rate da/dN is plotted against the maximum energy release rate, G_{max}. For the case of 914C laminate, the effect of stress ratio on growth behavior becomes smaller than that in Fig. 12. When compared at the same G_{max} in Figs. 14 and 15, da/dN decreases with increasing R value. It is interesting to note that the effect of R in the $da/dN-G_{max}$ relation is reversed in the $da/dN-\Delta K$ relation.

The growth rate da/dN is plotted against the range of energy release rate, $\Delta G = G_{max}(1 - R^2)$, in Figs. 16 and 17. Difference in rates due to the stress ratio is greatly reduced. For the case of 914C laminate, the effect of stress ratio on the $da/dN-\Delta G$ relations is almost negligible in the power law region. There is little dependence of R on the threshold values. For

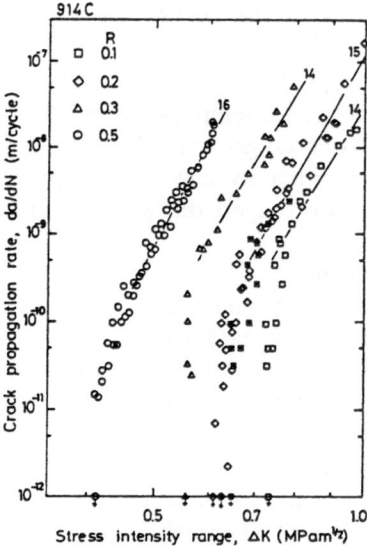

FIG. 12. Relation between da/dN and ΔK in 914C laminate.

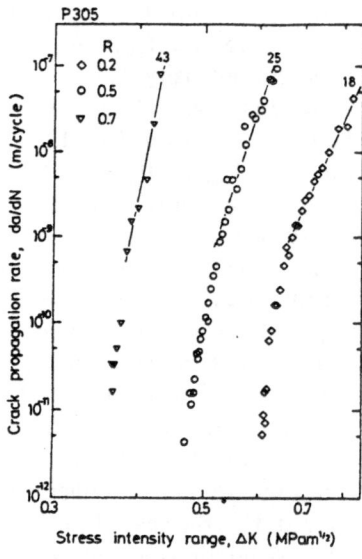

FIG. 13. Relation between da/dN and ΔK in P305 laminate.

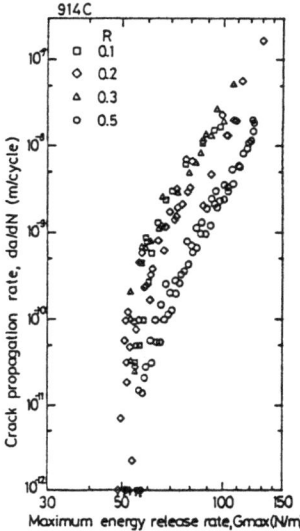

FIG. 14. Relation between da/dN and G_{max} in 914C laminate.

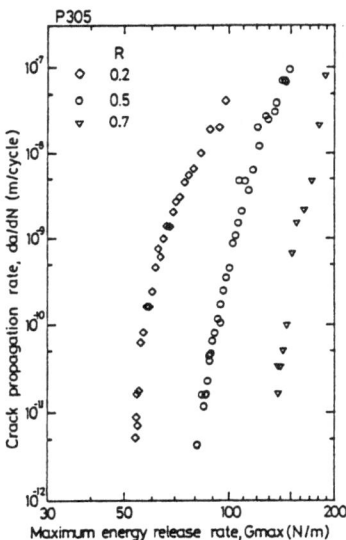

FIG. 15. Relation between da/dN and G_{max} in P305 laminate.

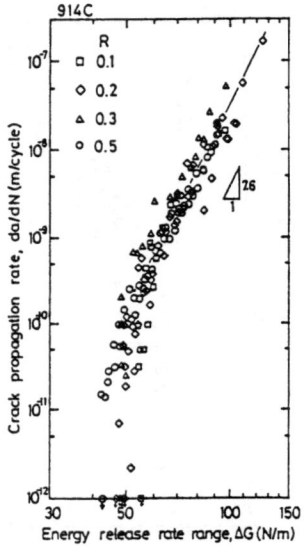

FɪG. '16. Relation between da/dN and ΔG in 914C laminate.

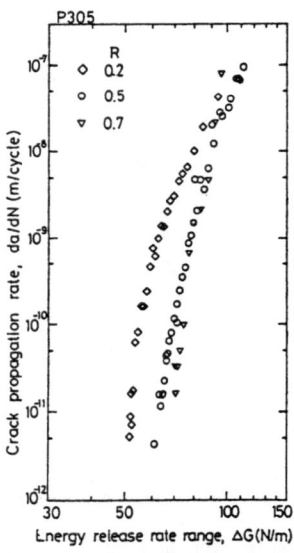

FɪG. 17. Relation between da/dN and ΔG in P305 laminate.

the case of P305 laminate, the data show smaller dependence of R in the da/dN–$\varDelta G$ relation than in the da/dN–$\varDelta K$ or da/dN–G_{max} relations. The threshold values of $\varDelta K$, K_{max}, $\varDelta G$ and G_{max} at the rate of about 10^{-11} m/cycle are summarized in Table 2. Although it can be concluded

TABLE 2
Stress intensity parameters at threshold condition

Material	R	$\varDelta K_I$ $(MPa\sqrt{m})$	K_{Imax}	$\varDelta G$ (N/m)	G_{max}
	0·1	0·73	0·81	54	55
	0·2	0·61	0·76	47	49
914C					
	0·3	0·55	0·79	48	52
	0·5	0·41	0·82	43	56
	0·2	0·61	0·76	52	54
P305	0·5	0·47	0·95	63	82
	0·7	0·37	1·22	70	137

that $\varDelta G$ is the best parameter among three parameters in correlating the fatigue crack growth rate under different R values, the physical meaning of $\varDelta G$ is not so clear as that of $\varDelta K$ or K_{max}. A better correlating parameter is necessary especially for P305 laminate. It will be a mixed parameter of $\varDelta K$ and K_{max} as discussed below.

3.3. Microscope Observation
Figure 18 shows a micrograph of a fatigue crack taken from the

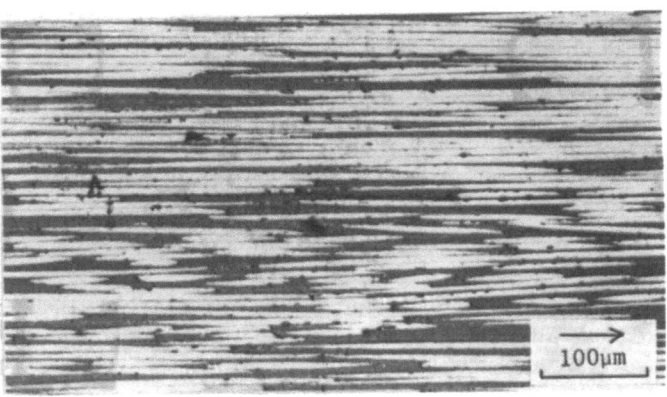

FIG. 18. Longitudinal section of crack in 914C laminate. $R = 0·5$, $da/dN = 1 \times 10^{-9}$ m/cycle.

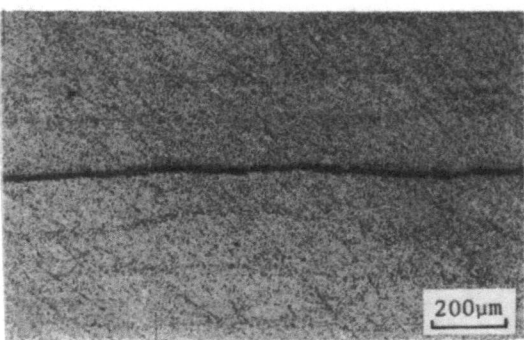

FIG. 19. Transverse section of crack in 914C laminate. $R = 0.3$, $da/dN = 1 \times 10^{-8}$ m/cycle.

longitudinal section of the specimen made of 914C laminate. Figure 19 is a micrograph of the transverse section of the specimen. The white parts in both figures are carbon fibers. There are resin rich layers corresponding to the prepreg interfaces. It is evident that the crack propagates on a single prepreg interface. No branches of cracks were observed. Moreover, no fiber breakage or fiber bridging was found by macroscopic observation of the fracture surface. Plastic deformation around the crack is supposed to be small.

Figure 20 presents scanning electron micrographs of the fracture surfaces of 914C laminate. The arrow indicates the growth direction. These three figures are characterized as matrix dominated surfaces. No significant difference was observed between the surface of fatigue fracture and that of static fracture. Moreover, there is no influence of the growth rate on the feature of the fracture surface. These figures suggest that the mechanism of fatigue crack propagation resembles that of static fracture.

3.4. Mechanism of Fatigue Fracture

In Fig. 21, the threshold value of ΔK and the ΔK value at $da/dN = 10^{-8}$ m/cycle are plotted against a stress ratio parameter. The fit of the data to the straight lines (eqn. 4) is fairly good.

$$\Delta K = \Delta K_0 (1 - R)^\gamma \qquad (4)$$

where ΔK_0 is the extrapolated value of ΔK to $R = 0$. The γ value is 0.86 at the threshold and 0.76 at $da/dN = 10^{-8}$ m/cycle in 914C laminate. In P305 laminate, $\gamma = 0.51$ at the threshold and 0.63 at $da/dN = 10^{-8}$ m/cycle. When ΔG is constant, ΔK is expressed as

$$\Delta K = \Delta K_0 ((1 - R)/(1 + R))^{1/2} \qquad (5)$$

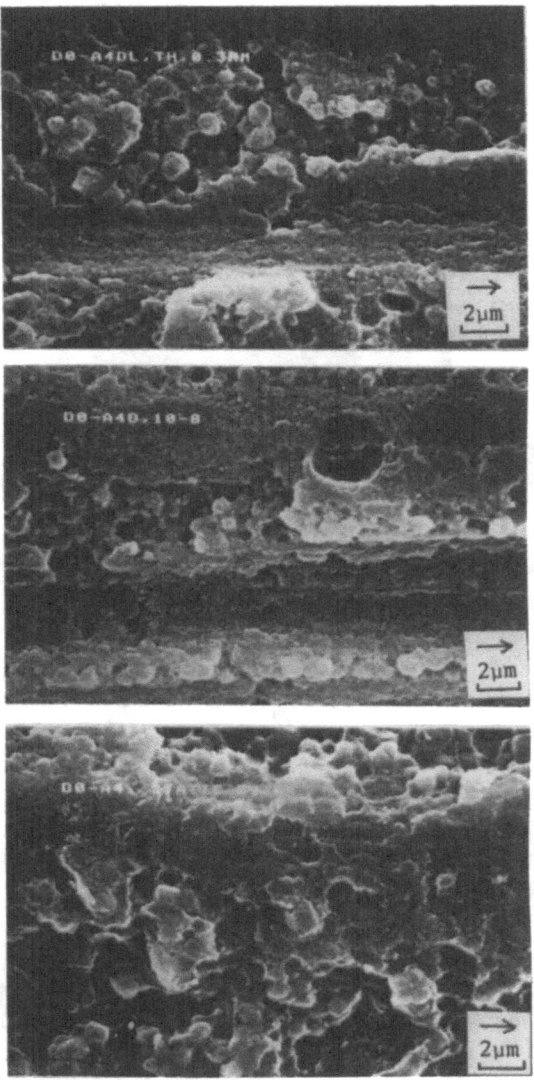

FIG. 20. Scanning electron micrographs of fracture surfaces in 914C laminate. (Top) Fatigue; $R = 0.5$, $\mathrm{d}a/\mathrm{d}N = 2 \times 10^{-10}$ m/cycle;   (middle)   fatigue; $R = 0.5$, $\mathrm{d}a/\mathrm{d}N = 1 \times 10^{-8}$ m/cycle; (bottom) static fracture.

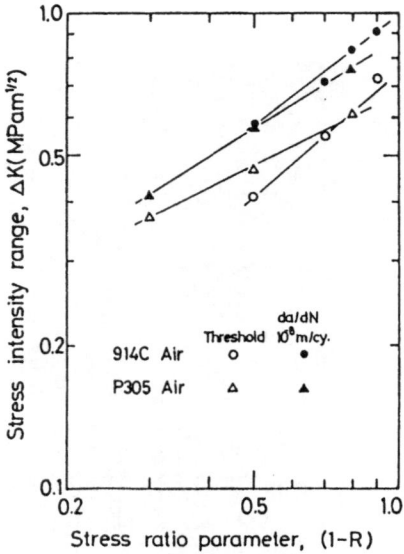

FIG. 21. Effect of R on ΔK.

The R-dependence on eqn. (5) is similar to that on eqn. (4) when $\gamma = 0.7$–0.8. This is the explanation why the rate is well correlated to ΔG for 914C laminate.

The expression of eqn. (4) is the same as that obtained for metallic materials by Klesnil and Lukas [13]. In metallic materials, crack closure causes the R-dependence on the rate near the threshold region, and the rate is correlated to ΔK_{eff} without respect to R [15]. On the other hand, the effect of crack closure in CFRP is negligible. Thus, the mechanism of fatigue crack growth in CFRP is different from that in metallic materials. Equation (4) gives that $\Delta K(1 - R)^{-\gamma}$ is constant. This parameter is expressed by ΔK and K_{max} as follows:

$$\Delta K(1 - R)^{-\gamma} = \Delta K^{1-\gamma} K_{max}^{\gamma} \tag{6}$$

In this equation, γ and $1 - \gamma$ indicate the contributions of the maximum stress and the cyclic stress on the crack growth rate. The high values of γ suggest that the mechanism of fatigue crack propagation is more or less controlled by the maximum stress. The results of SEM observation also support this fact. The values of γ in 914C laminate are larger than those in P305 laminate. When compared at the same R under high R values, ΔK of 914C laminate is lower than that of P305 laminate. This comes from the fact that epoxy resin becomes more brittle as the curing temperature is increased.

4. CONCLUSIONS

(1) The load-shedding condition, i.e. a normalized gradient of energy release rate below about -1 mm^{-1} does not influence the relation between the crack propagation rate and the energy release rate range from the threshold region to the high rate region. Moreover, the results of a load-increasing test agree well with those of a load-decreasing test.

(2) The crack propagation rate is expressed as a power function of the stress intensity range in the region of rates above about 5×10^{-10} m/cycle. Below this region, there exists the growth threshold of fatigue cracks. The exponent of the power functions are much higher than those for metallic materials.

(3) The effect of stress ratio is large in the crack propagation rate versus the stress intensity range relation and in the rate versus the maximum energy release rate relation from the threshold region to the power law region. The stress-ratio dependence on the rate becomes small in the relation between the rate and the energy release rate range.

(4) The ratio of contribution of the maximum stress to that of the cyclic stress is determined from the relation between the stress intensity range and the stress ratio at a certain rate. The degree of contribution of the maximum stress is high. This degree in 914C laminate is higher than that in P305 laminate.

(5) A fatigue crack propagates on a single prepreg interface without fiber breakage. Scanning electron micrographs showed no significant difference between the fracture surface of fatigue fracture and that of static fracture.

REFERENCES

1. O'BRIEN, T. K., ASTM STP775, 1982, p. 140.
2. GUSTAFSON, C. G. and SELDEN, R. B., ASTM STP876, 1986, p. 448.
3. PIPES, P. B. and PAGANO, N. J., *J. Comp. Mater.*, **4** (1970), 538.
4. WILKINS, D. J., EISENMANN, J. R., CAMIN, R. A., MARGOLIS, W. S. and BENSON, R. A., ASTM STP775, 1982, p. 168.
5. WANG, A. S., SLOMIANA, M. and BUCINELL, R. B., ASTM STP876, 1985, p. 135.
6. LEFEBVRE, D. and BATHIAS, C., ICCM-V, 1985, p. 331.
7. GUSTAFSON, C. G. and HOJO, M., *J. Reinforced Plastics and Composites*, **6** (1987).
8. GUSTAFSON, C. G., JILKEN, L. and GRADIN, A., ASTM STP876, 1985, p. 200.

9. KAGEYAMA, K. and KOBAYASHI, T., Preprint of 10th Symposium for Composite Materials, Japan Society for Composite Materials, 1985, p. 97.
10. SIH, G. C., PARIS, P. C. and IRWIN, G. R., *Int. J. Fract. Mech.*, **3** (1965), 189.
11. ASTM Committee E24, ASTM STP738, 1981, p. 340.
12. BATHIAS, C. and LAKSIMI, A., ASTM STP876, 1985, p. 217.
13. KLESNIL, M. and LUKAS, P., *Engng Fract. Mech.*, **4** (1972), 77.
14. OHTA, A. and SASAKI, E., *Engng Fract. Mech.*, **9** (1977), 307.
15. NAKAI, Y., TANAKA, K. and NAKANISHI, T., *Engng Fract. Mech.*, **15** (1981), 3–4, 291.

24

Dynamic and Cyclic Stress Corrosion Cracking Resistance of Metals

K. K OMAI and K. M INOSHIMA

Department of Mechanical Engineering, Kyoto University, Yoshida-honmachi, Sakyo-ku, Kyoto, 606, Japan

ABSTRACT

This paper deals with the influences of high-frequency small vibratory loads and low frequency variation on stress corrosion cracking (SCC) crack growth behavior of material/environment systems sensitive to active path corrosion (APC) and/or hydrogen embrittlement (HE) type SCC, and the concept of dynamic and cyclic SCC is proposed. The effects of vibratory stress range and stress cycle frequency on dynamic SCC and those of wave form and cathodic overprotection on cyclic SCC are summarized. And the importance of dynamic stresses on SCC crack growth is emphasized.

1. INTRODUCTION

With the progress of science and technology, there is a great demand for using high-strength materials. However, high-strength materials are sensitive to environments, and stress corrosion cracking (SCC) sometimes occurs. Machines and structures are frequently subjected to dynamic loads such as a sustained stress with high-frequency small vibratory stresses superimposed and/or low frequency variation. In the former case, superimposed small vibratory stresses are considered not only to cause damage of passive films formed at SCC crack tips, but also to affect the behavior of hydrogen in metals. Consequently, SCC crack growth under a sustained stress with high-frequency vibratory stresses superimposed (termed 'dynamic SCC' (DSCC)) and the threshold stress intensity factors for dynamic SCC, K_{DSCC}, are considered to be important problems. In the latter case, on the other hand, cyclic SCC is brought about

373

and the threshold values, K_{FSCC}, decrease from static SCC threshold, K_{ISCC} [1, 2].

In this paper, sustained load SCC tests with and without small vibratory loads superimposed and cyclic SCC tests under varying loads are performed on materials sensitive to active path corrosion (APC) type SCC and/or hydrogen embrittlement (HE) type SCC. Influences of stress ratios [3–5] and stress cycle frequency [4] on dynamic SCC crack growth and those of wave form [6] and cathodic overprotection [7] on cyclic SCC crack growth are investigated. And we discuss the SCC crack growth behavior under dynamic loads to elucidate the relation of static SCC under a sustained load (R being unity), cyclic SCC under a varying load (R being near zero), and dynamic SCC under high frequency vibratory loads (R being near unity).

2. DYNAMIC SCC CRACK GROWTH CHARACTERISTICS

2.1. Effects of Stress Ratios on Dynamic SCC Crack Growth

Figure 1 [3] illustrates the relation between crack growth rates, da/dt, and maximum stress intensity factors, K_{max}, in sensitized ZK 141 (0·14%–Cu, 0·19%–Fe, 0·06%–Si, 0·34%–Mn, 1·7%–Mg, 4·4%–Zn, 0·17%–Cr, 0·04%–Ti, 0·16%–Zr, Bal–Al, $\sigma_B = 416$ MPa) and 7075 (0·37%–Cu, 0·024%–Fe, 0·005%–Si, 2·11%–Mg, 6·19%–Cu, 0·001%–Ti, Bal–Al, $\sigma_B = 458$ MPa) alloys which belong to the 7XXX series of aluminum alloys sensitive to APC type SCC. A corrosive environment was 3·5% NaCl solution at 298 K, and the dynamic loads were sinusoidal waves with a stress cycle frequency f of 30 Hz. The threshold stress intensity factor at $R = 0.98$ of dynamic SCC, K_{DSCC}, was almost equal to that of static SCC, K_{ISCC}, at $R = 1.0$, At $R \leq 0.965$, however, K_{DSCC} was considerably lower than K_{ISCC}; the smaller the stress ratios, the lower the threshold values. Similarly, crack growth rates of dynamic SCC involving $R = 0.98$ in Region II, where subcritical crack growth is observed, were higher than those of static SCC; the smaller the stress ratios, the higher the crack growth rates.

SCC crack growth by an anodic dissolution mechanism is considered to be dominated by a process of damage (rupture) and restoration (repassivation) of passive films formed at crack tips, and the threshold condition of passive films not being damaged corresponds to the values of K_{ISCC}. For dynamic SCC, damage of the passive films was promoted by high frequency small vibratory stresses superimposed on a sustained stress, which resulted in an enhancement of da/dt and a fall of K_{DSCC}.

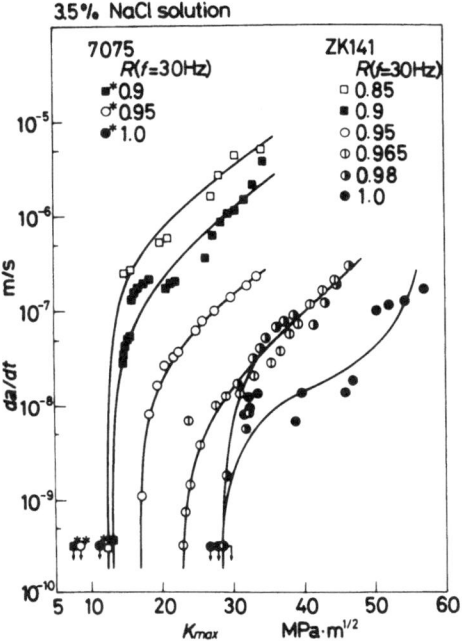

FIG. 1. da/dt–K_{max} plots; ZK 141 and 7075 alloys, 3·5% NaCl solution, $T = 298$ K.

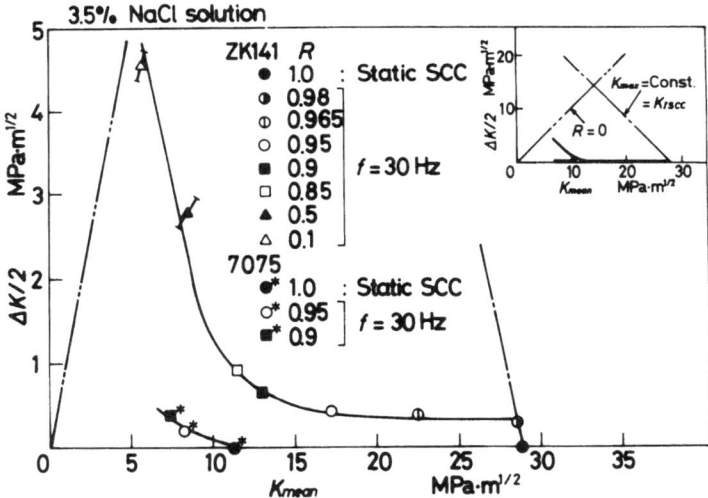

FIG. 2. Endurance limit diagram; ZK 141 and 7075 alloys, 3·5% NaCl solution, $T = 298$ K.

Figure 2 [3, 4] shows the endurance limit diagram, where the ordinate is an amplitude of stress intensity factor, $\Delta K/2$, and the abscissa a mean stress intensity factor, K_{mean}, taken from a crack growth rate of 3.0×10^{-10} m/s. The insert shown in the figure is a usual endurance limit diagram, where both scales of ordinate and abscissa are equalized.

It is clear that K_{DSCC} at $R = 0.98$ hardly decreased from K_{ISCC}. This may be due to the fact that the superimposed small vibratory stresses were lower than the fatigue strength of the passive films. In this case, the threshold stress intensity factor range against dynamic SCC crack growth, ΔK_{DSCC}, was approximated at $0.7 \text{MPa}\sqrt{m}$. At $0.95 \leq R < 0.98$, ΔK_{DSCC} values were unchanged and were almost equal to $0.7 \text{ MPa}\sqrt{m}$ corresponding to the fatigue strength of the films; the threshold condition of dynamic SCC crack growth at $R \geq 0.95$ was solely determined by ΔK_{DSCC}, below which integrity of the passive films was kept. Therefore, the threshold values of dynamic SCC crack growth $K_{DSCC} = \Delta K_{DSCC}/(1 - R)$ expressed in terms of K_{max} decreased with decrease in R. We must note that the value of ΔK_{DSCC} was smaller than the fatigue threshold value of $\Delta K_{th} = 1.19 \text{ MPa}\sqrt{m}$ in dry air at $R = 0.95$ [4].

At $R = 0.90$ and 0.85, the allowable vibratory stress range increased: ΔK taken from K_{DSCC} was greater than ΔK_{DSCC}, and gradually increased with decrease in R. In the region of $0.85 \leq R \leq 0.9$, even if the vibratory stress range ΔK greater than ΔK_{DSCC} caused damage of the passive films, restoration of the films predominated owing to K_{max} being small. In this situation, SCC crack growth is considered to be caused by dynamic SCC with cyclic SCC superimposed.

In the region of $K_{max} > K_{DSCC}$, da/dt of dynamic SCC inclusive of $R = 0.98$ was much enhanced compared with that of static SCC, because the passive films were ruptured by an increased vibratory stress associated with increase in K_{max}. Moreover, as compared with intergranular failures observed in static SCC, dynamic SCC failures were a mixed mode of intergranular and transgranular ones. This implies that an alternating mechanism of intergranular dissolution and subsequent transgranular hydrogen embrittlement caused an enhancement of dynamic SCC crack growth. On the other hand, a fall of K_{DSCC} from K_{ISCC} in 7075 alloy was also observed at $\Delta K = 0.4 \text{ MPa}\sqrt{m}$: see Fig. 2. The threshold values against the passive films, ΔK_{DSCC} might not exist, or have a very small value if present.

Similar results were also obtained in a high-strength steel sensitive to HE type SCC [5]. Figure 3 [5] shows the endurance limit diagram taken

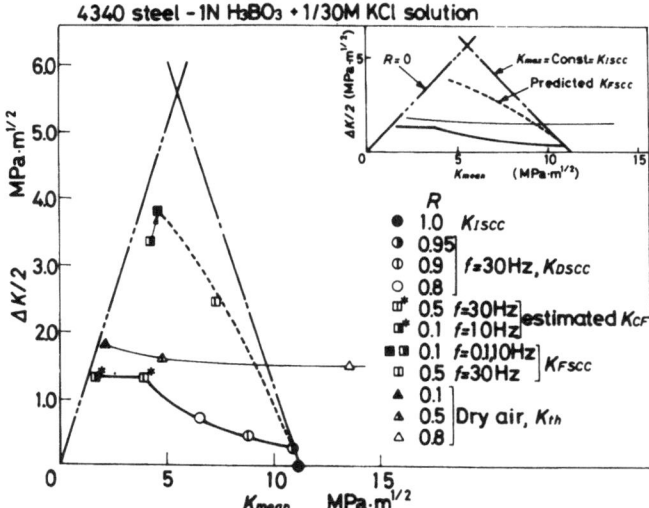

FIG. 3. Endurance limit diagram; 4340 steel, $1 \text{N } H_3 BO_3 + 1/30 \text{ M KCl}$ solution, $T = 298 \text{ K}$.

from a crack growth rate of $da/dt = 1.0 \times 10^{-9}$ m/s of 523 K-tempered 4340 steel (0·38%–C, 0·25%–Si, 0·78%–Mn, 0·019%–P, 0·014%–S, 1·77%–Ni, 0·80%–Cr, 0·24%–Mo, $\sigma_B = 1770$ MPa) in $1 \text{N } H_3 BO_3 + 1/30 \text{ M}$ KCl solution at 298 K. The stress wave forms were the same as the ones used in Fig. 1. The figure also illustrates the threshold values for fatigue crack growth in dry air. A remarkable fall of K_{DSCC} from K_{ISCC} due to superimposed small vibratory stresses was distinctly observed. A limiting value of ΔK_{DSCC}, below which no fall of K_{DSCC} from K_{ISCC} was observed, was approximated at $0.7 \text{ MPa}\sqrt{m}$. In dry air, fatigue crack growth was not observed at such a range of superimposed small vibratory stresses. Nevertheless in the SCC environment, dynamic SCC crack growth was brought about even at $K_{max} \leqq K_{ISCC}$ by the superimposed high frequency small vibratory stresses, because they promoted damage of the oxide films and increase in hydrogen content.

When R was further decreased, $R = 0.1$ and 0.5, the endurance limit was determined by the threshold values for corrosion fatigue; the endurance limit curve consisted of two regions, the one in the high R region, where dynamic SCC caused a decrease in threshold values from K_{ISCC}, and the other in the low R region, where corrosion fatigue predominated over the crack growth behavior. This transition from dynamic SCC to corrosion fatigue is also supported by the fractographical observations. In the high R region, intergranular failures were dominant, whereas transgranular failures predominated over fracture surfaces near threshold values in the low R region.

In the case of ZK141 alloy, on the other hand, the endurance limit line at lower R, i.e. $R = 0.1$ and 0.5, lay parallel to the line of $K_{max} = $ const. In this region, crack opening stress intensity factors, K_{op}, in 3.5% NaCl solution were higher than those in dry air, and the effective stress intensity factor ranges became too small to cause corrosion fatigue. That is to say, corrosion fatigue crack growth was suppressed and apparent threshold values, K_{CF}, increased with respect to K_{th} in dry air. This increase in K_{op} and K_{CF} was caused by corrosion products induced wedge effect [8]. In this case, a sudden enhancement of da/dN was brought about by cyclic SCC when K_{max} exceeded the threshold values for cyclic SCC, K_{FSCC}. In other words, the endurance limit was determined by the competition between the corrosion products-induced wedge effect and cyclic SCC, and therefore, it lay along the line of $K_{max} = K_{FSCC}$. In the case of 4340 steel at low R, cyclic SCC also occurred after K_{max} exceeded K_{FSCC}, which is shown in Fig. 3. In this case, the wedge effect was not so effective as ZK141 alloy, and da/dN increased from that of corrosion fatigue.

2.2. Effect of Stress Cycle Frequency on Dynamic SCC Crack Growth

As has been discussed before, dynamic SCC is caused by rupture of passive films. Therefore, the effects of stress cycle frequency are considered to be important problems. Figure 4 [4] shows the frequency effects ranged from 1 Hz to 120 Hz on dynamic SCC crack growth rates, da/dt, for ZK141 alloy, which was the same as the one used in Fig. 1. Experimental results at $f = 30$ Hz and $R = 0.95$ in dry air (dew point: 203 K) were converted into those at $f = 1$ Hz, which are plotted in the figure as solid triangles. Fatigue data at other frequencies are also shown by fine lines.

da/dt in Region II was enhanced with increase in frequency. At $f \geq 30$ Hz, however, little frequency effects were observed. In Region II, vibratory stresses caused degradation of the passive films at the crack tips, thereby bringing about an enhancement of da/dt from that of static SCC. The degradation of passive films and hydrogen embrittlement in a ligament zone were almost independent of stress cycle frequency above 30 Hz. At $f < 30$ Hz, however, decrease in frequency reduced da/dt down to that of static SCC, which was caused by decrease in anodic dissolution rates at grain boundaries, since restoration of the passive films became dominant.

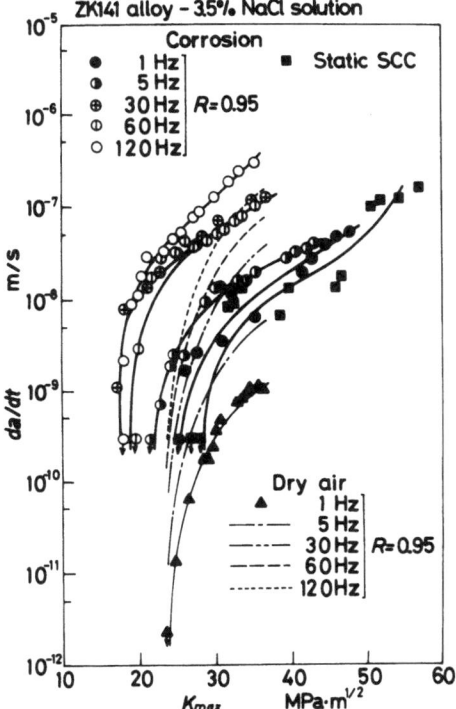

ZK141 alloy – 3.5% NaCl solution

F IG. 4. da/dt–K_{max} plots; ZK141 alloy, 3·5% NaCl solution, $T = 298$ K.

Now, let us consider the frequency effects on threshold values. At $f \geqq 30$ Hz K_{DSCC} values were almost constant, and were solely determined by the superimposed vibratory stress range, ΔK_{DSCC}, irrespective of stress cycle frequencies. At $f < 30$ Hz, on the other hand, K_{DSCC} increased with decrease in frequency, and approached up to K_{ISCC}. In this situation, even if ΔK taken from K_{DSCC} caused damage of the passive films at the crack tips, the restoration of passive films was more predominant. SCC crack growth never occurred until the vibratory stresses generated active slip steps at the crack tips. Therefore, SCC crack growth at $f < 30$ Hz was caused by dynamic SCC with cyclic SCC superimposed, resulting in an increase in K_{DSCC} values. As for a material sensitive to HE type SCC, similar results were also obtained in 300M steel [9]. These frequency effects are similarly explained from the viewpoints of repassivation and rupture of films at the SCC crack tip.

3. CYCLIC SCC CRACK GROWTH CHARACTERISTICS

3.1. Effects of Wave Form on Cyclic SCC Crack Growth

Previous sections have dealt with dynamic SCC crack growth behavior. However, low frequency varying loads also affect the SCC crack growth behavior, and cyclic SCC is brought about [1, 2]. In such cases, stress wave form is one of the important factors which dominate over crack growth mechanisms. Figure 5 [6] shows the relation between cyclic

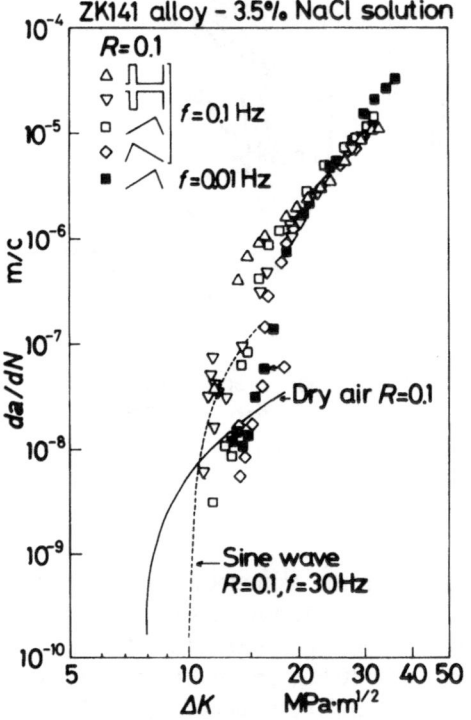

FIG. 5. da/dN–ΔK plots; ZK141 alloy, 3·5% NaCl solution, T = 298 K.

SCC crack growth rates da/dN and stress intensity factor ranges ΔK in various stress waves. The material tested was a high-strength aluminum alloy ZK141 which was the same as the one used in Fig. 1. Results of fatigue in dry air at R = 0·1 and in 3·5% NaCl solution at R = 0·1 and f = 30 Hz are also shown by the solid and the dotted lines in the figure, respectively.

At low ΔK, da/dN in 3·5% NaCl solution was lower than that in dry air except the positive pulse wave. This decrease in da/dN was caused by corrosion products-induced wedge effect. In fact, K_{op} in 3·5% NaCl solution in a low K_{max} region was higher than that in dry air. Above a certain ΔK, however, a sudden enhancement of da/dN from that in dry air was observed. At $f = 0·1$ Hz, the threshold stress intensity factor range of cyclic SCC, ΔK_{FSCC}, for the positive pulse wave was equal to the one for the negative pulse wave, and they had the smallest values. ΔK_{FSCC} increased in the order, the positive saw tooth wave and the negative saw tooth wave. ΔK_{FSCC} for the positive saw tooth wave at $f = 0·01$ Hz was higher than that for the same wave form at $f = 0·1$ Hz, and went up to the similar threshold value as the one for the negative saw tooth wave at $f = 0·1$ Hz.

As was mentioned before, ΔK_{FSCC} values or K_{FSCC} values in terms of maximum stress intensity factor were strongly dependent on the wave forms. To make clear the mechanisms, stress intensity factor rates, $\dot{K} = |dK/dt|$, at both loading and unloading periods were calculated (Fig. 6 [6]). Short lines (|) are added to the plots in the case of unloading. As

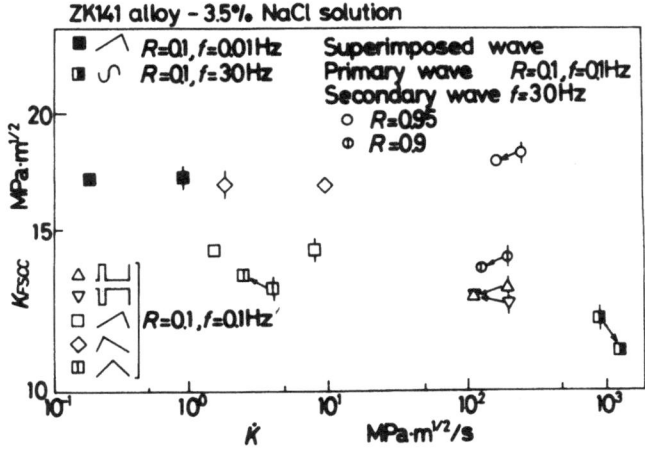

FIG. 6. K_{FSCC}–$\dot{K}$ plots; ZK141 alloy, 3·5% NaCl solution, $T = 298$ K.

was mentioned before, K_{FSCC} for the negative pulse wave was equal to that for the positive pulse wave, and an influence of hold time on K_{FSCC} was not observed. This indicates that passive films at the crack tip, which had been ruptured by the loading period, were restored within successive hold times ($< 1·35$ s; shorter hold time of the pulse wave form).

Moreover, it is clear that K_{FSCC} increased with decreasing $\dot{K}$. In this material/environment system, repassivation rates were very fast as was indicated by no influence of hold time at the maximum stress on K_{FSCC} and da/dN. Therefore, the restoration of passive films predominated at lower $\dot{K}$, i.e. positive saw tooth waves, resulting in higher K_{FSCC} with respect to that at higher $\dot{K}$, i.e. pulse waves. In the case of corrosion fatigue, crack growth rates were reported to be increased with decreasing $\dot{K}$ (i.e. longer rise time) [10]. One should note that the dependence of cyclic SCC thresholds on $\dot{K}$ is strongly opposed to that of corrosion fatigue.

The figure also illustrates K_{FSCC} values for superimposed waves. Here sinusoidal loads at a high frequency of 30 Hz were superimposed on the periods of the maximum loads of the negative pulse wave for 7·8 s (234 cycles). Stress ratios of secondary vibratory loads were 0·95 and 0·90. In

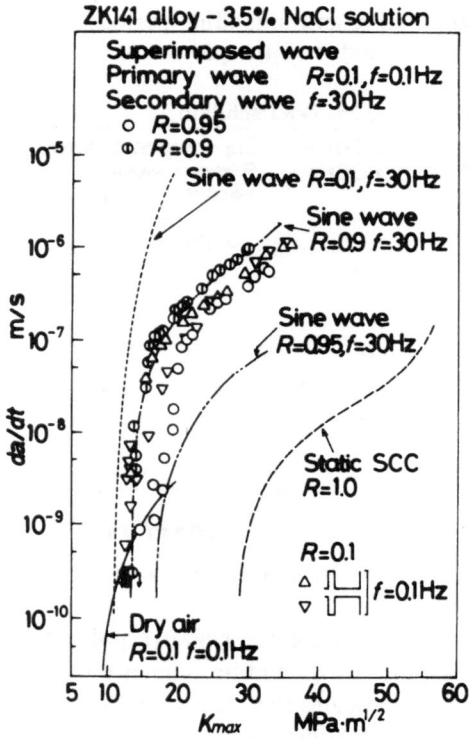

FIG. 7. da/dt–K_{max} plots; ZK141 alloy, 3·5% NaCl solution, $T = 298$ K.

this case, K_{FSCC} increased with respect to those for the positive and the negative pulse waves in spite of having similar $\dot{K}$ values. This increase of K_{FSCC} was especially remarkable for the $R = 0.95$ superimposed wave. Then da/dN values for the superimposed waves were converted into crack growth rates to be compared with dynamic SCC crack growth rates (see Fig. 7 [6]). It is clear that K_{FSCC} for the $R = 0.95$ superimposed wave was equal to K_{DSCC} at $R = 0.95$, and similarly K_{FSCC} for the $R = 0.9$ superimposed wave was to K_{DSCC} at $R = 0.9$. For superimposed waves, secondary vibratory stresses promoted corrosion products-induced wedge effect [8], and therefore, the primary varying loads could not bring about active slip steps effectively. This led to higher K_{FSCC} values for the superimposed waves than those for pulse waves. In this situation an abrupt crack growth enhancement from that in dry air never occurred until K_{max} exceeded K_{DSCC}, above which the secondary vibratory stresses caused dynamic SCC crack growth. Namely, K_{FSCC} values for the superimposed waves were determined by K_{DSCC} values corresponding to the secondary vibratory stress conditions.

3.2. Effect of Cathodic Overprotection on Cyclic SCC Crack Growth

Previous sections have dealt with stress corrosion cracking (SCC) crack growth under dynamic loading conditions in materials sensitive to static SCC. However, cyclic SCC sometimes occurs in materials insensitive to static SCC. In this section, cyclic SCC crack growth of a high-tensile strength steel under cathodic overprotection will be discussed. When cathodic protection is applied to machines and structures in corrosive environments, we must pay more attention to the uniformity of the cathodic current density. However, it is very difficult because of complicated shapes of machines and structures and local differences of environmental conditions including flow rates and oxygen concentration. And then the local potential of structures sometimes becomes more negative than that expected. So it is important to make clear the crack growth behavior under cathodic overprotection.

Figure 8 [7] shows the relation between crack growth rates, da/dN, and stress intensity factor ranges, ΔK, of high-tensile strength steel HT80. (0.12%–C, 0.29%–Si, 0.92%–Mn, 0.016%–P, 0.005%–S, 0.24%–Cu, 0.50%–Cr, 1.21%–Ni, 0.42%–Mo, 0.30%–V, 0.0013%–B, 0.06%–Al, $\sigma_B = 820$ MPa) in 3.5% NaCl solution at 298 K. The cathodic polarization of -1.2 V versus a saturated calomel electrode, SCE, was applied to specimens. The figure also shows the results in dry air and 1% NaCl solution [11]. Crack growth rates at $f = 0.1$ Hz and 0.01 Hz in 3.5% NaCl

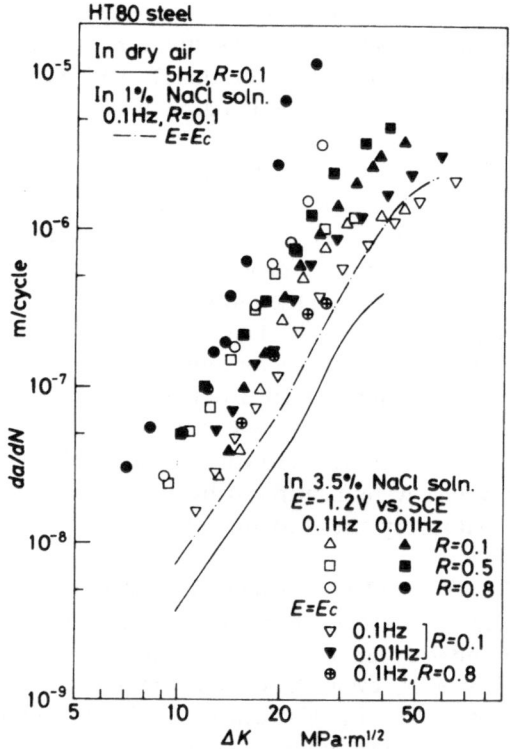

F IG. 8. da/dN–ΔK plots; HT80 steel, T = 298 K.

solution increased from that in dry air with increase in stress ratios. An
enhancement of da/dN was especially marked in a high ΔK region at
$R = 0.8$. Moreover, crack growth rates at $f = 0.01$ Hz were higher than
those at $f = 0.1$ Hz.

Figure 9 [7] shows the relation between da/dN and effective stress
intensity factor ranges, ΔK_{eff}. For the effects of cathodic protection in
3.5% NaCl solution at $R = 0.1$ and $f = 0.1$ Hz, da/dN was almost equal to
that at the free corrosion potential at $\Delta K_{eff} \leq 10$ MPa$\sqrt{m}$. At
$\Delta K_{eff} = 10$–44 MPa$\sqrt{m}$, however, da/dN under cathodic overprotection
was higher than that at the free corrosion potential. This enhancement of
crack growth rates was caused by both hydrogen embrittlement at such a
negative potential and caustic dissolution, which is suggested by the facts
of alkalization in the solution within the crack (pH values were about 10)
and the increase in K_{op} with respect to dry air values.

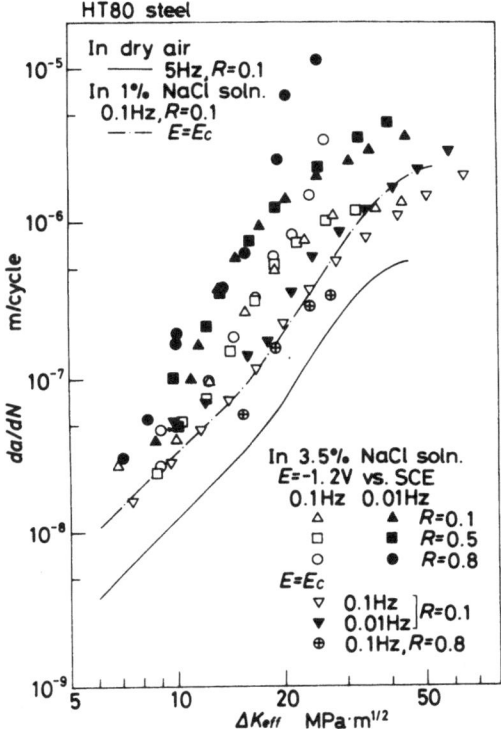

FIG. 9. da/dN–ΔK_{eff} plots; HT80 steel, $T = 298$ K.

Under cathodic overprotection, crack growth rates at $f = 0.1$ Hz and 0.01 Hz were independent of stress ratios at low ΔK_{eff}. However, crack growth rates at $R = 0.8$ were enhanced from those at $R = 0.1$ and 0.5 at $\Delta K_{\text{eff}} \geqq 21$ MPa$\sqrt{\text{m}}$ for $f = 0.1$ Hz and $\Delta K_{\text{eff}} > 19$ MPa$\sqrt{\text{m}}$ for $f = 0.01$ Hz. To make clear this enhancement, the acceleration ratio, R_{acc}, the ratio of $(\text{d}a/\text{d}N)_{\text{FSCC}}$ in 3.5% NaCl solution to $(\text{d}a/\text{d}N)_{\text{air}}$ in dry air was calculated (Fig. 10 [7]). At the free corrosion potential, R_{acc} in the whole range of ΔK_{eff} examined had constant values of about 2–4, whereas under cathodic overprotection R_{acc} had maxima at a certain ΔK_{eff} value, which increased with decrease in frequency. Especially under a high mean stress of $R = 0.8$, one must note that a dramatic enhancement of crack growth occurred, where R_{acc} at $\Delta K_{\text{eff}} = 20$–$25$ MPa$\sqrt{\text{m}}$ was 10–19 for $f = 0.1$ Hz and 40–90 for $f = 0.01$ Hz. This crack growth enhancement was due to maximum stress intensity factors being high

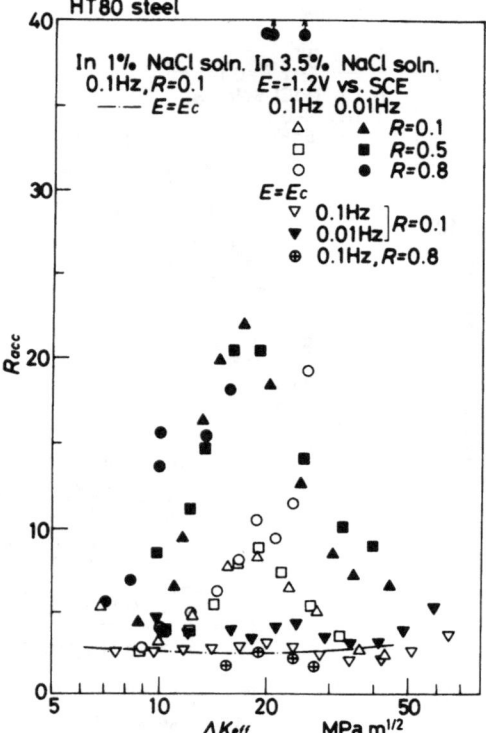

FIG. 10. R_{ace}–ΔK_{eff} plots; HT80 steel, $T = 298$ K.

values, and therefore, time dependent hydrogen embrittlement came to dominate over the crack growth kinetics, where the hydrogen content would be more than that at the free corrosion potential.

4. CONCLUSIONS

The influences of dynamic loads on stress corrosion cracking behavior, dynamic and cyclic SCC, have been investigated, and the following conclusions are summarized.

(1) Superimposed high-frequency vibratory stresses, where fatigue crack growth in inert environments does not occur, bring about a considerable fall in threshold values for dynamic SCC, K_{DSCC}, from

those for static SCC under a sustained load, K_{ISCC}. As for APC type dynamic SCC, this decrease in threshold values is caused by the fatigue rupture of passive films at the crack tip; for HE type dynamic SCC, that is due to increase in the hydrogen content and the damage of oxide films at the crack tip. For both types of SCC, K_{DSCC} decreases with decreasing stress ratios. However, there exists a limiting stress intensity factor range below which no fall of K_{DSCC} from K_{ISCC} is observed.

(2) The effects of vibratory stresses on SCC crack growth are also dependent on the stress cycle frequencies for both APC and HE type SCC; at higher frequencies a remarkable decrease in K_{DSCC} from K_{ISCC} is observed. However, this decrease of K_{DSCC} is saturated above a certain frequency. With decreasing stress cycle frequency, on the other hand, K_{DSCC} increases and becomes closer to K_{ISCC}. This frequency dependence is explained from the viewpoints of the repassivation and the rupture of films at the SCC crack tips.

(3) Low frequency varying loads also affect SCC crack growth behavior. Threshold values, K_{FSCC}, for APC type cyclic SCC decrease with increase in stress intensity factor rates per time, $\dot{K}$. This dependence upon $\dot{K}$ differs from that of corrosion fatigue in materials insensitive to SCC. In the case of the superimposed waves, that is, a low frequency trapezoidal varying stress with secondary high frequency vibratory stresses superimposed, the primary load repetitions become insignificant, and the threshold values, K_{FSCC}, are determined by the K_{DSCC} values according to the secondary loading conditions.

(4) Low frequency varying loads cause cyclic SCC even in materials insensitive to static SCC under a sustained load. Crack growth rates of high-tensile strength steel under cathodic overprotection are higher than those at the free corrosion potential; the enhancement is due to caustic dissolution and hydrogen embrittlement. Furthermore, one should note that the dramatic enhancement of crack growth under cathodic overprotection is brought about especially at a high stress ratio of 0·8.

REFERENCES

1. ENDO, K., KOMAI, K. and YAMAMOTO, I., *Bull. JSME*, **24** (1981), 1326.
2. ENDO, K., KOMAI, K. and MATSUDA, Y., *ibid.*, **24** (1981), 1885.

3. ENDO, K., KOMAI, K. and MINOSHIMA, K., *Proc. Fatigue '81*, SEE, 1981, p. 77.
4. KOMAI, K. and MINOSHIMA, K., *Proc. ICMC*, **1** (1984), 167.
5. ENDO, K., KOMAI, K. and MINOSHIMA, K., *Proc. 25th Japan Congr. Mater. Res.*, 1982, p. 56.
6. KOMAI, K. and MINOSHIMA, K., *Bull. JSME*, **28** (1985), 2211.
7. KOMAI, K., MINOSHIMA, K. and YAMAMOTO, K., *Proc. 25th Japan Congr. Mater. Res.*, 1985, p. 81.
8. ENDO, K., KOMAI, K. and SHIKIDA, T., ASTM STP 801, 1983, p. 81.
9. MINOSHIMA, K., D. Eng. Dissertation, Kyoto University, 1985.
10. SELINE, R. J. and PELLOUX, R. M., *Metall. Trans.*, **3** (1972), 2525.
11. KOMAI, K., KITA, S. and ENDO, K., *Bull. JSME*, **27** (1984), 847.

25

Local Corrosion of Welds of 60 kgf/mm² Class Steels for Ice-breaking Ships

K. Itoh, H. Mimura, T. Inoue, R. Todoroki, Y. Horii and H. Kihira

R & D Lab. II, Nippon Steel Corp., Fuchinobe 5-10-1, Sagamihara-shi 229, Kanagawa, Japan.

1. INTRODUCTION

It is difficult to maintain paint and cathodic protection of ice-breakers due to contact with ice. The result is corrosion problems with the steels. The most critical is local corrosion of welds which, together with residual stress and toughness deterioration, increases the risk of failure. With regard to the local corrosion of welds, studies on normalized HT-50 have been reported mainly by Finnish researchers, as shown in Fig. 1 [1].

Type A (local corrosion of W.M.)		Countermeasures; Use of Cu-Ni containing electrode(2) C Si Mn Cu Ni OK4823 (0.08 0.25 0.8 0.6 0.4)
Type B (local corrosion of HAZ)	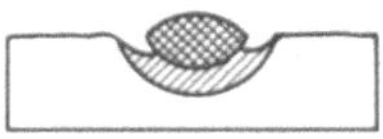	Countermeasures; Microstructure of HAZ must be Ferrite and Pearlite, same as base metal.(1) (1) Si < 0.3%, Mn < 1.1% (2) regulating heat input
Type C (knife edge corrosion at fusion line)	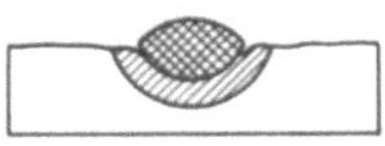	Causes; Continuous sulphide films at austenite grain boundaries. Countermeasures; Addition of Ce, Ti, Zr to B.M.

FIG. 1. Three types of local corrosion in welds [1].

389

Using the Finnish research data as reference, this report discusses local corrosion of a 60 kgf/mm² class high-strength steel whose base metal microstructure differs from F + P.

2. TEST STEELS AND WELDING MATERIALS

2.1. Test Steels

Table 1 shows test steels. Those in Table 1 are 60 kgf/mm² class steels selected from one of the following three viewpoints.

(1) Mn content must be as low as possible, as with 50 kgf/mm² class steels [1].
(2) Cr content must be less than 0·5%, because, while it generally increases corrosion resistance by forming protective films, it causes pitting corrosion in sea water to the detriment of the steel's corrosion resistance.
(3) To provide base metal with toughness and weldability (including toughness of HAZ), C content must be low while D_1 (D_1: parameter indicating quenching hardenability) must be high.

Figure 2 shows examples of base metal microstructure. These steels have a microstructure mainly consisting of fine grained upper bainite (Bu) which partially contains proeutectoid ferrite. With this microstructure, toughness is excellent.

2.2. Welding Materials

Table 2 shows the chemical compositions of tested welding materials. P, Q, R and S, shown in Table 2 are Cu–Ni–Cr containing materials for submerged arc welding (SAW) and shielded metal arc welding (SMAW), selected for their corrosion potential which is nobler than that of the test steels.

3. TEST PROCEDURES

For the corrosion test, the following immersion rotating testers were used.

(1) Low speed rotating tester

As shown in Fig. 3, this machine immerses test pieces in a corrosive solution in a 0·9 m³ stainless steel container and rotates them at a peripheral speed of 0·5 m/s. Ninety pieces 2 × 50 × 120 mm in size can be

TABLE 1
Chemical composition and heat treatment of tested HT-60 steels

Designation	C	Si	Mn	P	S	Cu	Ni	Cr	Mo	Nb	V	Remarks	Heat treatment
A	0·13	0·28	1·09	0·006	0·001	0·30	0·36	0·18	0·06	0·02	0·04	Ca treated	CLC-T[a]
B	0·06	0·20	1·42	0·005	0·002	0·90	0·73	—	—	0·01	—	Ca treated	CLC-T[a]
C	0·12	0·26	1·07	0·012	0·005	—	0·34	0·12	0·20	—	0·03		CLC-T[a]
D	0·15	0·24	1·14	0·005	0·003	0·17	0·55	0·05	0·09	—	0·02		CLC-T[a]
E	0·07	0·27	1·23	0·013	0·001	—	1·45	—	0·10	0·02	—	Ca treated	CLC-T[a]
F	0·10	0·17	1·35	0·006	0·003	0·35	0·31	—	0·16	—	—		CLC-T[a]
G	0·08	0·26	1·44	0·005	0·003	—	—	—	—	0·01	—	Ca treated	CLC-T[a]
H	0·07	0·21	0·79	0·009	0·005	0·30	1·04	0·47	0·26	—	—		Q.T[b]
I	0·06	0·22	0·85	0·011	0·005	0·64	1·10	0·49	0·25	—	—		Q.T[b]
J	0·07	0·22	0·85	0·008	0·005	1·04	1·07	0·50	0·25	—	—		Q.T[b]
K	0·06	0·20	0·85	0·008	0·005	—	1·07	0·50	0·25	—	—		Q.T[b]
L	0·13	0·24	1·38	0·007	0·001	—	—	—	—	—	0·04		Q.T[b]

[a]Continuous on line control process and tempered.
[b]Quenched and tempered.

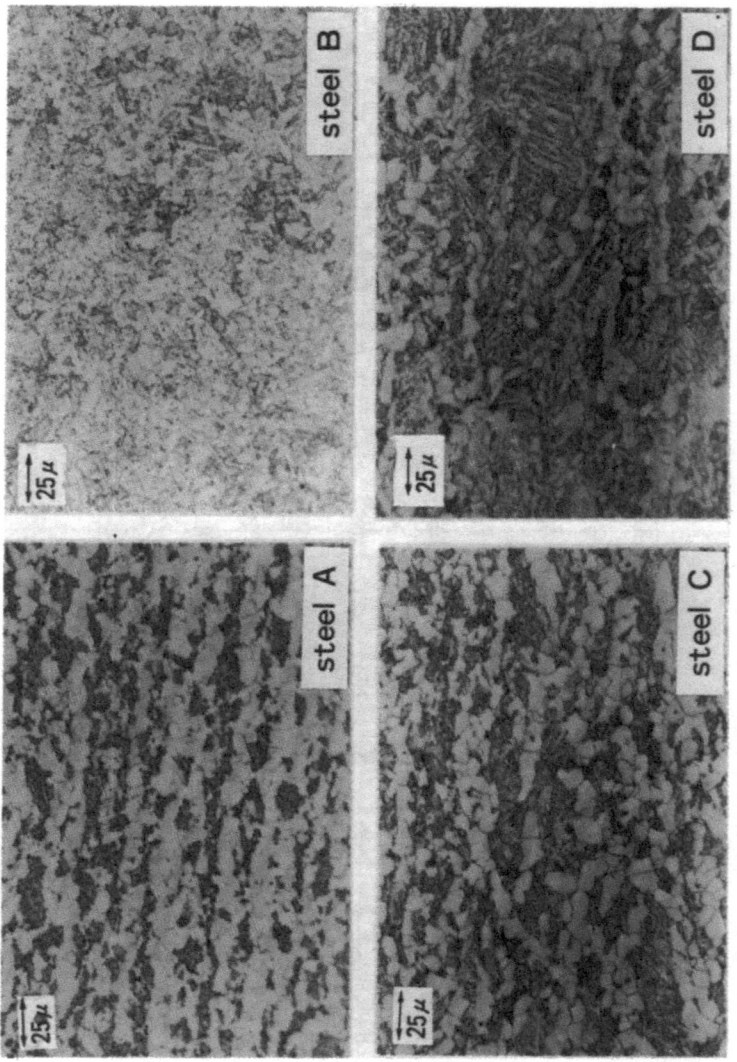

FIG. 2. Examples of microstructures of B.M.

TABLE 2
Chemical composition of tested welding materials

Designation	C	Si	Mn	P	S	Cu	Ni	Cr	Remarks
P	0·08	0·33	1·04	0·012	0·007	0·33	0·52	0·52	SAW
Q	0·07	0·40	0·70	0·011	0·004	0·28	0·49	0·15	SMAW ($3·2\phi$)
R	0·06	0·31	0·81	0·013	0·005	0·41	0·65	0·45	SMAW (4ϕ)
S	0·08	0·43	1·39	0·012	0·005	0·90	0·69	0·30	SMAW ($3·2\phi$)

( ): diameter of welding rod.

motor

dipping tank

FIG. 3. Low speed rotating corrosion test machine (0·5 m/s).

set at one time. Since it is capable of performing tests with a large number of small pieces, systematic testing was obtained by using welded test pieces consisting of various steels and welding materials combined.

(2) *High speed rotating tester*

This is an ESAB type test machine [2, 3]. As seen in Fig. 4, it is capable of handling 24 test pieces of 25 mm thickness, 70 mm upside, 240 mm downside and 305 mm height at a peripheral speed of 5 m/s.

FIG. 4. Setting of test pieces in high speed rotating tester.

Since the tester can simulate the service conditions encountered by an actual ice-breaker, it was used to confirm data obtained by the low-speed rotating tester.

3.1. Test for Influence of Chemical Composition of Steels and Welding Materials

(1) Preparation of welded specimen
  To observe the influence of the chemical composition of steels and welding materials on local corrosion, welded specimens were prepared by combining the steels shown in Table 1 and the welding materials in Table 2.
  A V-groove was machined in test pieces of 25 mm thickness, and one-pass welding was performed. Standard weld heat input was set at 15 kJ/cm for SAW and 7·5 kJ/cm for SMAW. This is in accordance with the theory that lower heat input places HT-50 [1] in more severe local corrosion conditions. A part of the test was carried out with high heat input welding and preheating.
  Test pieces of $2 \times 50 \times 120$ mm in size were taken from 1 mm under the surface in order to eliminate the toe part undercut. The width of weld is normally about 10 mm, and the test pieces were taken in a manner so as to place the weld at the center of its 120 mm length. This is in accordance

with the theory that welds with an area of less than 10% of the base metal are preferable in this sort of local corrosion test [4].

Prior to the corrosion test, machine polish was applied to all six surfaces of the test piece (surface roughness less than 4 μm), followed by degreasing and weighing.

(2) Conditions of corrosion test

Table 3 shows conditions of the corrosion test. A 3% NaCl solution was used as test liquid, which was changed once a month. The 3% NaCl solution was selected because it was thought to create a more severe corrosion condition than sea water. The test period was either 1 month or 2·5 months. After the test, the specimens were rinsed, pickled, weighed and visually examined for corrosion.

TABLE 3
Corrosion test conditions of low speed test

Solution	3% NaCl
Temperature	+50°C
Dissolved oxygen content	5 ppm (air saturated)
Velocity of test piece	0·5 m/s
Duration	1 month and 2·5 months
Dimensions of test piece	2 × 50 × 120

3.2. Confirmation Test

(1) Preparation of specimens

After gas cutting, scale was removed from the test piece by shot blasting. This was followed by machining a U groove in the center part and test welding.

For SMAW, weld heat input was selected for 7·5 kJ/cm, 15 kJ/cm and 30 kJ/cm. For SAW, it was fixed at 15 kJ/cm. Passes were repeated until the groove was filled. No preheating was performed and interpass temperature was 50°C or lower.

After welding, the center part of the test piece was subjected to a machine polish for 120 mm in the height direction. Finally, the test piece was fixed to a rotary disc with an acrylic plate sandwiched between the piece and the disc.

(2) Conditions of corrosion test

Specially prepared artificial sea water was used here as a solution, the composition of which is shown in Table 4. This was made from ordinary artificial sea water [4] from which the Na_2CO_3 was removed in order to avoid corrosion inhibition [5] due to deposits of $CaCO_3$ [3]. Solution temperature was set at 50°C for the accelerated test. The amount of dissolved oxygen was approximately 8 ppm.

TABLE 4
Composition of artificial seawater (g/litre)

NaCl	26·52
$MgCl_2$	2·45
$MgSO_4$	3·30
$CaCl_2$	1·14
KCl	0·73
NaBr	0·08

The solution was changed once a week. To eliminate rust, the test piece surface was brushed hard once a month with a stainless steel brush. Although superficial red rust came off, black rust could not be removed.

The test period was 1, 3 or 6 months. After the test, the specimens were rinsed, pickled and examined, in the same manner as in the low-speed rotating test.

3.3. Measurement of Corrosion Potential

Since the corrosion potential was thought to control local corrosion, the potential of welded test specimens was measured by a scanning reference electrode tester both in the as-polished, pre-corrosion state and after having been corroded.

3.4. Chemical Analysis of Corrosion Products by CMA

Since the corrosion products exerted a strong influence on local corrosion, corrosion products of specimens were subjected to chemical analysis by a computer-aided microanalyser (CMA). After rinsing and removal of superficial spongy rust, the test pieces, embedded in synthetic resin and polished, were analyzed by CMA.

4. TEST RESULTS

4.1. Results of Corrosion Test

(1) Influence of composition of steels and welding materials on local corrosion

Initially, the low-speed rotating test was carried out on the specimens made from the test steels shown in Table 1 and the welding materials shown in Table 2. Table 5 shows the results. Two test pieces from each

TABLE 5
Corrosion test results of low speed test

| | B.M. | Welding Material | Welding Method | Heat Input (kJ/cm) | Preheat Temp. (°C) | Test Results.* | | | | | | | |
| | | | | | | 1 month | | | | 2.5 months | | | |
						W.M.	F.L.	HAZ	Outside** of HAZ	W.M.	F.L.	HAZ	Outside of HAZ
1	B	P	SAW	15	R.T.	2	0	0	1	3	0	0	1
2	"	Q	SMAW	7.5	"	3	"	"	"	"	"	"	"
3	"	R	"	15	"	"	"	"	"	"	"	"	"
4	"	S	"	"	"	0	"	0.5	"	0.5	"	1	2
5	C	P	SAW	"	"	"	"	3	0	0	"	3	1
6	"	"	"	23	"	"	"	1	"	"	"	"	0
7	"	"	"	"	210	"	"	"	"	"	"	2	"
8	"	"	"	30	R.T.	"	"	"	"	"	"	3	"
9	"	Q	SMAW	7.5	"	"	"	3	"	"	"	3	1
10	D	P	SAW	15	"	"	"	1	"	0	"	2	0
11	"	Q	SMAW	7.5	"	"	"	"	"	"	"	"	1
12	E	P	SAW	15	"	0	"	"	"	"	"	"	"
13	"	Q	SMAW	7.5	"	1	"	0	"	0.5	"	0	"
14	F	P	SAW	15	"	2	"	"	"	2	"	1	"
15	"	Q	SMAW	7.5	"	3	"	"	"	"	"	0	"
16	G	P	SAW	15	"	0	"	3	"	0	"	3	0
17	"	Q	SMAW	7.5	"	"	"	"	"	"	"	"	"
18	H	P	SAW	23	"	0.5	"	0	"	0.5	"	0	"
19	I	"	"	"	"	"	"	"	"	"	"	"	"
20	"	"	"	15	"	"	"	"	"	"	"	"	"
21	J	"	"	23	"	"	"	"	"	"	"	"	"
22	K	"	"	"	"	"	"	"	"	"	"	"	"
23	"	"	"	15	"	"	"	"	"	"	"	"	"
24	L	"	"	"	"	0	"	2	"	0	"	1	1
25	"	Q	SMAW	7.5	"	"	"	1	"	"	"	2	0

*0, no corrosion; 1, slight corrosion; 2, clear corrosion; 3, strong corrosion;
0·5, no corrosion, slightly concaved side comparing between WM and HAZ.

** Outside of HAZ

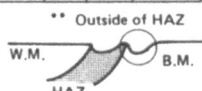

group were visually examined for corrosion, and graded in the manner described in the footnote of Table 5 [3]. Although the grading was qualitative, it was considered that the overall information had accurately been grasped, leaving few individual differences in judgement.

Since some of the test pieces showed local corrosion on the base metal side of the base metal/HAZ boundary (outside of HAZ), the same grading method was applied to this part.

Figure 5 shows examples of the section. From this test, the following became clear.

(1) Except for the B–S and F–P combinations after a 2·5 month test, only one side, either WM or HAZ of the test piece, suffered from local corrosion.

(2) In case of Steel E, local corrosion appeared on the HAZ of the E–P combination and on the WM of the E–Q combination. In addition, the WM of the B–P, B–Q and B–R combinations and the HAZ of the B–S combination were affected by local corrosion. In other words, the same kind of steel may or may not develop local corrosion of HAZ depending on the weld metal.

(3) Local corrosion of HAZ was observed also on Steel C, a low Mn type.

(4) The C–P combination, which had been subjected to a higher weld heat input or preheating, showed the effect of such treatments in a 1 month test, but not in a 2·5 month test.

(5) In any test case, no knife edge corrosion was observed at F.L.

Local corrosion appeared either in the WM or the HAZ. Even on the same base metal, however, WM local corrosion changed into HAZ local corrosion, or vice versa, depending on the welding materials. This phenomenon was also observed in case of the same welding material with different base metals.

Regression analysis was then made using as variables the difference in chemical composition between the steel and welding material, and as quality characteristics the corrosion degrees, which were defined as positive in case of WM local corrosion and negative in case of HAZ local corrosion. As a result, a significant difference was found in ΔCu and ΔNi whose coefficients were 3·8 and 1·1 respectively. Figure 6 shows the plotted result. The WM suffered from local corrosion when ΔE_c in formula (1) was positive, while the HAZ suffered when it was negative.

$$\Delta E_c = \Delta\text{Cu} \times 3\cdot8 + \Delta\text{Ni} \times 1\cdot1 + 0\cdot3 \qquad (1)$$

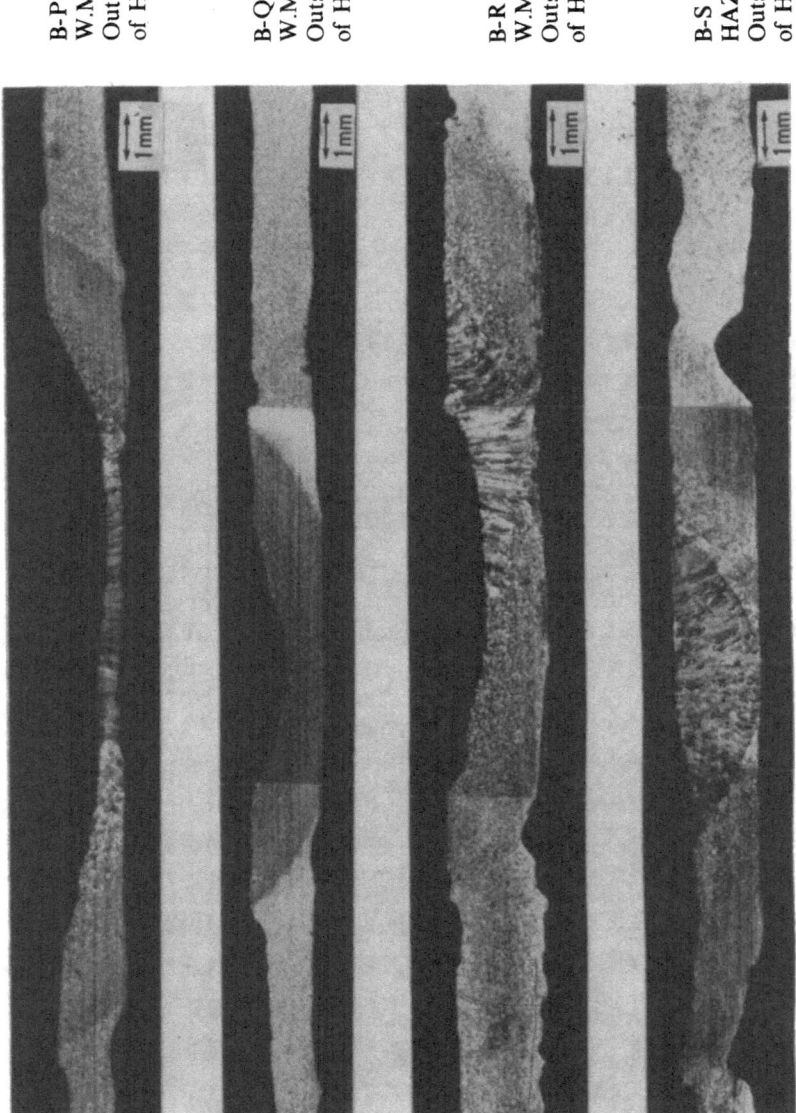

B-P
W.M. 3
Outside
of HAZ 1

B-Q
W.M. 3
Outside
of HAZ 1

B-R
W.M. 3
Outside
of HAZ 1

B-S
HAZ 1
Outside
of HAZ 2

Fig. 5. Sections of test pieces of low speed test after immersion of 2·5 months.

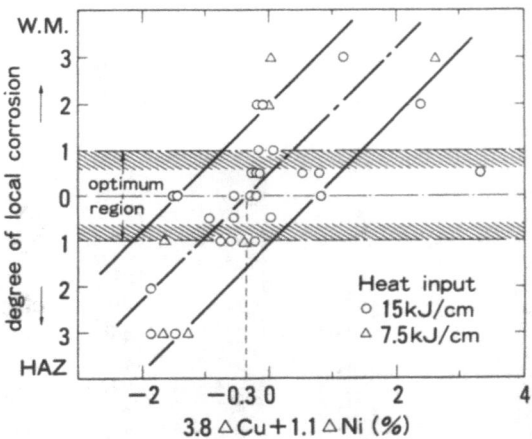

F IG. 6. The effects of the differences of alloying element between B.M. and WM to the
local corrosion.

ΔCu: difference in Cu content between B.M. and WM

ΔNi: difference in Ni content between B.M. and WM

(2) Result of Confirmation Test

In order to confirm the result of the low-speed rotating test, the specimens which had little local corrosion and those having distinctive local corrosion of the WM and of the HAZ were subjected to the confirmation test.

Figure 7 shows the result of the test. Figures 8 and 9 show examples of the surface and section of the specimens. The results indicated the following.

(1) When compared with the test result after one month, local corrosion was less than that in the low-speed rotating test with a 3% NaCl solution.

(2) No knife edge corrosion was observed at the fusion line.

(3) In the case of multi-pass welding, some of the HAZ in the WM showed local corrosion.

4.2. Result of Corrosion Potential Measurement

Using a scanning reference electrode tester, the corrosion potentials were measured in the as-polished state and after having been corroded. Figure 10 shows examples of the results. Although the specimen had shown local corrosion of the HAZ, the corrosion potential of the WM

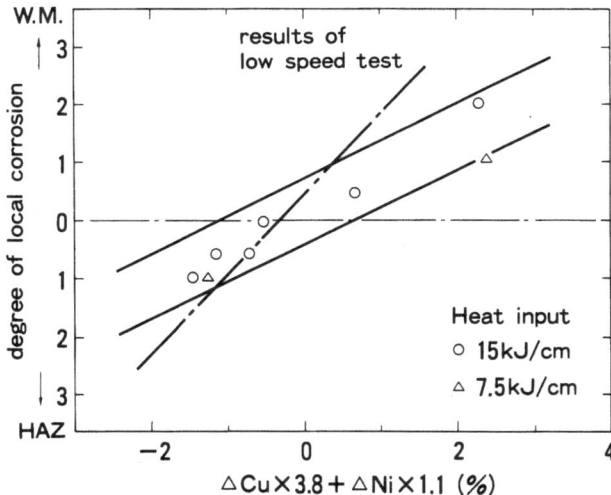

FIG. 7. Test results of high speed test.

was lower than that of the HAZ in the as-polished state while this trend was reversed when the potential was measured on corrosion products after a one-month corrosion test in artificial sea water. It can be concluded that the real corrosion behavior was closer to the test result with the corrosion product. In other words, the corrosion potential was influenced by corrosion products. Although corrosion potential should have been measured at the steel interface with the corrosion product, this was practically impossible.

4.3. Result of Chemical Analysis of Corrosion Products

Since it became clear from the above examination that corrosion products would exert a strong influence on local corrosion, chemical analysis was carried out on the corrosion products. It is seen that approximately 5–10 times more of Cu and 2·5–5 times more of Ni exist in uniform layers near the steel–rust border.

5. CONCLUSIONS

As reported in the above sections, local corrosion of welds in sea water was studied using the TMCP-produced 60 kgf/mm² class steel (whose microstructure mainly consisted of Bu), which shows excellent toughness

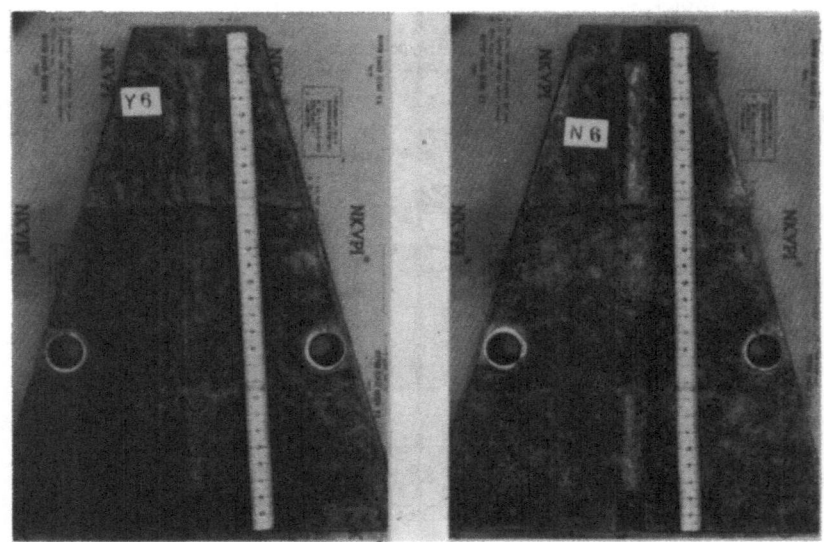

Y-6(A-Y13)   15kJ/cm   N-6(B-N1) 15kJ/cm
HAZ 0·5   Outside of HAZ 0·5
Outside of HAZ 1

FIG. 8. Appearances of specimens in high speed test after immersion of 6 months.

at low temperature. The result of the study can be summarized as follows.

(1) Local corrosion appears on WM when ΔE_c in formula (1) is positive, and appears on HAZ when it is negative. The local corrosion can be prevented by making ΔE_c almost 0.

$$\Delta E_c = \Delta Cu \times 3·8 + \Delta Ni \times 1·1 + 0·3 \tag{1}$$

ΔCu: difference in Cu content between B.M. and WM

ΔNi: difference in Ni content between B.M. and WM

This results from the following two factors.

(i) Cu and Ni concentrate in and stabilize corrosion products, making the steel noble.

(ii) In the 60 kgf/mm² class steels, the corrosion potential of the base metal becomes lower than that of the HAZ due to weld thermal cycle, and as a result, the corrosion potential of the WM and HAZ becomes dominant.

(2) Knife edge corrosion at fusion line can be prevented if the S content is less than 0·005%.

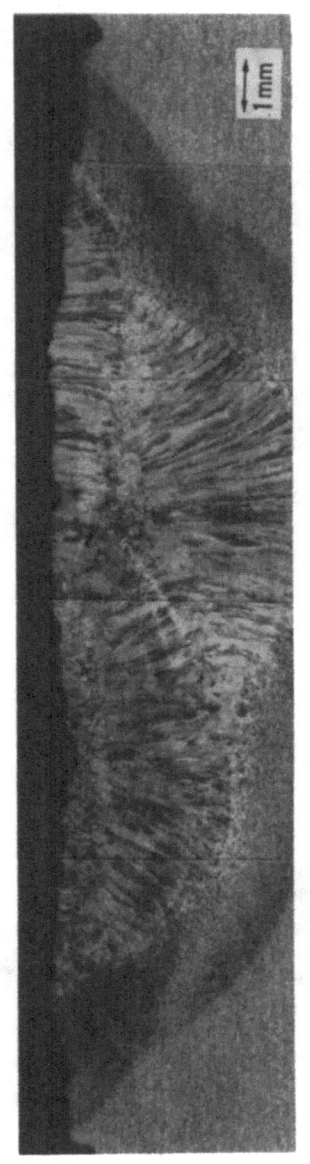

Y-6(A-Y13) 15kJ/cm; HAZ 0·5 Outside of HAZ 1

N-6(B-N1) 15kJ/cm; Outside of HAZ 0·5

FIG. 9. Sections of test pieces (6 months).

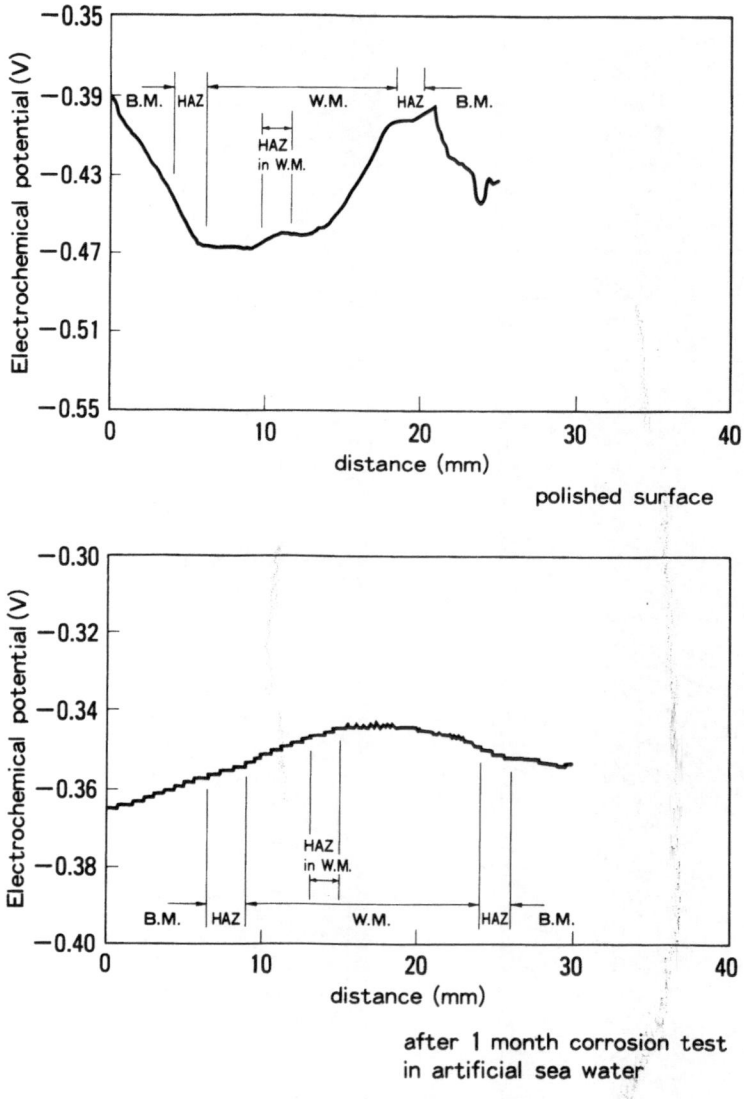

FIG. 10. Measured results of scanning reference electrochemical potential (versus Ag–AgCl electrode) of the specimen which had shown local corrosion of HAZ.

REFERENCES

1. RÄSÄNEN, E. and RELENDER, K., *Scand. J. Metallurgy*, **7** (1978), 11.
2. VALANTI, V. O., *Schiff und Hafen*, **H8** (1961), 695.
3. MICKWITY, B., private letter.
4. The U.S.S.R. Register of Shipping, Part XIV, 1978, 4, 3, 2, 6.
5. NANCOLLAS, G. H., KAGMIERCZAK, T. F. and SCHUTTRINGER, E., *Corrosion*, **37** (1981), 2, p. 76.

List of Participants

Mr T. Endo	Manager, Senior Research Engineer, Mitsubishi Heavy Industries Co., Ltd, Takasago R and D Center, Shinhama 2-1-1, Arai-cho, Takasago, Hyogo, Japan.
Dr A. G. Evans	Professor, University of California, Santa Barbara, College of Engineering, Dept. of Materials Science, Santa Barbara, CA 93106, USA.
Dr J. Fukakura	Chief, Senior Research Engineer, Toshiba Corp., Heavy Apparatus Engineering Laboratory, Suehiro-cho 2-4, Tsurumi-ku, Yokohama, Kanagawa, Japan.
Mr M. Hojo	Research Officer, Industrial Products Research Institute, Higashi 1-1-4, Yatabe-machi, Tsukuba-gun, Ibaraki, Japan.
Dr J. W. Hutchinson	Professor, Harvard University, Division of Applied Sciences, 316 Pierce Hall, 29 Oxford Street, Cambridge, MA 02138, USA.
Dr K. Iida	Professor, University of Tokyo, Faculty of Engineering, Dept. of Naval Architecture, Hongo 7-3-1, Bunkyo-ku, Tokyo, Japan.
Mr T. Inoue	Research Engineer, Nippon Steel Corp., R&D Lab. II, Fuchinobe 5-10-1, Sagamihara-shi, Kanagawa, Japan.
Dr K. Ishikawa	Group Leader (Senior Research Officer), National Research Institute for Metals, Sengen 1-2-1, Sakura-mura, Niihari-gun, Ibaraki, Japan.
Dr K. Itoh	Research Engineer, Nippon Steel Corp., R & D Lab. II, Fuchinobe S-10-1, Sagamihara-shi, Kanagawa, Japan.
Dr M. Jono	Professor, Osaka University, Faculty of Engineering, Dept. of Mechanical Engineering, Yamadaoka 2-1, Suita, Osaka, Japan.
Dr K. Kageyama	Senior Research Officer, Mechanical Engineering Laboratory, Namiki 1-2, Sakura-mura, Ibaraki, Japan.

Dr T. Kishi	Associate Professor, University of Tokyo, Faculty of Engineering, Institute of Interdisciplinary Research, Komaba 4-6-1, Meguro-ku, Tokyo, Japan.
Dr M. Kitagawa	Chief, Senior Research Engineer, Ishikawajima-Harima Heavy Industries Co., Ltd, Technical Institute, Toyosu 3-1-15, Koto-ku, Tokyo, Japan.
Dr A. S. Kobayashi	Professor, University of Washington, Dept. of Mechanical Engineering, FU-10, Seattle, WA 98195, USA.
Dr H. Kobayashi	Professor, Tokyo Institute of Technology, Faculty of Engineering, Dept. of Mechanical Engineering Science, Ohokayama 2-12-1, Meguro-ku, Tokyo, Japan.
Dr K. Komai	Professor, Kyoto University, Faculty of Engineering, Dept. of Mechanical Engineering, Yoshida-honmachi, Sakyo-ku, Kyoto, Japan.
Dr A. Komine	Research Engineer, Komatsu Ltd., Technical Research Center, Manda 1200, Hiratsuka, Kanagawa, Japan.
Dr F. Matsuda	Professor, Osaka University, Welding Research Institute, Mihogaoka 11-1, Ibaraki, Osaka, Japan.
Dr A. J. McEvily	Professor, University of Connecticut, Dept. of Metallurgy U-136, 97N. Eagleville Rd., Storrs, CT 06268, USA.
Dr K. Minakawa	Research Engineer, Nippon Kokan Corp., Technical Research Center, Minamiwatarida-cho 1-1, Kawasaki-ku, Kawasaki, Kanagawa, Japan.
Dr K. Miya	Professor, University of Tokyo, Faculty of Engineering, Nuclear Engineering Research Laboratory, Tohkai-mura, Naka-gun, Ibaraki, Japan.
Dr S. Miyazono	Chief Senior Researcher, Japan Atomic Energy Research Institute, Safety Engineering Division, Tohkai-mura, Naka-gun, Ibaraki, Japan.
Dr T. Mizoguchi	Chief Senior Research Engineer, Kobe Steel Ltd., Applied Mechanics Center, Wakinohama-cho 1-3-18, Chuo-ku, Kobe, Japan.
Dr H. Nakamura	Research Associate, Tokyo Institute of Technology, Faculty of Engineering, Dept. of Mechanical Engineering Science, Ohokayama 2-12-1, Meguro-ku, Tokyo, Japan.
Dr A. Narumoto	Senior Research Engineer, Kawasaki Steel Corp., Plate Laboratory, Kawasaki-cho 1, Chiba-shi, Chiba, Japan.
Dr M. Ohnami	Professor, Ritsumeikan University, Faculty of Science and Engineering, Dept. of Mechanical Engineering, Tojiin-Kitamachi, Kita-ku, Kyoto, Japan.
Dr F. S. Pettit	Professor, University of Pittsburgh, Dept. of Metallurgy and Materials Engineering, 848 Benedum Hall, Pittsburgh, PA 15261, USA.
Dr J. K. Tien	Professor, Columbia University, Henry Krumb School of Mines, 918 Mudd Building, New York, NY 10027, USA.
Dr S. Usami	Senior Research Engineer, Hitachi, Ltd, Mechanical Engineering Research Laboratory, Kandatsu-machi 502, Tsuchiura, Ibaraki, Japan.

Index